T0186526

BIOLOGICAL CLOCK
IN FISH

BIOLOGICAL CLOCK IN FISH

Editors:

Ewa Kulczykowska

Head of the Department of Genetics and Marine Biotechnology
Institute of Oceanology of Polish Academy of Sciences
Sopot, Poland

Włodzimierz Popek

Head of the Department of Ichthyology and Fisheries
University of Agriculture in Krakow
Krakow, Poland

B.G. Kapoor

Formerly Professor of Zoology
Jodhpur University
India

CRC Press
Taylor & Francis Group
Boca Raton London New York

CRC Press is an imprint of the
Taylor & Francis Group, an **informa** business
A SCIENCE PUBLISHERS BOOK

CRC Press
Taylor & Francis Group
6000 Broken Sound Parkway NW, Suite 300
Boca Raton, FL 33487-2742

First issued in paperback 2019

Copyright reserved © 2010 by Taylor & Francis Group, LLC
CRC Press is an imprint of Taylor & Francis Group, an Informa business

No claim to original U.S. Government works

ISBN-13: 978-1-57808-675-7 (hbk)
ISBN-13: 978-0-367-38395-4 (pbk)

This book contains information obtained from authentic and highly regarded
sources. Reasonable efforts have been made to publish reliable data and
information, but the author and publisher can- not assume responsibility for the
validity of all materials or the consequences of their use. The authors and publishers
have attempted to trace the copyright holders of all material reproduced in this
publication and apologize to copyright holders if permission to publish in this form
has not been obtained. If any copyright material has not been acknowledged please
write and let us know so we may rectify in any future reprint.

Except as permitted under U.S. Copyright Law, no part of this book may be reprinted,
reproduced, transmitted, or utilized in any form by any electronic, mechanical, or
other means, now known or hereafter invented, including photocopying,
microfilming, and recording, or in any information storage or retrieval system,
without written permission from the publishers.

For permission to photocopy or use material electronically from this work, please
access www.copy- right.com (http://www.copyright.com/) or contact the Copyright
Clearance Center, Inc. (CCC), 222 Rosewood Drive, Danvers, MA 01923, 978-750-8400.
CCC is a not-for-profit organization that pro- vides licenses and registration for a
variety of users. For organizations that have been granted a photocopy license by the
CCC, a separate system of payment has been arranged.

Trademark Notice: Product or corporate names may be trademarks or registered
trademarks, and are used only for identification and explanation without intent to
infringe.

        Library of Congress Cataloging-in-Publication Data

Biological clock in fish / editors, Ewa Kulczykowska,
Wlodzimierz Popek,
B.G. Kapoor.
     p. cm.
  Includes bibliographical references and index.
  ISBN 978-1-57808-675-7 (hardcover)
1. Fishes--Reproduction. 2. Sexual cycle. I.
Kulczykowska, Ewa. II. Popek, Wlodzimierz. III. Kapoor, B. G.
  QL639.2.B56 2010
  571.8'17--dc22

                                          2009054198

Visit the Taylor & Francis Web site at
http://www.taylorandfrancis.com

and the CRC Press Web site at
http://www.crcpress.com

Contents

List of Contributors

A.L. Alonso-Gomez
Department of Physiology (Animal Physiology II), Faculty of Biology, Complutense University of Madrid, 28040 Madrid, Spain.

Laurence Besseau
CNRS, FRE 3247 et GDR 2821, Modèles en Biologie cellulaire et évolutive, Avenue Fontaulé, BP 44, F-66651 Banyuls-sur-Mer, Cedex, France.

Université Pierre et Marie Curie (UPMC Univ. Paris 6), FRE 3247 et GDR 2821, Laboratoire Arago, Avenue Fontaulé, BP 44, F-66651 Banyuls-sur-Mer, Cedex, France.

Gilles Boeuf
CNRS, FRE 3247 et GDR 2821, Modèles en Biologie cellulaire et évolutive, Avenue Fontaulé, BP 44, F-66651 Banyuls-sur-Mer, Cedex, France.

Université Pierre et Marie Curie (UPMC Univ. Paris 6), FRE 3247 et GDR 2821, Laboratoire Arago, Avenue Fontaulé, BP 44, F-66651 Banyuls-sur-Mer, Cedex, France.

Museum National d'Histoire Naturelle, 43 rue Cuvier, 75005 Paris, France.

Elżbieta Ćwioro
Department of Ichthyobiology and Fisheries, University of Agriculture in Krakow, Krakow, Poland.

Bronisław Cymborowski
Department of Animal Physiology, University of Warsaw, Miecznikowa 1, 02-096 Warsaw, Poland.
E-mail: _bron@biol.uw.edu.pl_

M.J. Delgado
Department of Physiology (Animal Physiology II), Faculty of Biology, Complutense University of Madrid, 28040 Madrid, Spain.

Peter Ekström
Department of Cell and Organism Biology, Lund University, Helgonavägen 3, S-22362 Lund, Sweden.
E-mail: *peter.ekstrom@cob.lu.se*

Jack Falcón
CNRS, FRE 3247 et GDR 2821, Modèles en Biologie cellulaire et évolutive, Avenue Fontaulé, BP 44, F-66651 Banyuls-sur-Mer, Cedex, France.
Université Pierre et Marie Curie (UPMC Univ. Paris 6), FRE 3247 et GDR 2821, Laboratoire Arago, Avenue Fontaulé, BP 44, F-66651 Banyuls-sur-Mer, Cedex, France.
E-mail: *falcon@obs-banyuls.fr*

Nicholas S. Foulkes
Institute of Toxicology and Genetics, Karlsruhe Institute of Technology, Hermann-von-Helmholtz Platz 1, Eggenstein-Leopoldshafen 76344, Germany.
E-mail: *nicholas.foulkes@itg.fzk.de*

Michael Fuentès
CNRS, FRE 3247 et GDR 2821, Modèles en Biologie cellulaire et évolutive, Avenue Fontaulé, BP 44, F-66651 Banyuls-sur-Mer, Cedex, France.
Université Pierre et Marie Curie (UPMC Univ. Paris 6), FRE 3247 et GDR 2821, Laboratoire Arago, Avenue Fontaulé, BP 44, F-66651 Banyuls-sur-Mer, Cedex, France.

R. Haque
Departments of Ophthalmology and Pharmacology, Emory University School of Medicine, Atlanta, GA 30322, USA.

P.M. Iuvone
Departments of Ophthalmology and Pharmacology, Emory University School of Medicine, Atlanta, GA 30322, USA.
E-mail: *miuvone@emory.edu*

Ewa Drąg-Kozak
Department of Ichthyobiology and Fisheries, University of Agriculture in Krakow, Krakow, Poland.

Kajori Lahiri
Institute of Toxicology and Genetics, Karlsruhe Institute of Technology, Hermann-von-Helmholtz Platz 1, Eggenstein-Leopoldshafen 76344, Germany.

Jose Fernando López-Olmeda
Department of Physiology, Faculty of Biology, University of Murcia, 30100 Murcia, Spain.
E-mail: *jflopez@um.es*

Ewa Łuszczek-Trojnar
Department of Ichthyobiology and Fisheries, University of Agriculture in Krakow, Krakow, Poland.

Elodie Magnanou
CNRS, FRE 3247 et GDR 2821, Modèles en Biologie cellulaire et évolutive, Avenue Fontaulé, BP 44, F-66651 Banyuls-sur-Mer, Cedex, France.

Université Pierre et Marie Curie (UPMC Univ. Paris 6), FRE 3247 et GDR 2821, Laboratoire Arago, Avenue Fontaulé, BP 44, F-66651 Banyuls-sur-Mer, Cedex, France.

Lucien C. Manchester
Department of Cellular and Structural Biology, University of Texas Health Science Center, San Antonio, Texas, USA.

Hilmar Meissl
Max Planck Institute for Brain Research, Deutschordenstr. 46, 60528 Frankfurt/Main, Germany.
E-mail: *meissl@mpih-frankfurt.mpg.de*

Hanna Natanek
Department of Ichthyobiology and Fisheries, University of Agriculture in Krakow, Krakow, Poland.

Catarina Oliveira
Department of Physiology, Faculty of Biology, University of Murcia, 30100 Murcia, Spain.
E-mail: *oliveira@um.es*

Włodzimierz Popek
Department of Ichthyobiology and Fisheries, University of Agriculture in Krakow, Krakow, Poland.
E-mail: *rzpopek@cyf-kr.edu.pl*

Russel J. Reiter
Department of Cellular and Structural Biology, University of Texas Health Science Center, San Antonio, Texas, USA.
E-mail: *reiter@uthscsa.edu*

Francisco Javier Sánchez-Vázquez
Department of Physiology, Faculty of Biology, University of Murcia, 30100 Murcia, Spain.
E-mail: *javisan@um.es*

Sandrine Sauzet
CNRS, FRE 3247 et GDR 2821, Modèles en Biologie cellulaire et évolutive, Avenue Fontaulé, BP 44, F-66651 Banyuls-sur-Mer, Cedex, France.

Université Pierre et Marie Curie (UPMC Univ. Paris 6), FRE 3247 et GDR 2821, Laboratoire Arago, Avenue Fontaulé, BP 44, F-66651 Banyuls-sur-Mer, Cedex, France.

Dan-Xian Tan
Department of Cellular and Structural Biology, University of Texas Health Science Center, San Antonio, Texas, USA.

E. Velarde
Department of Physiology (Animal Physiology II), Faculty of Biology, Complutense University of Madrid, 28040 Madrid, Spain.

David Whitmore
University College London, Dept. of Cell and Developmental Biology, Centre for Cell and Molecular Dynamics, Rockefeller Building, 21 University Street, London, WC1E6DE, UK.
E-mail: *d.whitmore@ucl.ac.uk*

1

INTRODUCTION TO CIRCADIAN RHYTHMS

Bronisław Cymborowski

Since appearance of life on our planet it has been subjected to daily rhythm of light and darkness, and also to seasonal cycles of climate changes, caused by the rotation of the Earth around its axis and around the sun. Therefore, from the earliest times the alternation of day and night has left its very deep marks on the leaving organisms. In addition, marine organisms have been subjected to tidal and lunar cycles. Only animals lived in the depth of the oceans, or in caves, have avoided these environment fluctuations. The majority of animals show daily and annual rhythms of activity. They may be nocturnal, diurnal or crepuscular. They may diapose, hibernate or estivate. Plants may produce leaves or flowers only at certain seasons, and flowers may open and close at particular times of the day. Even if the flowers are kept in constant darkness, they still open around the same time. These innate rhythms has been observed for the first time by the French astronomer De Mairan who discovered in 1729 that the daily leaf movements of *Mimosa* plant would persist also in constant darkness. Later on Carolus Linnaeus had an idea that it should be possible to take advantage of several plants that open or close their flowers at particular times of the day to accurately predict the time, and form kind of *Horologium Florae* (flower clock). He proposed this concept in the 1751 publication Philosophia Botanica. In the early 19th century, several botanical gardens have succeeded in "construction" of such a flower clock. The study of photoperiodic responses of organisms has revealed other very interesting examples of time measurement. It was first discovered in plants, and later in animals, that

Department of Animal Physiology, University of Warsaw, Miecznikowa 1, 02-096 Warsaw, Poland.
e-mail: bron@biol.uw.edu.pl

some developmental processes are controlled by the length of day. For example, in animal's day length can control the annual cycle of reproduction, or the beginning of rest period, e.g., winter dormancy of plant buds, hibernation in mammals, the diapose and estivation in insects, etc. These photoperiodic reactions are actually caused by day length and not by the quantity of light. The migration of animals, change in pelage (fur or hair) or feather colors, are among the more common events readily observed in nature. The importance of time measurement by animals was also recognized during the study of their ability to orientate by the use of a sun compass. It was found that animals could compensate for changes in the position of the sun with the progress of time during the day.

Now we know that organisms not only can indicate the time of the day, but that they also make use of their clock for actual time measurement. The plants and animals do not use the hour glass principle for determination of the most suitable time of the day for a given physiological process to be performed; they rather measure time by means of oscillations, which has distinct selective advantage. The oscillations underlying life functions are now known to provide a temporal organization for physiological and behavioral activities in practically every group of organisms, including prokaryotes such as cyanobacteria. The use of oscillations makes it possible to "predict" environmental changes even for a several days ahead.

GENERAL PROPERTIES OF BIOLOGICAL RHYTHMS

The basic characteristic of biological rhythms is that they persist under constant environmental conditions, in absence of any direct external influence. The rhythms, therefore, are termed endogenous; they arise within organism itself and are not imposed by the environment. However, it does not mean that their natural period is exactly the same as that of the environment's cycle. The periods of the endogenous rhythms actually differ from those in the nature, but exogenous synchronizing cues, commonly called Zeitgeber ("timegiver" in German), adjust or "entrain" the biological rhythms, so that the organisms remain in phase with their surroundings, e.g. they are synchronized to environmental cycles. The daily alternation of light and dark (day and night) is a very powerful entrainer factor for plants and animals. The slight difference in period of the biological rhythms from the geophysical cycles has led to their description as free running or circadian rhythms. The term *circadian* derives from the Greek words *circa* (about) and *dian* (day). In the absence of temporal cues (in noncycling conditions) from the environment (i.e., in darkness and constant temperature) the rhythms reveal their own natural period which is close to, but significantly different (longer or shorter—depending on species) from that of the solar day. Under these conditions the term *circadian time* (CT) is commonly used, which is the time of the internal clock. Therefore, the term "circadian" classifies the daily rhythms as approximately a day or 24 hour, and circannual as about a year, or seasonal rhythms. Jurgen Aschoff noted that the periodicity of the same organism might differ under different conditions, depending on the functional state of the organism. He proposed a circadian rule according to which diurnal animals have shorter than 24-hour rhythms in constant light (LL) and longer than 24-hour rhythms in constant darkness (DD). The reverse would be true for nocturnal animals. There are some exceptions to this rule.

The time at which the rhythmic process occurs defines the phase of that particular rhythm. For example, the time at which animal sleeps, or rests, can be used as a measure of phase. Therefore, the nocturnal and diurnal animals display opposite phases in their sleep-wake rhythms. By exposing an organism to pulses of light while in constant conditions it is possible to obtain the phase shift of a given rhythm. In such situation, light can have three differing effects; it may cause no effect, a phase advance or a phase delay. Typically no effect is obtained during most of the subjective day, phase delays are obtained in late subjective night extending into early subjective day. The light phase response curve (PRC) appears to be quite similar among wide range of organisms. PRCs may be obtained in constant conditions as well from stimuli other than light (so called non-photic stimuli). The phase response curve describes very well the phenomena of rhythms' entrainment. The best analogy for a phase shift can be found in the natural day-night cycle. The dawn occurring earlier each day in the spring would lead to earlier onset of the next day's phase of a given rhythm (e.g., activity, feeding, etc.).

Since most of rhythms are synchronized to environmental cycles, chronobiologists refer to time as defined by these environmental cycles, calling it Zeitgeber Time (ZT). When animal is kept in a 24-hour cycle that consists of 12 hours of light and 12-hours of darkness (LD 12:12), zeitgeber time or ZT0 corresponds to "light on", and ZT12 to "lights off". Thus, all time points between 0 and 12 refer to daytime hours, while those between 12 and 24 (the same as "0") refer to nighttime hours. Since the periodicity of the internal clock is generally different from 24 hours, CT is not the same as ZT. This terminology will be used throughout this book.

An important feature of circadian rhythms is temperature-compensation of their periods, as already mentioned above. For the first time, it was demonstrated in some classical experiments with *Drosophila pseudoobscura* by Colin Pittendrigh in early 1950s. This property is an absolute functional prerequisite for any clock mechanism. It is also essential for effective entrainment by a natural (24 hours) Zeitgebers. The rate of most physiological processes doubles with a 10°C rise in temperature. The fact that this does not applies to circadian rhythms because their circadian periods remain constant over a wide temperature range. This indicates that there must be a mechanism compensating temperature-induced changes.

The manner in which temperature-compensation is achieved in biological clocks is still not understood. The observation that the period of rhythm is temperature-compensated, justifies the use of the term "biological clock" in analogy to man-made time-measuring devices.

GENERALIZED SCHEMATIC MODEL FOR BIOLOGICAL CLOCK

A simplified time-measuring system (biological clock) functioning in organism is depicted in Fig. 1.1. This very basic circadian system includes environmental receptors (e.g., light-absorbing pigments), biochemical and/or biophysical input pathway leading to the clock (primary oscillator or pacemaker), and output pathways leading to the oscillation of overt rhythm. In practice, there can be more than one external input that may be more specific to a given organism or system, such as water or humidity, feed/fast

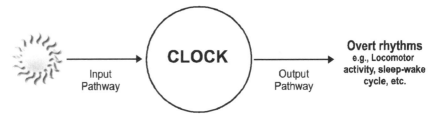

Fig. 1.1 A simplified time-measuring system (biological clock) in organism.

regimens, social cues, noise, etc. The output pathway might include a regulatory center for a given physiological/behavioral rhythmic processes. In addition also outputs can feedback on the clock and perhaps even on input. At this point should be emphasize that external factors do not "drive" the oscillator, but serve as synchronizers of the overt rhythms ("hands" of clock). The effect of light that is independent of the endogenous clock is term "masking effect". It is called so, because it can mask the manifestation of the endogenous rhythm. In principle, it can apply not only to light effect, but also to any influence that obscures a rhythmic process. In some cases the oscillator that accounts for amplitude of the rhythm may disappear (dampen) when agents of entrainment are absent.

THE MULTIOSCILLATORY CIRCADIAN SYSTEM

It has now become clear that multicellular organisms have more than one clock. It appears that with increasing complexity of organism, there is an increasing need to have clocks specific for different physiological functions. Apparently independent or semi-independent, light-entrainable circadian clocks exist at all levels of organization from cells, through tissues to organs. Such systems have been discovered in nervous tissue, endocrine glands, gonads, etc. The question arises whether they all are truly independent or may serve as part of a physiological hierarchy? Among animals there is usually a master or central clock which is located in the brain. Clocks located in other parts of the body are referred to as peripheral oscillators (clocks) to separate them from central clock (Fig. 1.2A, B). The degree of autonomy of such oscillators depends very much on species. In mammals, the

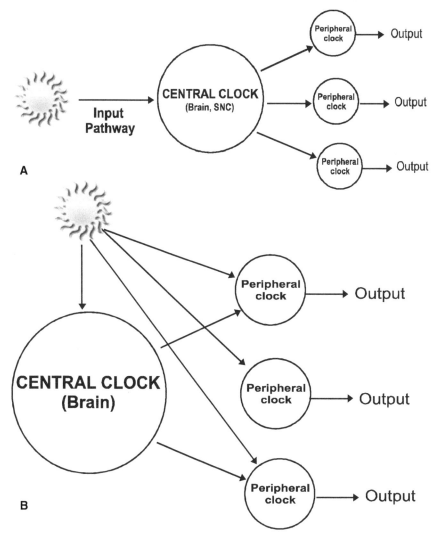

Fig. 1.2A, B Central clock and peripheral oscillators (clocks) in organism.

light-sensitive central clock is located in the brain and controls the peripheral clocks. They are not photosensitive but they can be entrained by other signals originated within the organism (e.g., hormones). In other organisms (e.g., insects), peripheral clocks can be photosensitive (*via* extraretinal photoreceptors) and independent from the central clock. In the nonmammalian vertebrate, zebrafish most oscillators are autonomous and have their own photoreceptors, although with some degree of dependence from central clock. In mammals there is a central (master) clock is located in the hypothalamus in the suprachiasmaticus nucleus (SCN). The SCN clock can sustain its rhythmic activity for a number of days *in vitro*, and can be synchronized by light signals. In organism it is entrained by the photic information coming from photoreceptors. At this point it is worth mentioning that in some vertebrates a functioning clock was also found in the eyes. As it was mentioned above, the mammalian peripheral clock can not be directly stimulated by light, but they might be directly entrained by various hormones which some of them might be controlled by CNS. It is obvious that within the organism there are many interactions between central and peripheral clock, which will be discuss in different chapters of this book.

Special attention will be focused on the role of melatonin in fish circadian system. It was found that SCN clock controls rhythmic production of melatonin by the pineal glands, and is suppressed by light—no other circadian rhythm is so immediately and completely suppressed by exposure to environmental light. Previously it was postulated that SCN clock is "on" at night (stimulating melatonin production) and "off" during the day, because lessoning the CNS resulted in a decrease N-acetyltransferase activity, a key enzyme in melatonin synthesis pathway. However, another studies showed that within SCN there are both inhibitory and stimulatory neurons effecting pineal activity in response to light.

MOLECULAR BASIS OF BIOLOGICAL CLOCK

The time-measuring system in both prokaryotes (cyanobacteria) and eukaryotes (fungi, plants and animals) has a number of key molecular components. This includes, among others, clock genes, cycling proteins, and photoreceptors. As it was mentioned above, photoreceptors are part of an input pathway that receives and transmits signals from external environmental synchronizers to a central clock that generates rhythmicity. In turn, an output pathway transmits temporal signals from the clock to biological variables that oscillate. These oscillations are the overt rhythms of variables ("hands of the clock") such as activity cycle, leaf movements, enzyme activity, etc. In all organisms examined, the basic clock mechanism includes transcription/translation loops with positive and negative feedback that produce delays and generate rhythmicity (Fig. 1.3). In other

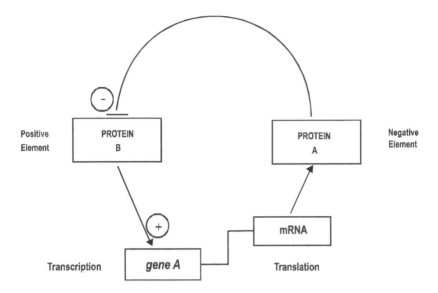

Fig. 1.3 A feedback loop with oscillating products of specific clock genes regulating their own expression.

words the basic clock mechanism is a feedback loop in which oscillating products of specific clock genes regulate their own expression. Each complete turn of the loop takes about 24 hours to accomplish, resulting in circadian oscillations of RNA and protein levels. The proteins that function in this manner differ from one species to another, although some of them are conserved from cyanobacteria to mammals. At present there are more and more information on how the clocks transmit their temporal signals to the rest of the organism. It is clear now, that one of the ways clock transmit output signals is by driving rhythms of the clock controlled gene expression by rhythmically regulating of its promotor activity in various tissues and organs. Genetic and molecular approaches have been very successful at unraveling functional properties of the clock. At present, various organisms subjected to random mutagenesis are widely screened for circadian phenotypes. Such approach has already allowed identifying genes involved in many aspects of the circadian clock system. Much more work is needed in order to understand all aspects of temporal organization within the organism.

2

THE PINEAL ORGAN OF FISH

Jack Falcón,[1,2,CA] *Laurence Besseau,*[1,2] *Elodie Magnanou,*[1,2]
Sandrine Sauzet,[1,2] *Michael Fuentès*[1,2] *and Gilles Boeuf*[1,2,3]

INTRODUCTION

The first studies dealing with the teleost pineal organ emphasized its photoreceptive characteristics, and it soon appeared it was involved in the control of functions and behaviors displaying daily rhythms (for reviews see: Zachmann et al., 1992a; Ekström and Meissl, 1997; Falcón, 1999; Boeuf and Falcón, 2001). Indeed, the pineal epithelium contains photoreceptors cells that resemble the retinal cone photoreceptors, both, on a structural and functional point of view. The cone-like photoreceptors establish synaptic contacts with ganglion cells (or second order neurons) that send their axons to the brain. Interstitial cells make the third main cell type of the pineal epithelium, which so resembles to a simplified retina.

Photoreceptors release at night an excitatory neurotransmitter at the synaptic junctions with the ganglion cells. The latter transmit in turn and immediately the information to brain centers. This nervous information mediates short latency responses to changes in ambient illumination. A second message—melatonin—is produced at night by the photoreceptors and released into the cerebro-spinal fluid and blood. The melatonin signal provides the organism periodic information of longer latency (24 h basis) than the nervous information. Temperature and internal factors (neurotransmitters, hormones) may modulate the melatonin pattern of production.

---

[1] CNRS, FRE 3247 et GDR 2821, Modèles en Biologie cellulaire et évolutive, Avenue Fontaulé, BP 44, F-66651 Banyuls-sur-Mer, Cedex, France.
[2] Université Pierre et Marie Curie (UPMC Univ. Paris 6), FRE 3247 et GDR 2821, Laboratoire Arago, Avenue Fontaulé, BP 44, F-66651 Banyuls-sur-Mer, Cedex, France.
[3] Museum National d'Histoire Naturelle, 43 rue Cuvier, 75005 Paris, France.
[CA] Corresponding author: e-mail: falcon@obs-banyuls.fr.

In most species, the rhythmic production of melatonin involves a circadian pacemaker located within the photoreceptor cells, so that photoperiod synchronizes the clock, which in turn drives the rhythmic production of melatonin. Melatonin thus represents a hormonal hand of the pineal circadian clocks that helps synchronizing functions and behaviors to variations in the external cues. If it is today well admitted that the pineal organ is a key component in the circadian organization of fish, little is known yet on its modes of action. We review here our current knowledge on the mechanisms underlying the rhythmic control of melatonin production by the pineal organ of fish.

FUNCTIONAL ORGANISATION OF THE PINEAL

Anatomy

The pineal organ appears generally as an end-vesicle connected to the dorsal epithalamus by a stalk (Ekström and Meissl, 1997; Falcón, 1999; Falcón *et al.*, 2007). In adult fish, the end vesicle is located in a "window" below the skull, which appears often thinner in this area, and the skin covering it is less pigmented. The end vesicle may be large enough to cover the whole cerebral hemispheres and olfactory bulbs. The lumen of the organ communicates with the 3rd ventricle, although this might not be always the case (Omura and Oguri, 1969). The organ is surrounded by fenestrated blood vessels that do not enter the epithelium (Fig. 2.1). It lacks a blood-brain barrier, so that the epithelium is exposed to the haemal environment in its basal part, and to the cerebrospinal fluid in its apical part (Falcón, 1979a; Omura *et al.*, 1985; reviewed *in* Ekström and Meissl, 1997).

Fig. 2.1 The pineal epithelium. 1, Cone type photoreceptors; 1', modified photoreceptors; 2, Interstitial cells; 3, Second order neurons; 4, Macrophages. For details see text.

The pineal cell types

Three main cell types make the pineal epithelium, which is generally pseudo-stratified and more or less enfolded. The photoreceptor cells resemble to the cone photoreceptors of the retina. They display a polarized organization, with four distinct cell compartments: the outer and inner segments, cell body and synaptic pedicle(s) (Fig. 2.1). The outer segment, which protrudes into the pineal lumen, is composed of flattened stacks (20–70) made by folds of the plasma membrane. The outer segment covers the apical part of the inner segment from which it arises, and which contains the centrosome (formed by a distal and a proximal centriole (a 9×2+0 and a 9×3+0 cilia, respectively; Falcón *et al.*, 2007). Mitochondria are particularly abundant in the inner segment. The cell body contains the nucleus surrounded by the Golgi apparatus and rough endoplasmic reticulum. One or several synaptic pedicles with synaptic buttons constitute the basal part of the cell. The synaptic pedicles make synaptic contacts with dendrites from the ganglion cells (or second-order neurons) through ribbon type synapses (Fig. 2.1). They accumulate synaptic vesicles and few dense-core (secretory-like) vesicles. In some fish species the outer segment displays some kind of more or less pronounced disorganization, or even is absent; in addition, ganglion cells may become scarce or disappear; when this the case, then the photoreceptor end pedicles just contact other photoreceptors or the basal lamina (Fig. 2.1; Falcón *et al.*, 2007). As observed in pike, cone-type photoreceptors and "modified" photoreceptors can be observed within the same pineal gland, and display regional distribution.

The ganglion cells constitute the second-order neurons of the pineal epithelium (Fig. 2.1). Details on their morphology and distribution have been provided elsewhere (Falcón, 1999; Falcón *et al.*, 2007). Except for some inter-neurons, most of them send their axons to the brain and thus contribute to forming the pineal tract that runs dorsally along the pineal stalk. This tract enters the brain at the level of the subcommissural organ. In the species investigated, the axons exit the pineal stalk to reach a number of structures including the *habenula*, ventral and dorsal thalamus, posterior commissure, periventricular *pretectum*, pretectal area, posterior *tuberculum*, paraventricular organ, posterior tuberal nucleus, dorsal synencephalon and *tegmentum* (Falcón *et al.*, 2009). Pinealofugal terminals have also been observed in the preoptic area (POA) of some but not all species; it is noteworthy that the POA is also a retino-recipient area. The functional significance of the pinealofugal innervations remains enigmatic.

The fish pineal organ also receives axon terminals originating from cells in the thalamic *eminentia, habenula,* dorsal thalamus, ventromedial thalamus, periventricular pretectum, posterior commissure, posterior *tuberculum* and dorsal synencephalon (Ekström *et al.*, 1994; Jiménez *et al.*,

1995; Yáñez and Anadón, 1996; Pombal *et al.*, 1999; Mandado *et al.*, 2001). Some of these brain areas overlap with brain regions that also appear connected with the retina (ventral and dorsal thalamus, pretectal area, and posterior *tuberculum*). This highlights the importance of these areas in the integration of photoperiod information, which constitute a possible pathway for the exchange of information between the retina and pineal organ. The nature and role of these "pinealopetal" innervations is unclear; it consists of neurons containing catecholamines, FMRF-amide, neuropeptide Y, growth hormone releasing hormone (GHRH) or gonadotropin releasing hormone (GnRH) (Ekström and Meissl, 1989; Rao *et al.*, 1996; Subhedar *et al.*, 1996).

The interstitial (glial) cells constitute the third main cell type of the pineal epithelium, where they occupy the whole height (Fig. 2.1). They are thus in contact with the cerebrospinal fluid in their apical part and the vascular spaces in their basal part. In addition to a supposed supportive role, there is indication that these cells are involved in an intense synthetic activity (Falcón, 1999; Falcón *et al.*, 2007). One of their possible functions could be to provide photoreceptors with nutrients and differentiation factors. Indeed, dissociated fish pineal cells rebuild, *in vitro*, the general architecture of a pineal epithelium, including differentiated and polarized photoreceptors, a central lumen, and a peripheral network of collagen fibers, whereas isolated photoreceptors are unable to do so when cultured alone (Bégay *et al.*, 1992; Bolliet *et al.*, 1997). As a matter of fact, the retinal Muller cells, express growth factors, neurotransmitter transporters and antioxidant agents, which mediate protective effects from various neurological insults (Garcia *et al.*, 2002; Garcia and Vecino, 2003; Harada *et al.*, 2003). Other possible functions of glial cells include recycling of photoreceptor outer segments and visual chromophore or production of paracrine modulators (Falcón *et al.*, 2007).

Finally, macrophages are abundant in the pineal lumen. Their strong acid phosphatase activity, suggests they play a role in the breakdown and digestion cellular debris including phagocytic photoreceptor outer segments.

LIGHT SENSITIVITY AND PRODUCTION OF A NERVOUS MESSAGE

The pineal photoreceptors are analogous to the retinal cones not only in terms of structure but also function (O'Brien and Klein, 1986; Falcón *et al.*, 2007). The molecules of the phototransduction cascade are present and functional. These include chromophores and different types of opsins, transducin, arrestin, and a cyclic nucleotide-gated (CNG) channel operated by cyclic GMP (cGMP). Also, the lipid environment is very similar, and different from other neuronal and non-neuronal structures (Henderson *et al.*, 1994). Both, the pineal and retinal photoreceptors are particularly rich in polyunsaturated fatty acids (PUFA), which are critical to photoreceptor structure, function, and development (Bell *et al.*, 1995; Boesze-Battaglia and Allen, 1998). Among the PUFA of interest, docosahexaenoïc acid (DHA) is one major constituent of the outer segment membranes (Bazan *et al.*, 1992; Chen and Anderson, 1993; Brown, 1994; Henderson *et al.*, 1994; Falcón and Henderson, 2001). DHA containing phospholipids confer special fluidity to the membrane, and could be required for the proper localization of rhodopsin in the outer segment (Weisinger *et al.*, 1999).

After the pioneer studies of E. Dodt (1962, 1963) many studies have investigated the electrical responses from the photoreceptor and ganglion cells of the pineal organ. Early receptor potentials (ERP), electropinealograms (EPG) and variations in membrane potential can be recorded from the photoreceptor cells. Upon illumination of dark adapted photoreceptors, ERP first appear with very short latency (100 µs); they are generated by charge redistribution in the photopigment molecule (Falcón and Tanabe, 1983). The ERP response is followed by membrane hyperpolarization, which can reach –60 to –90 mV (in the dark, the resting potential is at –20/–30 mV) (Meissl and Ekström, 1997; Falcón *et al.*, 2007). The pineal photoreceptor cannot discriminate between rapid changes in light stimuli and, during prolonged stimulation it maintains the same response amplitude. In other words, the membrane potential reflects the ambient level of illumination (Meissl and Ekström, 1997). Finally, mass responses of cells, the EPG, can be recorded, which are analogous to the electroretinogram (ERG) (Meissl and Dodt, 1981). Latency (± 50 ms) and amplitude of the EPG depend on both the duration and intensity of the stimulus at a given wavelength. In the species investigated, the ERP and EPG spectral sensitivity curves were similar.

In the dark the photoreceptor cell releases an excitatory neurotransmitter (glutamate and/or aspartate) (Meissl and Ekström, 1997). This stimulates ganglion cells activity, which can be monitored through recording of their spike discharges (Ekström and Meissl, 1997). There is a linear relationship between the logarithm of light intensity and the frequency of the discharges over a range of 5 to 6 log units (Falcón and Meissl, 1981). This provides a wide range of sensitivity because only 1 to 2 log units are absorbed by the skull (Meissl and Dodt, 1981). Two types of responses have been recorded, chromatic and achromatic. The chromatic response consists of a long lasting inhibition upon exposure to UV light and stimulation by green or red light. The mechanisms underlying the chromatic response have been discussed and reviewed elsewhere (Meissl and Dodt, 1981; Solessio and Engbretson, 1993; Falcón *et al.*, 2007). The achromatic (or luminance) response is the most common and is characterized by an inhibitory effect of light at all wavelengths from UV to red light.

LIGHT SENSITIVITY AND PRODUCTION OF A HORMONAL MESSAGE: MELATONIN

The fish pineal organ produces melatonin in a rhythmic manner (Falcón *et al.*, 2007, 2009). The synthesis and release of melatonin is controlled by the environmental LD cycle in such a manner that it increases in the dark and decreases upon illumination. In a general manner, this rhythm is of high amplitude/short duration under long photoperiod, and of low amplitude/ long duration under short photoperiod (Fig. 2.2) (Kezuka *et al.*, 1988; Masuda *et al.*, 2003a). Thus, the melatonin signal, which reflects the prevailing photoperiod, provides the animal with accurate information on daily and calendar time, as is also the case for the nervous signal sent to brain centers (see above).

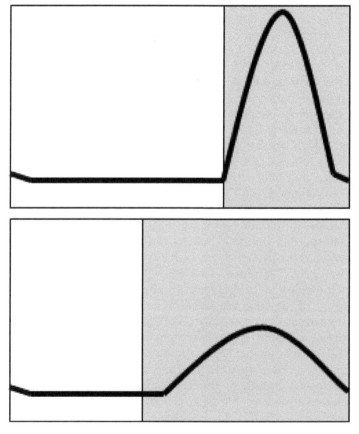

Fig. 2.2 Schematic presentation of the daily profiles of melatonin production and release by the pineal organ under long (above) and short (below) photoperiod. Gray boxes correspond to the scotophase.

The melatonin biosynthesis pathway operates in the pineal photoreceptor cells (Bolliet *et al.*, 1997). It involves four enzymatic steps (Fig. 2.3; Klein *et al.*, 1997; Falcón *et al.*, 2007): 1) tryptophan hydroxylase (TPOH) catalyzes the conversion of tryptophan into 5-hydroxytryptophan; 2) 5-hydroxytryptophan is decarboxylated by the aromatic amino-acid decarboxylase to produce serotonin; 3) the arylalkylamine *N*-acetyltransferase (AANAT) converts serotonin to *N*-acetylserotonin; 4) *N*-acetylserotonin is *O*-methylated by the action of the hydroxyindole-*O*-methyltransferase (HIOMT) to produce melatonin. Other indole

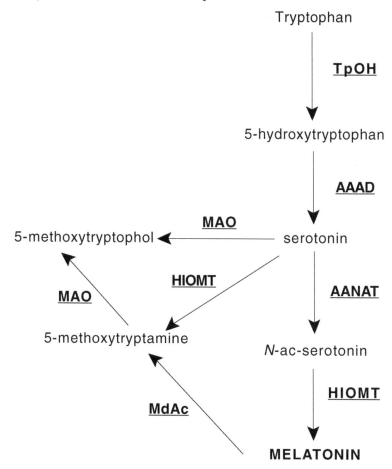

Fig. 2.3 Melatonin biosynthesis pathways. Thick arrows correspond to the main pathway. AAAD: aromatic amino acid decarboxylase; AANAT: arylalkylamine-*N*-acetyltransferase; *N*-ac-serotonin: *N*-acetylserotonin; HIOMT: hydroxyindole-*O*-methyltransferase; MdAc: melatonin decarboxylase; MAO: monoamine oxydase; TpOH: tryptophan hydroxylase. Modified from Falcón *et al.* (2007).

compounds (5-hydroxyindole acetic acid and 5-hydroxytryptophol) are produced by the fish pineal organ after oxidative deamination of serotonin, which is catalyzed by monoamine oxidase (MAO). These compounds as well as serotonin are also substrates for HIOMT. In addition to being released in the blood or cerebrospinal fluid (CSF), melatonin can be deacetylated *in situ* to produce 5-methoxytryptamine and 5-methoxytryptophol (Fig. 2.3; Falcón *et al.*, 1985; Yañez, 1996).

Melatonin is highly lipophilic, and thus crosses easily the cell membrane. In the teleost fish investigated only pineal melatonin is released into the blood stream and CSF as soon as it is synthesized (Falcón *et al.*, 2007). Although the retina also produces melatonin in significant amounts, it seems not to be released but rather to be acting in an autocrine/paracrine manner. And, the presence of a strong melatonin deacetylase activity in retinal tissues (Grace *et al.*, 1991), prevents melatonin from being released into the blood stream and conversely, of melatonin from other sources to reach the retina.

In vitro, the release of melatonin by fish pineal glands or cells cultured under a LD cycle reflects the pattern observed *in vivo* in the pineal organ and plasma (Falcón *et al.*, 2007). And, unexpected light at night decreases pineal melatonin content and release both *in vivo* and *in vitro*. This indicates that the regulation of melatonin production by light is a property of the pineal photoreceptor cells. The amplitude of the response to light is proportional to the intensity of the stimulus (Max and Menaker, 1992; Zachmann *et al.*, 1992b; Bolliet *et al.*, 1995; Bayarri *et al.*, 2002). Spectral sensitivity curves indicate melatonin production is highly sensitive to blue and green wavelengths (Max and Menaker, 1992; Bayarri *et al.*, 2002). Interestingly, it appears that the light-dependent inhibition of the nervous and neurohormonal messages display the same spectral sensitivity curve. Beside this general scheme, recent studies indicate that in some species the control by light may also rely on light perceived through retina, either partially of even exclusively (Migaud *et al.*, 2007). The modalities of the regulation by the retina remain to be elucidated. In mammals, the photic information perceived through the eyes is conveyed through a retino-hypothalalamic tract (RHT), to the suprachiasmatic nuclei of the hypothalamus (SCN), where the master circadian clocks reside; from there, a multisynaptic pathway (hypothalamic paraventricular nuclei [PVN] → preganglionic neurons of the sympathetic nervous system → superior cervical ganglion [SCG]) connects the SCN to the melatonin producing units of the pineal gland (Simonneaux and Ribelayga, 2003). In birds, both the direct (pineal) and indirect (retinal) pathways mediate the effects of light on the pineal melatonin secretion. It is possible that the situation observed in some teleost fish species reflects the existence of a convergent evolution with tetrapods. The large anatomical diversity observed in the pineal organ of fish as well as the presence, in some species, of more or less "rudimentary" photoreceptors, pinealopetal innervations or catecholaminergic control of melatonin secretion (see below), might be pieces of this "evolution puzzle" (Falcón *et al.*, 2009).

THE ARYLALKYLAMINE *N*-ACETYLTRANSFERASE (AANAT), THE ENZYME OF THE MELATONIN RHYTHM

In the last two decades much attention has focused on the AANAT enzyme because it soon appeared that the rhythm in melatonin biosynthesis results from the rhythmic activity of the enzyme, as is the case in mammals and birds (reviewed *in*: Klein *et al.*, 1981; 1997; Falcón, 1999; Falcón *et al.*, 2007, 2009). The daily rhythms of AANAT activity and melatonin production can be superimposed both *in vivo* and *in vitro*. In contrast, HIOMT activity remains rather constant throughout the LD cycle (Falcón *et al.*, 1987; Morton and Forbes, 1988).

The cloning of fish AANATs soon indicated that teleost fish are unique among vertebrates because they express two or even three AANAT genes that define two families (Falcón *et al.*, 2007, 2009). It is believed that a first round of whole genome duplication at the base of the teleost fish lineage is at the origin of the AANAT1 and AANAT2 families of enzymes. A second round of duplication at some point of the teleost fish evolution, resulted in the appearance of two AANAT1, AANAT1a and AANAT1b (Coon and Klein, 2006). This situation is unique among vertebrates. Most interestingly, the different AANAT families exhibit tissue specific expression. Thus, AANAT2 is specifically expressed in the pineal organ; it has no equivalent in other vertebrate classes. The AANAT1a and AANAT1b enzymes are expressed in the retina and discrete brain areas; AANAT1a and AANAT1b are more similar to the AANATs found in other vertebrates. Studies in the zebrafish indicated two regions of the *Aanat2* gene determine pineal specific expression; one is located in the 5'-flanking region, and the other 6 kb downstream the transcribed region (Appelbaum *et al.*, 2004). The encoded AANAT1 and AANAT2 proteins have distinct affinities for serotonin and differ in their relative affinities for indole-ethylamines versus phenyl-ethylamines (Falcón and Bolliet, 1996, Coon *et al.*, 1999; Benyassi *et al.*, 2000; Zilberman-Peled *et al.*, 2004).

Regulation of pineal AANAT2 by light

Regulation of pineal AANAT2 operates at the transcriptional, traductional and post-traductional levels, although differences exist from one species to another. In the few fish species investigated, AANAT2 transcripts abundance varies in some species, not in others (Fig. 2.4). Variations have been observed in species where an intra-pineal circadian control of melatonin secretion operates (pike, trout; see below). In this case, the rhythm in transcripts abundance is 3–4 h phase advanced compared to the rhythms in AANAT protein amount and activity. In species where an intra-pineal circadian control does not operate (trout), AANAT2 transcript levels remain constant throughout the LD cycle (Fig. 2.4; Bégay *et al.*, 1998; Coon *et al.*, 1999; Gothilf *et al.*, 1999). It is interesting that TpOH mRNA abundance exhibits similar

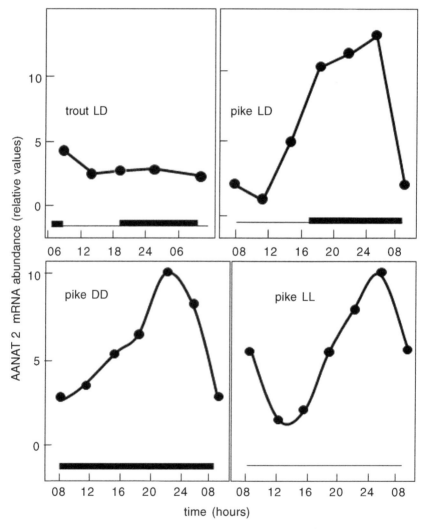

Fig. 2.4 LD variations in AANAT2 mRNA abundance in the pineal organs of the trout (without circadian clock control) and pike (with circadian clock control). Adapted from Falcón *et al.*, 2007.

variations than AANAT transcripts abundance, when the latter are observed only, suggesting a similar intracellular mechanism regulates expression of *Tpoh* and *Aanat2* (Bégay *et al.*, 1998). In the case AANAT2 transcripts abundance is regulated, transcription inhibitors prevent the nocturnal increase in AANAT2 activity (Falcón *et al.*, 1998), while they are without effect if no variation in AANAT2 mRNA abundance is seen (Falcón *et al.*, 1998). In this latter situation, melatonin release may however be inhibited suggesting another component of the melatonin biosynthesis pathway is transcriptionnally regulated (Misuzawa *et al.*, 2001). Despite the differences observed at the transcriptional level, a constant is that night results in *de novo* synthesis of AANAT protein, which does not occur in the presence of traduction inhibitors. The increase in AANAT protein amount is paralleled by an increase in activity and a rise in melatonin production; the three rhythms (AANAT protein amount, AANAT activity and melatonin production) can be superimposed. And, light induces AANAT protein degradation *via* the proteasome, with concomitant decreases in AANAT activity and melatonin secretion (Falcón *et al.*, 2001b).

Photoperiodic or circadian control of *Aanat2* expression

The pineal organs of salmonids and rabbitfish are simple light sensors with regard to the production of melatonin because *in vivo* (rabbitfish) or *in vitro* (salmonids), AANAT2 activity and/or melatonin release remain arrhythmic under constant conditions, low under constant light (LL), and high under constant darkness (DD). This is because in either case, AANAT2 transcript levels are constitutively expressed, as is the case in LD (Fig. 2.4). Thus, the variations in melatonin secretion rely exclusively on the stability of AANAT2, higher in the dark than during day (Falcón *et al.*, 2007). In other fish, *Aanat2* expression is driven by a circadian clock system, which free runs under constant LL or DD (Falcón, 1999; Falcón *et al.*, 2007; Martinez-Chavez *et al.*, 2009). Thus, under LL or DD, AANAT2 mRNA abundance (as well as TpOH mRNA abundance) continues to be rhythmic with a period that approximates 24 h (Fig. 2.4); so is the case with AANAT2 protein amount and activity but only under DD, because AANAT2 protein is degraded under LL. Studies in the pike have shown that one pineal organ is made of multi-oscillatory cellular units, each located in a photoreceptor cell (Bolliet *et al.*, 1997). In culture, the shape of the circadian oscillations depends very much on the cell density in the wells, because the endogenous melatonin rhythm dampens faster at low or high cell densities, than at intermediate densities (Bolliet *et al.*, 1997). The damping observed at low or high densities could result from uncoupling between the cellular circadian units. Uncoupling could be due to lack of cellular communication at low density, or accumulation of some uncoupling agent at high density. Whatever it may be, the photoreceptor constitutes a cellular circadian system by itself because it contains the core of the clock machinery as well as the input to the clock, the photosensitive unit, and the output of the clock, the melatonin producing unit.

Extensive discussion on how the circadian time-keeping mechanism operates and is entrained by light is available elsewhere (Cahill, 2002; Reppert and Weaver, 2002; Korf *et al.*, 2003; Sato *et al.*, 2004). Briefly, the circadian clock machinery is based on a molecular feed-back loop consisting of two heterodimers, PER/CRY acting as repressors, and BMAL/CLOCK acting as activators, and additional interlocking loops. The situation is complex in fish because several copies of each actor may exist; thus, in zebrafish four *Per*, six *Cry*, three *Clock* and three *Bmal* genes have been cloned (Cahill, 2002; Hirayama *et al.*, 2003; Vallone *et al.*, 2004). If the clock mechanisms seem very similar between fish and mammals (Cahill, 2002), it is not yet known if all the clock genes operate in the same cell or whether there is a tissue-specific expression of the different homolog genes expressed (as is the case for the AANATs for example).

The CLOCK-BMAL heterodimer could be the link through which *Aanat2* expression is driven because its promoter contains several consensus binding sites for the CLOCK-BMAL heterodimer (E-box responsive elements). In zebrafish, *zfAanat2* expression is enhanced as a result of the synergistic action of the BMAL/CLOCK dimer and the photoreceptor specific homeogene OTX5; the dimer binds one E-box element and the OTX5 binds 3 photoreceptor conserved elements (PCE) all four located in the 3'-end of the *zfAanat2*, the so called PRDM (photoreceptor restrictive downstream modulator) (Appelbaum *et al.*, 2004, 2005; Appelbaum and Gothilf, 2006; Zilberman-Peled *et al.*, 2007). *Aanat2* also contains REV-ERBα/RORα responsive elements, which are part of the interlocking loops of the clock. Thus Aanat2 would appear as a direct target of the circadian clock. Light resets the circadian clock, and thus synchronizes the rhythm to the 24 h cycle, through morning induction of zfPER2, the only clock gene the expression of which remains constant under DD (Ziv *et al.*, 2005). The light-induced expression of *zfPer2* determines the phase of the oscillations (Vuilleumier *et al.*, 2006). Indeed, in zebrafish embryos maintained under DD, light pulses applied at the same circadian time induce the same phase shift in *zAanat2* expression, independent of their duration; conversely, under LL, dark pulses of different durations induce different phase shifts, even if they are applied at the same circadian time. This indicates the phase of the rhythm is determined by the time at which the dark-to-light transition occurs, i.e., when *zPer2* is induced. The sub-sequent later induction of the CLOCK/ BMAL heterodimer would initiate transcription. The accumulation of AANAT mRNA late in the afternoon and early at night makes AANAT production possible as soon as night starts. Morning light resets the clock (Ziv *et al.*, 2005) and inhibits AANAT activity and melatonin secretion. The presence of such a circadian clock allows the system to anticipate changes in the LD conditions (Falcón, 1999; Falcón *et al.,* 2007b; Martinez-Chavez *et al.,* 2008).

Another pathway through which the clock may modulate *Aanat2* expression might be the cAMP/PKA/CREB pathway. Indeed, as shown in the pike, pineal cAMP content and extrusion are under circadian control, and studies in the chicken have demonstrated a causal link between the circadian variations in cAMP content and AANAT activity (Ivanova and Iuvone, 2003). In view of the broad range of functions under cAMP control, one can assume that the clock may also regulate a number of other metabolic pathways in photoreceptor cells.

Photoperiodic control of AANAT2 protein amount and activity

In all species, whether there is or not a circadian clock, light acts directly on AANAT2 protein amount. In fish, it is believed that both Ca^{2+} and cAMP intracellular concentrations, which are controlled by light, contribute to the intracellular regulation of AANAT activity and melatonin secretion (Falcón, 1999; Falcón *et al.,* 2007). As mentioned above, the phototransduction process is similar in the retinal and pineal photoreceptors. The photoreceptors are depolarized in the dark because of an intracellular accumulation of cGMP, which allows entry of cations (Na^+, Ca^{2+}) through the cGMP-gated channel; intracellular Ca^{2+} accumulation is further enhanced by opening of the voltage-gated L-type Ca^{2+} channel at the depolarized state (Bégay *et al.,* 1994a; Meissl *et al.,* 1996; Kroeber *et al.,* 2000). Evidence indicates this nocturnal entry of Ca^{2+} contributes to increasing melatonin secretion both directly (through Ca^{2+}-binding proteins) and indirectly through stimulation of cAMP production (Bégay *et al.,* 1994a,b). Cyclic AMP induces AANAT2 protein accumulation and AANAT2 activity increase (Falcón *et al.,* 1992a; Thibault *et al.,* 1993a,b; Kroeber *et al.,* 2000; Falcón *et al.,* 2001). Illumination results in the opposite effects, i.e., cGMP degradation, closure of the cGMP-gated channel, photoreceptor hyperpolarization, and closure of the voltage-gated L-type Ca^{2+} channel, which eventually results in decreasing intracellular Ca^{2+} and cAMP contents (Falcón and Gaildrat, 1997; Falcón *et al.,* 1990, 1992). In mammals, a decrease in cAMP induces AANAT dephosphorylation and dissociation from the 14-3-3, a chaperone protein, which eventually results in AANAT protein degradation through the proteasome (Klein *et al.,* 1997; Ganguly *et al.,* 2002). A similar situation probably operates in the fish pineal organ because AANAT2 possesses the required phosphorylation sites; and, inhibitors of the proteasome prevent the light-induced degradation of AANAT2 protein and AANAT2 activity (Falcón *et al.,* 2001).

NON PHOTIC REGULATION OF ARYLALKYLAMINE N-ACETYLTRANSFERASE (AANAT)

Temperature

Most fish are ectotherms and as such directly influenced by the external temperature, which fluctuates on a daily and seasonal basis. Temperature acts directly on the pineal organ because in cultured isolated organs, cAMP accumulation, AANAT2 activity and melatonin release depend on ambient temperature (Zachmann *et al.*, 1992; Falcón *et al.*, 1994, 1996; Coon *et al.*, 1999; Falcón, 1999; Benyassi *et al.*, 2000). Interestingly, there is a good correlation between the peak of activity and the fish optimal physiological temperature. The mechanisms that operate are not known. They might involve specific membrane sensors (Seebacher and Murray, 2007; Myers *et al.*, 2009). Another possible mechanism includes the AANAT2 enzyme protein itself; indeed, studies on recombinant enzymes have shown distinct responses of the AANAT2 expressed proteins to temperature (trout, *Oncorhyncus mykiss*: 12°C, pike, *Esox lucius*: 25°C, seabream, *Sparus aurata*: 27°C; zebrafish, *Danio rerio*: 30°C) (Falcón *et al.*, 1996; Coon *et al.*, 1999; Benyassi *et al.*, 2000; Zilberman-Peled *et al.*, 2004). This contrasts with the AANAT1 family of enzymes, which respond similarly to temperature, showing all maximum activity at 37°C. It is possible that species specific amino-acid components in the AANAT2 sequence, that affect protein 3-D environment and properties (accessibility of the catalytic site, flexibility, hydrophobicity, etc …), are crucial in determining the response to temperature. Temperature does not seem to have an impact on the period of the melatonin oscillations (Falcón *et al.*, 1994). It is believed that photoperiod determines the duration of the melatonin signal, whereas temperature determines its amplitude; this provides accurate definitions of both the daily and annual cycles. Any change in temperature, related to husbandry conditions or global warming, may thus have dramatic consequences on the time-keeping system of fish.

Internal neurotransmitters, neuromodulators and hormones

The role played by internal factors in the control of melatonin production has received little attention. The data available are limited to few fish species, so that no general rule can be extrapolated (review in Falcón *et al.*, 2007a). These factors include hormones, neuromodulators and neurotransmitters.

The hormones that modulate melatonin secretion are melatonin itself and steroids. Melatonin inhibits is own production as seen in cultured trout pineal organs *in vitro* (Yañez and Meissl, 1995). This explains that production by pineal photoreceptor cells in culture depends on the frequency of medium renewal or seeding density (Bégay *et al.*, 1992; Bolliet *et al.*, 1997). The functional significance of this effect and mechanisms of action remain unknown. Melatonin receptors have not been identified in the fish pineal, and the effects might result from an interference with some metabolic pathway (e.g., inhibition of HIOMT activity).

The steroids that modulate melatonin secretion include 17-β-estradiol and glucocorticoids (Bégay *et al.*, 1993, 1994c; Benyassi *et al.*, 2001). In trout, specific probes detected mRNA corresponding to the 17-β-estradiol receptor, and specific antibodies directed against trout glucocorticoid receptor labeled a single band of proteins in pineal extracts. In isolated and cultured pineal cells, physiological concentrations of 17-β-estradiol and glucocorticoids inhibit the nocturnal rise in melatonin secretion (Bégay *et al.*, 1993; 1994b; Benyassi *et al.*, 2001). These effects could involve modulation of *Aanat2* expression because several putative binding sequences for both 17-β-estradiol and glucocorticoid receptors are present in the *Aanat2* promoter. At least for the glucocorticoids, there is direct inhibition of AANAT2 activity (HIOMT activity is not affected).

Adenosine and GABA are two neuromodulators produced locally by the fish pineal organ (Ekström *et al.*, 1987a; Falcón *et al.*, 1988). Adenosine is a ubiquitous local neuromodulator produced in the cells or in the extracellular spaces. There are two potential sources: the first is the catabolism of extracellular ATP or extruded cAMP, *via* ecto-nucleotidases and 5'-nucleotidases; the second is the degradation of both cAMP and S-adenosylmethionine (SAM); SAM is the co-substrate of methyltransferases including HIOMT, the last enzyme in the melatonin pathway. Adenosine may cross the plasma membrane through specific transporters. In trout and pike pineal organ, local adenosine modulates AANAT2 activity and melatonin release through a cAMP dependent mechanism involving cell surface adenosine receptors of high (A_1) and low (A_2) affinity, which mediate inhibition and stimulation on cAMP production, respectively (Falcón *et al.*, 1991, 1992b). The control by adenosine offers a fine-tuning of cAMP levels, depending on the balance between the intra- and extra-cellular concentrations of the nucleoside. This balance might vary during the LD cycle, providing the pineal with a continuous feed-back modulation of melatonin secretion.

Two populations of intra-pineal neurons produce GABA in trout (Ekström *et al.*, 1987a): one is located in the rostral pineal end-vesicle, and presumably constitutes a population of inter-neurons; the other is found in the pineal stalk, and their axons are sent to the brain. GABA is also present in glial cells in the pineal end-vesicle. GABA reduces melatonin secretion through $GABA_A$ receptors (Meissl *et al.*, 1993) but benzodiazepines, which are active at the $GABA_A$ receptors, have opposite effects. They increase melatonin production in the mesopic and partly in the photopic range of illumination, without showing clear effects in the dark-adapted organ (Meissl *et al.*, 1994).

Finally, catecholamines are to be added to the spectrum as neurotransmitters affecting melatonin secretion. And, they deserves special attention; indeed, in mammals norepinephrine is the final link in the pathway that brings light information from the eyes to the pineal gland, through the circadian clocks of the SCN (Klein *et al.*, 1997): the nocturnal release of the neurotransmitter synergistically activates α1- and β1-adrenergic receptors to stimulate nocturnal melatonin secretion. An intermediate situation is seen in birds, which possess both a direct and an indirect photosensitivity, and where norepinephrine acting through α2-adrenergic receptors and light, concomitantly inhibit daytime melatonin secretion. Arguments favor the view that catecholamines impact on the pineal organ in some fish species. Dopamine inhibits AANAT2 activity in zebrafish and stimulates cAMP accumulation in trout pineal organ (Martin *et al.*, 1991; Cahill, 1997). In pike, low concentrations (10^{-8} M) of norepinephrine potentiate the nocturnal increase in AANAT2 activity, whereas high concentrations (10^{-6} M) are reversing this effect (Falcón *et al.*, 1991). Stimulation could be mediated by β-adrenergic receptors, while inhibition would be mediated by α-adrenergic receptors (α_1 and/or α_2). This latter point remains however questionable because the pharmacological tools used for the classification of receptors have been characterized in mammals and may not exhibit the same characteristics in fish (Johansson, 1984; Fabbri *et al.*, 1998).

Whatever it may be, the existence of a catecholaminergic control in the pineal gland of some fish indicates there is diversity in the strategies developed by teleost to control melatonin secretion. One could speculate that in some cases there is convergent evolution with tetrapods concerning the involvement of the retina (as mentioned above for the structure of photoreceptor cells): (i) Migaud *et al.* (2007) have recently shown that the light control of melatonin production relies either partially or totally on the eyes in some fish species. (ii) Pineal serotonin and melatonin contents oscillate on a circadian basis in trout maintained under DD, in marked contrasts with the situation observed *ex vivo* in culture (see above) (Ceinos *et al.*, 2008). This would favor the idea that an extra-pineal circadian clock is operating as is the case in mammals, e.g., *via* a retinal/brain pathway. (iii) The fish pineal organ receives innervations from the brain (Ekström *et al.*, 1994; Jiménez *et al.*, 1995; Yáñez and Anadón, 1998; Pombal *et al.*, 1999; Mandado *et al.*, 2001; Servili *et al.*, 2005), and some of the brain areas involved (ventral and dorsal thalamus, pretectal area, posterior *tuberculum*), also receive information from the retina. Whether catecholamines and other neurotransmitters are the final link in a retina/brain/pineal connection awaits further investigations in fish.

It is interesting that many of the compounds that affect melatonin secretion also modulate the nervous message. These include melatonin itself, which inhibits the impulse frequency of the pineal neurons through yet unidentified mechanisms (Meissl *et al.*, 1990), and GABA (or GABA agonists), which may inhibit or increase the spontaneous discharges of the pinealofugal neurons, depending on the state of light- or dark-adaptation (Meissl and Ekström, 1991); the effects of GABA are mediated through GABA$_A$ receptors.

CONCLUSIONS AND PERSPECTIVES

Fish have two photoreceptive organs: the retina and the pineal organ, which share many common properties. Many of the properties established for the retinal cones have been extended to the pineal photoreceptors, and conversely (Collin and Oksche, 1981; O'Brien and Klein, 1986; Falcón, 1999; Klein, 2004). Actually, the more we know on the structural and functional genes express in each of these organs, the more it appears they are similar but distinct. This is the case for genes encoding proteins of the phototransduction and of the melatonin biosynthesis pathway. The genes encoding AANAT give a remarkable example. The presence of several AANATs in fish is a unique situation among vertebrates. It has recently been suggested that whole genome duplication occurred close to the origin of the teleost fish lineage (Amores *et al.*, 1998; Hoegg *et al.*, 2004; Jaillon *et al.*, 2004). The observation that other vertebrates possess only one AANAT (related to fish retinal AANAT1) would suggest that this whole genome duplication, which occurred ~230 million years ago, is responsible for the appearance of AANAT2 in teleost fish. The presence of two AANAT1 in some but not all teleost could result from another genome duplication that occurred later during evolution of teleot fish (Coon and Klein, 2006). The search for AANAT in early chordates should help clarifying this point and determining when the melatonin signal appeared for the first time. One hypothesis suggests AANAT and HIOMT genes could have been acquired through horizontal gene transfer from bacteria to chordates (Iyer *et al.*, 2004). It is noteworthy that the genome of the urochordate *Ciona* does not contain any AANAT encoding gene, whereas the genome of the cephalochordate *amphioxus* expresses several AANATs but of a different subtype both structurally and functionally (Pavlicek *et al.*, in preparation).

The main function of the fish pineal organ is to integrate light information and elaborate messages that will impact on animals' physiology. The photoreceptor cells occupy a key position because they are at the interface between the environment and the organism providing rapid (time scale of seconds) and less rapid (time scale of several hours) responses to environmental light. The downstream nervous information transmitted to brain centers *via* the ganglion cells provides the appropriate response to rapid changes. The hormonal -melatonin- response is slower and rhythmic, and synchronized by the 24 h LD cycle. Fish have developed several strategies, which all result in a nocturnal surge in melatonin secretion. The photoreceptor cells involved in this production may or may not express a functional circadian clock, and may or may not receive input from the retina. More species need to be investigated in order to get a precise idea of the different existing schemes. It is interesting that when a clock is present AANAT2 appears as a direct target of the clock genes and as such can be used as a monitor of clock function (Vuilleumier *et al.*, 2006). With or without a clock AANAT2 is the light regulated target in the photoreceptor cells. The latter appear to integrate other information from the environment (temperature) or the body (hormones, neurotransmitters and neuromodulators). These factors do not seem to affect the period of the circadian clocks that drive (when present) the melatonin rhythm; but they shape the amplitude of the melatonin oscillations, though regulation of AANAT2 activity. The pineal photoreceptor must be viewed as a multi-effectors' cell, which activity is modulated by hormones and neurotransmitters acting through specific receptors. More enigmatic is the case of temperature. Studies in other ectotherms indicate membrane channels might be involved (Seebacher and Murray, 2007; Myers *et al.*, 2009). But even more, evidence suggests the structure of AANAT2 confers a specific response to temperature. Also, although some information accumulates questions remain regarding the pathways that link the phototransduction unit to the circadian clock unit, and these two units to the melatonin producing machinery. Another interesting question regards the tissue-specificity of gene expression. Studies in the zebrafish have shown that selective *Aanat2* pineal expression is determined by specific sequences located in the 5'- and 3'-flanking regions of the gene (Appelbaum *et al.*, 2004). This tissue-specific expression allows using the *Aanat* genes as markers of photoreceptor and circadian clock functions (Gothilf *et al.*, 1999). This might also be of interest for studies related to photoreceptor differentiation and development. As was the case in the past, comparative studies between the retina and pineal will benefit from each other, keeping in mind that photoreceptor cells of the pineal organ are only of the cone type. The observation that melatonin secretion and AANAT1 activity are higher during day, not night, in the fish retina (Besseau *et al.*, 2005) or that AANAT1 and HIOMT are expressed in discrete brain areas (authors' unpublished results) further enhances the interest in such comparative studies and opens new lines of investigations regarding the role of these enzymes in the central nervous system.

References

Amores, A., A. Force, Y.L. Yan, L. Joly, C. Amemiya, A. Fritz, R.K. Ho, J. Langeland, V. Prince, Y.L. Wang, M. Westerfield, M. Ekker and J.H. Postlethwait. 1998. Zebrafish hox clusters and vertebrate genome evolution. *Science* 282(5394): 1711–1714.

Appelbaum, L. and Y. Gothilf. 2006. Mechanism of pineal-specific gene expression: the role of E-box and photoreceptor conserved elements. *Molecular and Cell Endocrinology* 27(252): 27–33.

Appelbaum, L., A. Anzulovich, R. Baler and Y. Gothilf. 2005. Homeobox-clock protein interaction in zebrafish. A shared mechanism for pineal-specific and circadian gene expression. *Journal Biological Chemistry* 280(12): 11544–11551.

Appelbaum, L., R. Toyama, I.B. Dawid, D.C. Klein, R. Baler and Y. Gothilf. 2004. Zebrafish serotonin-N-acetyltransferase-2 gene regulation: pineal-restrictive downstream module contains a functional E-box and three photoreceptor conserved elements. *Molecular Endocrinology* 18: 1210–1221.

Bayarri, M.J., J.A. Madrid and F.J. Sanchez-Vazquez. 2002. Influence of light intensity, spectrum and orientation on sea bass plasma and ocular melatonin. *Journal of Pineal Research* 32: 34–40.

Bazan, N.G., W.C. Gordon and E.B. Rodriguez de Turco. 1992. Docosahexaenoic acid uptake and metabolism in photoreceptors: retinal conservation by an efficient retinal pigment epithelial cell-mediated recycling process. *Advances in Experimental Medicine and Biology* 318: 295–306.

Bégay, V., J. Falcón, C. Thibault, J.P. Ravault and J.P. Collin. 1992. Pineal photoreceptor cells: photoperiodic control of melatonin production after cell dissociation and culture. *Journal of Neuroendocrinology* 4: 337–345.

Bégay, V., Y. Valotaire, J.P. Ravault, J.P. Collin and J. Falcón. 1993. Photoreceptor cells of the pineal body in culture: effect of 17 beta-estradiol on the production of melatonin. *Comptes Rendus des Séances de la Société de Biologie* 187: 77–86.

Bégay, V., P. Bois, J.P. Collin, J. Lenfant and J. Falcón. 1994a. Calcium and melatonin production in dissociated trout pineal photoreceptor cells in culture. *Cell Calcium* 16: 37–46.

Bégay, V., J.P. Collin and J. Falcón. 1994b. Calciproteins regulate cyclic AMP content and melatonin secretion in trout pineal photoreceptors. *NeuroReport* 5: 2019–2022.

Bégay, V., Y. Valotaire, J.P. Ravault, J.P. Collin and J. Falcón. 1994c. Detection of estrogen receptor mRNA in trout pineal and retina: estradiol-17 beta modulates melatonin production by cultured pineal photoreceptor cells. *General and Comparative Endocrinology* 93: 61–69.

Bégay, V., J. Falcón, G.M. Cahill, D.C. Klein and S.L. Coon. 1998. Transcripts encoding two melatonin synthesis enzymes in the teleost pineal organ: circadian regulation in pike and zebrafish, but not in trout. *Endocrinology* 139: 905–912.

Bell, MV., R.S. Batty, J.R. Dick, K. Fretwell, J.C. Navarro and J.R. Sargent. 1995. Dietary deficiency of docosahexaenoic acid impairs vision at low light intensities in juvenile herring (*Clupea harengus* L.). *Lipids* 30: 443–449.

Benyassi, A., C. Schwartz, B. Ducouret and J. Falcón. 2001. Glucocorticoid receptors and serotonin N-acetyltransferase activity in the fish pineal organ. *NeuroReport* 12: 889–892.

Benyassi, A., C. Schwartz, S.L. Coon, D.C. Klein and J. Falcón. 2000. Melatonin synthesis: arylalkylamine N-acetyltransferases in trout retina and pineal organ are different. *NeuroReport* 11: 255–258.

Besseau, L., A. Benyassi, M. Møller, S.L. Coon, J.L. Weller, G. Boeuf, D.C. Klein and J. Falcón. 2006. Melatonin Pathway: Breaking the "High-at-Night" Rule in Trout Retina. *Experimental Eye Research* 82: 620–627.

Boesze-Battaglia, K. and C. Allen. 1998. Differential rhodopsin regeneration in photoreceptor membranes is correlated with variations in membrane properties. *Bioscience Reports* 18: 29–38.

Boeuf, G. and J. Falcón. 2001. Photoperiod and growth in fish. *Vie et Milieu* 51: 237–246.

Bolliet, V., J. Falcón and M.A. Ali. 1995. Regulation of melatonin secretion by light in the isolated pineal organ of the white sucker (*Catostomus commersoni*). *Journal of Neuroendocrinology* 7: 535–542.

Bolliet, V., V. Bégay, C. Taragnat, J.P. Ravault, J.P. Collin and J. Falcón. 1997. Photoreceptor cells of the pike pineal organ as cellular circadian oscillators. *European Journal of Neuroscience* 9: 643–653.

Brandstätter, R. and A. Hermann. 1996. Gamma-Aminobutyric acid enhances the light response of ganglion cells in the trout pineal organ. *Neuroscience Letters* 210 : 173–176.

Brandstätter, R., E. Fait and A. Hermann. 1995. Acetylcholine modulates ganglion cell activity in the trout pineal organ. *NeuroReport* 6: 1553–1556.

Brown, M.F. 1994. Modulation of rhodopsin function by properties of the membrane bilayer. *Chemistry and Physics of Lipids* 73: 159–180.

Cahill, G.M. 1997. Circadian melatonin rhythms in cultured zebrafish pineals are not affected by catecholamine receptor agonists. *General and Comparative Endocrinology* 105: 270–275.

Cahill, G.M. 2002. Clock mechanisms in zebrafish. *Cell and Tissue Research* 309: 27–34.

Ceinos, R.M., S. Polakof, A.R. Illamola, J.L. Soengas and J.M. Míguez. 2008. Food deprivation and refeeding effects on pineal indoles metabolism and melatonin synthesis in the rainbow trout *Oncorhynchus mykiss*. *General and Comparative Endocrinology* 156(2): 410–417.

Chen, H. and R.E. Anderson. 1993. Comparison of uptake and incorporation of docosahexaenoic and arachidonic acids by frog retinas. *Current Eye Research* 12: 851–860.

Collin, J.P. and A. Oksche. 1981. Structural and functional relationships in the nonmammalian pineal organ. In: *The Pineal Gland*. Vol. I. *Anatomy and Biochemistry*. R.J. Reiter (ed.). CRC Press, Boca Raton. pp. 27–67.

Coon, S.L. and D.C. Klein. 2006. Evolution of arylalkylamine *n*-acetyltransferase: emergence and divergence. *Molecular and Cellular Endocrinology* 252: 2–10.

Coon, S.L., V. Bégay, D. Deurloo, J. Falcón and D.C. Klein. 1999. Two arylalkylamine N-acetyltransferase genes mediate melatonin synthesis in fish. *Journal of Biological Chemistry* 274: 9076–9082.

Dodt, E. 1963. Photosensitivity of the pineal organ in the teleost, *Salmo irideus* (Gibbons). *Experientia* 19: 642–643.

Dodt, E. and E. Heerd.1962. Mode of action of pineal nerve fibers in frogs. *Journal of Neurophysiology* 25: 405–429.

Ekström, P. 1994. Developmental changes in the brain-stem serotonergic nuclei of teleost fish and neural plasticity. *Cellular and Molecular Neurobiology* 14(4): 381–393.

Ekström, P. and H. Meissl. 1989. Signal processing in a simple vertebrate photoreceptor system: the teleost pineal organ. *Physiology Bohemoslov* 38: 311–326.

Ekström, P. and H. Meissl. 1997. The pineal organ of teleost fishes. *Reviews in Fish Biology and Fisheries* 7: 284.

Ekström, P. and H. Meissl. 2003. Evolution of photosensory pineal organs in new light: the fate of neuroendocrine photoreceptors. *Philosophical Transactions of the Royal Society London B* 358: 1679–1700.

Ekström, P., T. Östholm and B.I. Holmqvist. 1994. Primary visual projections and pineal neural connections in fishes, amphibians and reptiles. *Advances in Pineal Research* 8: 1–18.

Ekström, P., T. van Veen, A. Bruun and B. Ehinger. 1987. GABA-immunoreactive neurons in the photosensory pineal organ of the rainbow trout: two distinct neuronal populations. *Cell and Tissue Research* 250: 87–92.

Fabbri, E., A. Capuzzo and T.W. Moon. 1998. The role of circulating catecholamines in the regulation of fish metabolism: an overview. *Comparative Biochemistry and Physiology C: Pharmacology, Toxicology and Endocrinology* 120: 177–192.

Falcón, J. 1979. L'organe pinéal du Brochet (*Esox lucius* L.) I. Etude anatomique et cytologique. *Reproduction Nutrition Development* 19: 445–465.

Falcón, J. 1999. Cellular circadian clocks in the pineal. *Progress in Neurobiology* 58: 121–162.

Falcón, J. and H. Meissl. 1981. The photosensory function of the pineal organ of the pike (*Esox lucius*, L.). Correlation between structure and function. *Journal of Comparative Physiology* 144: 127–137.

Falcón, J. and J. Tanabe. 1983. Early receptor potential of pineal organ and lateral eye of the pike. *Naturwissenschaften* 70: 149–150.

Falcón, J. and P. Gaildrat. 1997. Variations in cyclic adenosine 3',5'-monophosphate and cyclic guanosine 3',5'-monophosphate content and efflux from the photosensitive pineal organ of the pike in culture. *Pflügers Archiv European Journal of Physiology* 433: 336–342.

Falcón, J. and R.J. Henderson. 2001. Incorporation, distribution, and metabolism of polyunsaturated fatty acids in the pineal gland of rainbow trout (*Oncorhynchus mykiss*) *in vitro*. *Journal of Pineal Research* 31: 127–137.

Falcón, J., V. Bolliet and J.P. Collin. 1996. Partial characterization of serotonin N-acetyltransferases from northern pike (*Esox lucius* L.) pineal organ and retina: effects of temperature. *Pflügers Archiv European Journal of Physiology* 432: 386–393.

Falcón, J., L. Besseau and G. Boeuf. 2007. Molecular and cellular regulation of pineal organ responses. In: *Sensory Systems Neuroscience—Fish Physiology*, T. Hara and B. Zielinski (eds.). Academic Press, N.Y., Elsevier, Amsterdam, pp. 203–406.

Falcón, J., M.G. Balemans, J. van Benthem and J.P Collin. 1985. In vitro uptake and metabolism of [^{3}H]indole compounds in the pineal organ of the pike. I. A radiochromatographic study. *Journal of Pineal Research* 2: 341–356.

Falcón, J., C. Besse, J. Guerlotte and J.P. Collin. 1988. 5'-Nucleotidase activity in the pineal organ of the pike. An electron-microscopic study. *Cell and Tissue Research* 251: 495–502.

Falcón, J., V. Bégay, C. Besse, J.P. Ravault and J.P. Collin. 1992a. Pineal photoreceptor cells in culture: fine structure and light control of cyclic nucleotide levels. *Journal of Neuroendocrinology* 4: 641–651.

Falcón, J., C. Thibault, V. Bégay, A. Zachmann and J.P. Collin. 1992b. Regulation of the rhythmic melatonin secretion by the fish pineal photoreceptor cells. Plenum press, New York.

Falcón, J., S. Barraud, C. Thibault and V. Begay. 1998. Inhibitors of messenger RNA and protein synthesis affect differently serotonin arylalkylamine N-acetyltransferase activity in clock-controlled and non-clock-controlled fish pineal. *Brain Research* 797: 109–117.

Falcón, J., C. Thibault, J.L. Blazquez, H. Vaudry, N. Lin and J.P. Collin. 1990. Atrial natriuretic factor increases cyclic GMP and cyclic AMP levels in a directly photosensitive fish pineal organ. *Pflügers Archiv European Journal of Physiology* 417: 243–245.

Falcón, J., C. Thibault, C. Martin, J. Brun-Marmillon, B. Claustrat and J.P. Collin. 1991. Regulation of melatonin production by catecholamines and adenosine in a photoreceptive pineal organ. An *in vitro* study in the pike and the trout. *Journal of Pineal Research* 11: 123–134.

Falcón, J., V. Bolliet, J.P. Ravault, D. Chesneau, M.A. Ali and J.P. Collin. 1994. Rhythmic secretion of melatonin by the superfused pike pineal organ: thermo- and photoperiod interaction. *Neuroendocrinology* 60: 535–543.

Falcón, J., L. Besseau, M. Fuentès, S. Sauzet, E. Magnanou and G. Boeuf. 2009. Structural and functional evolution of the pineal melatonin system in vertebrates. *Annals of the New York Academy of Sciences* 1163: 101–111.

Falcón, J., K.M. Galarneau, J.L. Weller, B. Ron, G. Chen, S.L. Coon and D.C. Klein. 2001. Regulation of arylalkylamine N-acetyltransferase-2 (AANAT2, EC 2.3.1.87) in the fish pineal organ: evidence for a role of proteasomal proteolysis. *Endocrinology* 142: 1804–1813.

Ganguly, S., S.L. Coon and D.C. Klein. 2002. Control of melatonin synthesis in the mammalian pineal gland: the critical role of serotonin acetylation. *Cell and Tissue Research* 309: 127–137.

Garcia, M. and E. Vecino. 2003. Role of Muller glia in neuroprotection and regeneration in the retina. *Histology and Histopathology* 18: 1205–1218.

Garcia, M., V. Forster, D. Hicks and E. Vecino. 2002. Effects of muller glia on cell survival and neuritogenesis in adult porcine retina *in vitro*. *Investigative Ophthalmology and Visual Science* 43: 3735–3743.

Gothilf, Y., S.L. Coon, R. Toyama, A. Chitnis, M.A. Namboodiri and D.C. Klein. 1999. Zebrafish serotonin N-acetyltransferase-2: marker for development of pineal photoreceptors and circadian clock function. *Endocrinology* 140: 4895–4903.

Grace, M.S., G.M. Cahill and J.C. Besharse. 1991. Melatonin deacetylation: retinal vertebrate class distribution and *Xenopus laevis* tissue distribution. *Brain Research* 559: 56–63.

Harada, C., T. Harada, H.M. Quah, F. Maekawa, K. Yoshida, S. Ohno, K. Wada, L.F. Parada and K. Tanaka. 2003. Potential role of glial cell line-derived neurotrophic factor receptors in Muller glial cells during light-induced retinal degeneration. *Neuroscience* 122: 229–235.

Henderson, R.J., M.V. Bell, M.T. Park , J.R. Sargent and J. Falcón. 1994. Lipid composition of the pineal organ from rainbow trout (*Oncorhynchus mykiss*). *Lipids* 29: 311–317.

Hirayama, J., I. Fukuda, T. Ishikawa, Y. Kobayashi and T. Todo. 2003. New role of zCRY and zPER2 as regulators of sub-cellular distributions of zCLOCK and zBMAL proteins. *Nucleic Acids Research* 31: 935–943.

Hoegg, S., H. Brinkmann, J.S. Taylor and A. Meyer. 2004. Phylogenetic timing of the fish-specific genome duplication correlates with the diversification of Teleost fish. *Journal of Molecular Evolution* 59: 190–203.

Isorna, E., L. Besseau, G. Boeuf, Y. Desdevises, R. Vuilleumier, A.L. Alonso-Gómez, M.J. Delgado and J. Falcón. 2006. Retinal, pineal and diencephalic expression of frog arylalkylamine N-acetyltransferase-1. *Molecular and Cellular Endocrinology* 252: 11–18.

Ivanova, T.N. and P.M. Iuvone. 2003. Circadian rhythm and photic control of cAMP level in chick retinal cell cultures: a mechanism for coupling the circadian oscillator to the melatonin-synthesizing enzyme, arylalkylamine N-acetyltransferase, in photoreceptor cells. *Brain Research* 991: 96–103.

Iyer, L.M., L. Aravind, S.L. Coon, D.C. Klein and E.V. Koonin. 2004. Evolution of cell-cell signaling in animals: did late horizontal gene transfer from bacteria have a role? *Trends in Genetics* 20: 292–299.

Jaillon, O., J.M. Aury, F. Brunet, J.L. Petit, N. Stange-Thomann, E. Mauceli, L. Bouneau, C. Fischer, C. Ozouf-Costaz, A. Bernot, S. Nicaud, D. Jaffe, S. Fisher, G. Lutfalla, C. Dossat, B. Segurens, C. Dasilva, M. Salanoubat, M. Levy, N. Boudet, S. Castellano, V. Anthouard, C. Jubin, V. Castelli, M. Katinka, B. Vacherie, C. Biemont, Z. Skalli, L. Cattolico, J. Poulain, V. De Berardinis, C. Cruaud, S. Duprat, P. Brottier, J.P. Coutanceau, J. Gouzy, G. Parra, G. Lardier, C. Chapple, K.J. McKernan, P. McEwan,

S. Bosak, M. Kellis, J.N. Volff, R. Guigo, M.C. Zody, J. Mesirov, K. Lindblad-Toh, B. Birren, C. Nusbaum, D. Kahn, M. Robinson-Rechavi, V. Laudet, V. Schachter, F. Quetier, W. Saurin, C. Scarpelli, P. Wincker, E.S. Lander, J. Weissenbach and H. Roest Crollius. 2004. Genome duplication in the teleost fish *Tetraodon nigroviridis* reveals the early vertebrate proto-karyotype. *Nature* 431: 946–957.

Jimenez, A.J., P. Fernandez-Llebrez and J.M. Perez-Figares. 1995. Central projections from the goldfish pineal organ traced by HRP-immunocytochemistry. *Histology Histopathology* 10: 847–852.

Johansson, P. 1984. Alpha-adrenoceptors: recent development and some comparative aspects. *Comparative Biochemistry and Physiology—Part C: Toxicology & Pharmacology* 78: 253–261.

Kezuka, H., K. Furukawa, K. Aida and I. Hanyu. 1988. Daily cycles in plasma melatonin levels under long or short photoperiod in the common carp, *Cyprinus carpio*. *General and Comparative Endocrinology* 72: 296–302.

Klein, D.C. 2004. The 2004 Aschoff/Pittendrigh lecture: Theory of the origin of the pineal gland—a tale of conflict and resolution. *Journal of Biological Rhythms* 19: 264–279.

Klein, D.C., D.A. Auerbach, M.A. Namboodiri and G.H.T. Wheler. 1981. Indole metabolism in the mammalian pineal gland, In: *The Pineal Gland*, R.J. Reiter (ed.). CRC Press, Boca Raton, pp. 199–227.

Klein, DC., S.L. Coon, P.H. Roseboom, J.L. Weller, M. Bernard, J.A. Gastel, M. Zatz, P.M. Iuvone, I.R. Rodriguez, V. Bégay, J. Falcón, G.M. Cahill, V.M. Cassone and R. Baler. 1997. The melatonin rhythm-generating enzyme: molecular regulation of serotonin N-acetyltransferase in the pineal gland. *Recent Progress in Hormone Research* 52: 307–357.

Korf, H.W., C. Von Gall and J. Stehle. 2003. The circadian system and melatonin: lessons from rats and mice. *Chronobiology International* 20: 697–710.

Kroeber, S., H. Meissl, E. Maronde and H.W. Korf. 2000. Analyses of signal transduction cascades reveal an essential role of calcium ions for regulation of melatonin biosynthesis in the light-sensitive pineal organ of the rainbow trout (*Oncorhynchus mykiss*). *Journal of Neurochemistry* 74: 2478–2489.

Mandado, M., P. Molist, R. Anadón and J. Yáñez . 2001. A DiI-tracing study of the neural connections of the pineal organ in two elasmobranchs (*Scyliorhinus caniculo* and *Raja montagui*) suggests a pineal projection to the midbrain GnRH-immunoreactive nucleus. *Cell and Tissue Research* 303: 391–401.

Martin, C., J. Falcón and J.P. Collin. 1991. Catecholamines regulate cAMP levels in the photosensitive trout pineal organ. *Advances in Pineal Research* 5: 137–140.

Martinez-Chavez, C.C. and H. Migaud. 2009. Retinal light input is required to sustain plasma melatonin rhythms in Nile tilapia *Oreochromis niloticus niloticus*. *Brain Research* 1269: 61–67.

Martinez-Chavez, C.C., S. Al-Khamees, A. Campos-Mendoza, D.J. Penman and H. Migaud. 2008. Clock-controlled endogenous melatonin rhythms in Nile tilapia (*Oreochromis niloticus niloticus*) and African catfish (*Clarias gariepinus*). *Chronobiology International* 25: 31–49.

Masuda, T., M. Iigo, K. Mizusawa, M. Naruse, T. Oishi, K. Aida and M. Tabata. 2003a Variations in plasma melatonin levels of the rainbow trout (*Oncorhynchus mykiss*) under various light and temperature conditions. *Zoological Science* 20: 1011–1016.

Max, M. and M. Menaker. 1992. Regulation of melatonin production by light, darkness, and temperature in the trout pineal. *Journal of Comparative Physiology* A 170: 479–489.

Meissl, H. and E. Dodt. 1981. Comparative physiology of pineal photoreceptor organs. In: *The Pineal Organ: Photobiology—Biochronometry—Endocrinology*, A. Oksche and P. Pévet (eds). Elsevier, Amsterdam, pp. 61–80.

Meissl, H. and S.R. George. 1984. Electrophysiological studies on neuronal transmission in the frog's photosensory pineal organ. The effect of amino acids and biogenic amines. *Vision Research* 24: 1727–1734.

Meissl, H. and P. Ekström. 1991. Action of gamma-aminobutyric acid (GABA) in the isolated photosensory pineal organ. *Brain Research* 562: 71–78.

Meissl, H., C. Martin and M. Tabata. 1990. Melatonin modulates the neural activity in photosensory pineal organ of the trout: evidence for endocrine-neuronal interactions. *Journal of Comparative Physiology* A 167: 641–648.

Meissl, H., M. Anzelius, T. Östholm and P. Ekström. 1993. Interaction of GABA, benzodiazepines and melatonin in the photosensory pineal organ of salmonid fish. In: *Melatonin and the Pineal Gland: From Basic Science to Clinical Applications.* Y. Touitou, J. Arendt and P. Pévet (eds), Elsevier, Amsterdam, pp. 95–98.

Meissl, H., J. Yanez, P. Ekström and E. Grossmann. 1994. Benzodiazepines influence melatonin secretion of the pineal organ of the trout *in vitro. Journal of Pineal Research* 17: 69–78.

Meissl, H., S. Kroeber, J. Yanez and H.W. Korf. 1996. Regulation of melatonin production and intracellular calcium concentrations in the trout pineal organ. *Cell and Tissue Research* 286: 315–323.

Myers, B.R., Y.M. Sigal and D. Julius. 2009. Evolution of Thermal Response Properties in a Cold-Activated TRP Channel. *PloS ONE* 4: e5741.

Migaud, H., A. Davie, C.C. Martinez Chavez and S. Al-Khamees. 2007 Evidence for differential photic regulation of pineal melatonin synthesis in teleosts. *Journal of Pineal Research* 43: 327–335.

Mizusawa, K., M. Iigo, T. Masuda and K. Aida. 2001. Inhibition of RNA synthesis differentially affects in vitro melatonin release from the pineal organs of ayu (*Plecoglossus altivelis*) and rainbow trout (*Oncorhynchus mykiss*). *Neuroscience Letters* 309: 72–76.

Morton, D.J. and H.J. Forbes. 1988. Pineal gland N-acetyltransferase and hydroxyindole-O- methyltransferase activity in the rainbow trout (*Salmo gairdneri*): seasonal variation linked to photoperiod. *Neuroscience Letters* 94: 333–337.

O'Brien, P.J. and D.C. Klein. 1986. *Pineal and retinal relationships.* Academic Press, Orlando, FL.

Omura, Y., H.W. Korf and A. Oksche. 1985. Vascular permeability (problem of the blood-brain barrier) in the pineal organ of the rainbow trout, *Salmo gairdneri. Cell and Tissue Research* 239: 599–610.

Pombal, M.A., J. Yáñez, O. Marín, A. González and R. Anadón. 1999. Cholinergic and GABAergic neuronal elements in the pineal organ of lampreys, and tract-tracing observations of differential connections of pinealofugal neurons. *Cell and Tissue Research* 295: 215–223.

Rao, S.D., P.D. Rao and R.E. Peter. 1996. Growth hormone-releasing hormone immunoreactivity in the brain, pituitary, and pineal of the goldfish, *Carassius auratus. General and Comparative Endocrinology* 102: 210–220.

Reppert, S.M. and D.R. Weaver. 2002. Coordination of circadian timing in mammals. *Nature* 418(6901): 935–941.

Sato, T.K., S. Panda, L.J. Miraglia, T.M. Reyes, R.D. Rudic, P. McNamara, K.A. Naik, G.A. FitzGerald, S.A. Kay and J.B. Hogenesch. 2004. A functional genomics strategy reveals Rora as a component of the mammalian circadian clock. *Neuron* 43: 527–537.

Schmitz, Y. and P. Witkovsky. 1996. Glutamate release by the intact light-responsive photoreceptor layer of the *Xenopus* retina. *Journal of Neuroscience Methods* 68: 55–60.

Schmitz, Y. and P. Witkovsky. 1997. Dependence of photoreceptor glutamate release on a dihydropyridine-sensitive calcium channel. *Neuroscience* 78: 1209–1216.

Seebacher, F. and S.A. Murray. 2007. Transient receptor potential ion channels control thermoregulatory behaviour in reptiles. *PLoS ONE* 2: e281.

Simonneaux, V. and C. Ribelayga. 2003. Generation of the melatonin endocrine message in mammals: a review of the complex regulation of melatonin synthesis by norepinephrine, peptides, and other pineal transmitters. *Pharmacological Reviews* 55: 325–395.

Solessio, E. and G.A. Engbretson. 1993. Antagonistic chromatic mechanisms in photoreceptors of the parietal eye of lizards. *Nature (London)* 364: 442–445.

Subhedar, N., J. Cerda and R.A. Wallace. 1996. Neuropeptide Y in the forebrain and retina of the killifish, *Fundulus heteroclitus. Cell and Tissue Research* 283: 313–323.

Taylor, J.S. and J. Raes. 2004. Duplication and divergence: the evolution of new genes and old ideas. *Annual Review of Genetics* 38: 615–643.

Thibault, C., J.P. Collin and J. Falcón. 1993a. Intrapineal circadian oscillator(s), cyclic nucleotides and melatnin production in pike pineal photoreceptor cells. In: *Melatonin and the Pineal Gland: From Basic Science to Clinical Application,* Y. Touitou (ed.). Elsevier, Amsterdam, pp. 11–18.

Thibault, C., J. Falcón, S.S. Greenhouse, C.A. Lowery, W.A. Gern and J.P. Collin. 1993b. Regulation of melatonin production by pineal photoreceptor cells: role of cyclic nucleotides in the trout (*Oncorhynchus mykiss*). *Journal of Neurochemistry* 61: 332–339.

Vallone, D., S.B. Gondi, D. Whitmore and N.S. Foulkes. 2004. E-box function in a period gene repressed by light. *Proceedings of the National Academy of Sciences USA* 101: 4106–4111.

Yañez, J. and H. Meissl. 1995. Secretion of methoxyindoles from trout pineal organs in vitro: indication for a paracrine melatonin feedback. *Neurochemistry International* 27: 195–200.

Yañez, J. and R. Añadon. 1996. Afferent and efferent connections of the habenula in the rainbow trout (*Oncorhynchus mykiss*): an indocarbocyanine dye (DiI) study. *Journal of Comparative Neurology* 372: 529–543.

Zachmann, A., M.A. Ali and J. Falcón. 1992a. Melatonin and its effects in fishes: an overview. In: *Rhythms in Fishes,* M.A. Ali (ed.). Plenum Press, New York, pp. 149–165.

Zachmann, A., S.C.M. Knijff, M.A. Ali and M. Anctil. 1992b. Effects of photoperiod and different intensities of light exposure on melatonin levels in the blood, pineal organ and retina of the brook trout (*Salvelinus frontinalis* Mitchill). *Canadian Journal of Zoology* 70: 25–29.

Zachmann, A., J. Falcón, S.C. Knijff, V. Bolliet and M.A. Ali. 1992c. Effects of photoperiod and temperature on rhythmic melatonin secretion from the pineal organ of the white sucker (*Catostomus commersoni*) *in vitro. General and Comparative Endocrinology* 86: 26–33.

Zilberman-Peled, B., I. Benhar, S.L. Coon, B. Ron and Y. Gothilf. 2004. Duality of serotonin-N-acetyltransferase in the gilthead seabream (*Sparus aurata*): molecular cloning and characterization of recombinant enzymes. *General and Comparative Endocrinology* 138: 139–147.

Zilberman-Peled, B., L. Appelbaum, D. Vallone, N.S. Foulkes, S. Anava, A. Anzulovich, S.L. Coon, D.C. Klein, J. Falcón, B. Ron and Y. Gothilf. 2007 Transcriptional regulation of arylalkylamine-N-acetyltransferase-2 gene in the pineal gland of the gilthead seabream. *Journal of Neuroendocrinology* 19: 46–53.

Ziv, L., S. Levkovitz, R. Toyama, J. Falcón and Y. Gothilf. 2005. Functional development of the zebrafish pineal gland: light-induced expression of period2 is required for onset of the circadian clock. *Journal of Neuroendocrinology* 17: 314–320.

3

PINEAL PHOTORECEPTION AND TEMPORAL PHYSIOLOGY IN FISH

Peter Ekström[1] and *Hilmar Meissl*[2]

STRUCTURE AND SPECIALIZATIONS OF THE PINEAL ORGAN—PINEAL PHOTORECEPTORS AND NEURONS

The pineal complex of non-mammalian vertebrates is a directly photoreceptive structure that contains photoreceptor cells similar to those of the retina. In lampreys and most bony fish the pineal complex has two components: a pineal organ and a parapineal organ. Elasmobranchs, as most other vertebrates, generally possess only a pineal organ (epiphysis cerebri). A few vertebrate taxa, including hagfishes and electric rays, have apparently lost both components during evolution (Oksche, 1965; Vollrath, 1981).

Lampreys possess the two components of the pineal complex: the pineal organ and the parapineal organ (Fig. 3.1). Both parts contain different types of photosensory cells and neurons that transmit information about the photic environment as perceived by the photoreceptor cells (Ekström and Meissl, 2003). The pineal complex of lampreys probably acts primarily as a luminance detector, although it is capable of some spectral discrimination. Photic information is relayed to the brain *via* a pineal tract. Melatonin biosynthesis is regulated by an endogenous intrapineal circadian oscillator, and by direct light perception (Vollrath, 1981).

[1]Department of Cell and Organism Biology, Lund University, Helgonavägen 3, S-22362 Lund, Sweden.
e-mail: peter.ekstrom@cob.lu.se
[2]Max Planck Institute for Brain Research, Deutschordenstr. 46, 60528 Frankfurt/Main, Germany.
e-mail: meissl@mpih-frankfurt.mpg.de

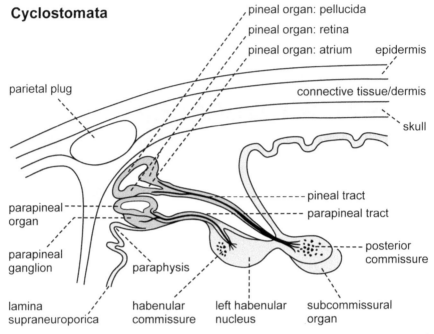

Cyclostomata

pineal organ: pellucida

pineal organ: retina

pineal organ: atrium   epidermis

parietal plug   connective tissue/dermis

skull

pineal tract

parapineal organ   parapineal tract

parapineal ganglion   posterior commissure

paraphysis

lamina supraneuroporica   habenular commissure   left habenular nucleus   subcommissural organ

Fig. 3.1 Diagrammatic representation of a near-midsagittal section through the pineal region of a lamprey. The elongated structure of the pineal organ consists of a thickened distal end-vesicle and a thin proximal pineal stalk that consists mainly of nerve fibers of the pineal tract. The pineal end-vesicle can be subdivided into a dorsal wall, the pellucida, and a ventral wall, the retina, and an atrium. The parapineal organ is located ventral to the pineal organ; its caudal portion is intimately attached to the parapineal ganglion.

Elasmobranchs possess only a pineal organ, which contains photoreceptor cells and neurons for transmission of photic information to the brain. The pineal organ of sharks is usually very long, and consists of a long stalk region that ends in a small end-vesicle (Vollrath, 1981). The end-vesicle may be attached to the frontal or prefrontal cartilages of the skull, or reside in a pineal foramen (Fig. 3.2). The pineal organ acts as a luminance detector, and photic information is relayed to the brain *via* a pineal tract (Hamasaki and Streck, 1971).

Most bony fish possess both a pineal organ and a parapineal organ (Fig. 3.3). When present, the parapineal organ is always very small and remains located close to the habenular nucleus, whereas the pineal organ is larger and of very variable morphology. The parapineal organ becomes laterally displaced during early development, usually to the left side (Ekström and Meissl, 1997).

The pineal organ comes in many morphological varieties in teleosts, ranging from simple tubular, saccular, and extensively folded saccular, to compact parenchymal structures (Omura and Oguri, 1969; Hafeez, 1971).

Elasmobranchii

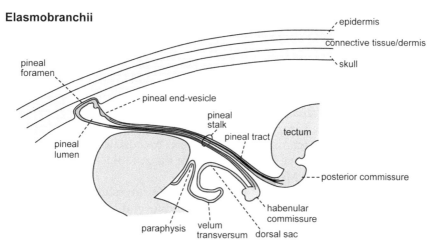

Fig. 3.2 Schematic drawing of the pineal region of an elasmobranch. The pineal organ often resides in a pineal foramen. It consists of a long tubular structure with a end-vesicle and a long pineal stalk with a narrow lumen. A parapineal is missing in elasmobranchs.

Actinopterygii

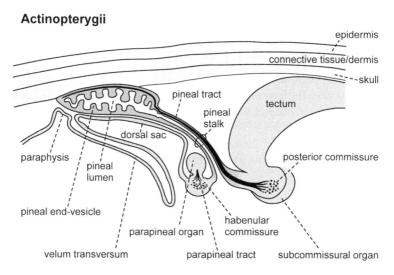

Fig. 3.3 Overview, in a near-midsagittal section, of the pineal region of a teleost. Most bony fish possess both a pineal organ and a parapineal organ. The parapineal organ becomes laterally displaced during early development, usually to the left side. The pineal organ comes in many morphological varieties in teleosts, ranging from simple tubular, saccular, and extensively folded saccular, to compact parenchymal. There may be a distinct subdivision between a pineal end-vesicle and a stalk division, but in some species there is no clear gross morphological differentiation.

There may be a distinct subdivision between a pineal end-vesicle and a stalk division, but in some species there is no clear gross morphological differentiation.

The pineal organ of bony fishes is clearly a photosensory organ, with photoreceptor cells and neurons that transmit photic information via the pineal tract to various brain centres. The parapineal organ, in contrast, consists mainly of small neurons that project to the habenular nucleus on the same side as the parapineal organ. However, a photoreceptive function of the parapineal organ seems to be likely, because a small number of photoreceptor cells have been observed in the parapineal organ of some species (Ekström and Meissl, 1997).

The pineal organ of most bony fish species investigated contains an endogenous circadian oscillator. Melatonin biosynthesis is controlled by direct light perception by pineal photoreceptors, and the circadian oscillator (Ekström and Meissl, 1997; Falcón, 1999). An interesting exception to this mechanism is found in salmonids, which lack an intrapineal oscillator. In these fishes, melatonin biosynthesis is directly controlled by the ambient light intensity (Gern and Greenhouse, 1988; Iigo *et al.*, 2007).

CELLULAR COMPONENTS OF THE PINEAL ORGAN

Three types of cells comprise the pineal parenchyma of lampreys, cartilaginous fish, and bony fish: photoreceptor cells, neurons, and supportive cells. The fine details of their morphology have been extensively reviewed previously (e.g., Oksche, 1965; Collin, 1969; Vollrath, 1981; Ekström and Meissl, 1997; Falcón, 1999) and it is beyond the scope of this review to give a detailed account of all morphological varieties. Here, we shall concentrate on the main features of the cellular components from a functional perspective.

PINEAL PHOTORECEPTORS

Photic stimuli perceived by pineal photoreceptors are transduced into either a neural signal (graded potentials) or a neuroendocrine signal (melatonin). The neural signal is transmitted by the photoreceptors to second-order neurons, which innervate various centers in the brain. The neuroendocrine signal—melatonin—diffuses out of the photoreceptor after synthesis, and is transported by the blood to the systemic circulation. Melatonin biosynthesis is regulated directly by light, as light always suppresses melatonin production. In general, the daily melatonin rhythm is controlled by endogenous circadian oscillators located in the photoreceptor cells, which are entrained by the ambient light-dark cycle (Falcón, 1999).

In lampreys, cartilaginous fish and bony fish most photoreceptors have morphology similar to that of retinal photoreceptors. They have a "sensory pole" with a mitochondria-rich ellipsoid and a well-developed photoreceptor (cone-like) outer segment, and an "output pole" consisting of a basal axon with synaptic specializations at the terminals. This type of photoreceptors was named "neural-mode photoreceptor" (Ekström and Meissl, 2003).

The typical presynaptic specialization is the synaptic ribbon. In most instances, reports of photoreceptor synaptic ribbon synapses on dendrites describe monad synapses, i.e., one presynaptic and one postsynaptic element (Collin, 1969; Rüdeberg, 1968; McNulty, 1984), although dyad synapses have also been observed (Pu and Dowling, 1981; Falcón, 1979). Several photoreceptors may converge on one neuron (Collin, 1969; McNulty, 1984), and a single photoreceptor cell with ramifying basal processes may innervate several dendrites (Collin, 1969; Cole and Youson, 1982).

However, some photoreceptors have less well-developed outer segments and do not form synaptic contacts with intrapineal neurons. In some cases the axon extends to contact the basal lamina of the pineal epithelium, and synaptic ribbon-like structures may be aligned along the zone of contact with the basal lamina (Collin and Meiniel, 1971). In lampreys, indoleamine-storing "dense bodies" are located in the cytoplasm (Meiniel, 1980). As these photoreceptors obviously do not have a neural output, they were designated "neuroendocrine-mode photoreceptors" (Ekström and Meissl, 2003). They correspond to the "rudimentary photoreceptor cells" of Collin and Oksche (1981).

It is important, though, that both types of photoreceptors have an active melatonin biosynthesis, and it may be assumed that both types possess an endogenous circadian oscillator. As discussed below several photopigments are often present in the pineal organ. It is not known whether specific photopigments are associated with either type of photoreceptor.

PINEAL NEURONS

In addition to melatonin production and release during darkness, pineal photoreceptors transmit information about the ambient illumination to the brain with neural signals. This information is conveyed primarily by second-order neurons with centrally projecting axons. Although directly post-synaptic to the photoreceptor cells, these neurons are analogous with the ganglion cells of the retina, and have often been named "pineal ganglion cells".

Pineal neurons are present in all vertebrates that possess photoreceptors. Lampreys, chondrichthyans and bony fish all show a variety of morphological types of neurons, as well as a well-developed bundle of afferent axons, the pineal tract.

The predominant pineal neuron is of a pseudounipolar morphology, reminiscent of dorsal root sensory neurons and embryonic primary neurons (Fig. 3.4). In pineal organs with a clear distinction between a distal end-vesicle and a proximal stalk region, the pseudounipolar neuron is the by far most numerous neuronal type in the end-vesicle, while bipolar and multipolar neurons occur in smaller numbers. In some species there is a cluster of large neurons (or even a bilateral pair of clusters) in the most distal portion of the end-vesicle. Such clusters consist mainly of large multipolar and pseudounipolar neurons, with occasional large bipolar neurons (Fig. 3.4). Small, regularly spaced multipolar neurons with basket-like dendrite trees have been observed in the pineal stalk of the minnow (Ekström and Korf, 1986).

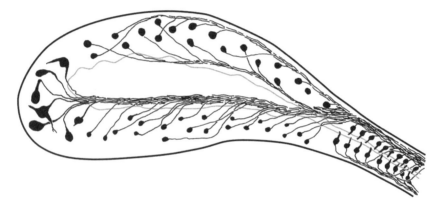

Fig. 3.4 Schematic drawing of the localization of pineal neurons (ganglion cells) in a teleost (rainbow trout). The predominant pineal neuron is of a pseudounipolar morphology that is the by far most numerous neuronal type in the end-vesicle, while bipolar and multipolar neurons occur in smaller numbers. In some species there is a cluster of large multipolar and pseudounipolar neurons in the distal portion of the end-vesicle. In addition, the pineal organ contains numerous cerebrospinal fluid (CSF) contacting neurons. The large majority of pineal neurons constitute centrally projecting neurons, i.e. pineal "ganglion cells".

In addition, the pineal organ contains numerous cerebrospinal fluid (CSF) contacting neurons. A CSF contacting neuron is a bipolar neuron with an apical pole that carries one or more cilia, which may be structurally modified (cf. the photopigment-carrying outer segments of photoreceptor cells), and a basal axon (Fig. 3.4). Pineal photoreceptors are thus structurally similar to CSF contacting neurons (Vigh *et al.*, 1975; Vigh-Teichmann and Vigh, 1983). Centrally projecting CSF contacting neurons have been described in trout pineal organs (Ekström and Korf, 1985). It is presently not known whether these latter cells possess a photoreceptive capacity and thus could represent modified photoreceptors with long axons.

The large majority of pineal neurons constitute centrally projecting neurons, i.e. pineal "ganglion cells", which receive direct synaptic input from photoreceptor cells. Thus, photic information is largely transmitted to the brain via a bineuronal (photoreceptor-ganglion cell) chain (See below; Fig. 3.6). There is no counterpart to retinal bipolar cells, and interneurons— if present—are assumed to play only a minor role in modulating ganglion cell signaling. The distal (rostral) clusters of large neurons were originally believed to constitute a group of interneurons (Wake, 1973; Korf, 1974), but it was later shown that these large neurons are centrally projecting neurons (Ekström and Korf, 1985). However, pineal interneurons have been unequivocally identified by intracellular recording and dye filling (Ekström and Meissl, 1988). It is presently not known whether they constitute part of the AChE-positive population, or if they belong to another, yet unidentified neuronal population. The GABA-immunoreactive neurons demonstrated in the pineal complex of trout are plausible candidates for a population of pineal interneurons (Meissl and Ekström, 1991).

Lampreys and bony fish possess a bipartite pineal complex. The neuronal types in the parapineal and pineal organs of lampreys have not been systematically compared, but available evidence suggests that the predominant types of neurons may be different in the two parts of the pineal complex. In lampreys, the centrally projecting neurons of the pineal organ comprise basally located neurons ("ganglion cells") and bipolar photoreceptor-like cells, whereas in the parapineal organ only the latter type has been observed (Pombal *et al.*, 1999; Yáñez *et al.*, 1999). The parapineal organ of bony fish consists mainly of small, densely packed pseudounipolar neurons (Rüdeberg, 1969; van Veen, 1982; Borg *et al.*,1983) that give rise to an afferent pathway, the parapineal tract, whereas the pineal organ contains several types of neurons (cf. above).

Conventional synapses between neurons have been identified in teleosts (Omura, 1984). The identities of the pre- and postsynaptic neurons have never been characterized, and it is thus not known whether they represent interactions between ganglion cells, between interneurons and ganglion cells, or efferent axons contacting intrapineal neurons.

In contrast, direct input from pineal neurons to photoreceptor cell basal processes has been observed in teleosts (Omura and Ali, 1981). It has been speculated that this innervation serves to optimize secretory activity at night, and inhibit photoreceptor-to-neuron signaling during light exposure.

Available data on the synaptology of the pineal organ indicates that the main purpose of the photoreceptor-to-neuron signaling is to provide an "average signal" reflecting ambient illumination. The absence of evidence for discrete signaling channels with high convergence of specific photoreceptor parts, together with consistent indications of signal divergence (each photoreceptor innervates several neurons), fits well with physiological evidence that the high light sensitivity of the pineal organ does not depend on a high rate of convergence.

EVOLUTION OF THE PINEAL ORGAN

Evolutionary loss of photosensory function

The pineal organ of vertebrates is a neuroendocrine organ that produces the "darkness hormone" melatonin, but there exist considerable differences in the light-dependent control of melatonin production. In non-mammalian species the pineal organ is a directly photosensory organ, containing cells that directly respond to changes in the environmental light conditions. In contrast, the pineal organ of mammals, although lacking intrinsic photosensory cells, is indirectly photosensitive and so also under the control of the ambient light-dark cycle. Light information reaches the pineal through a pathway originating in the retina, and then transmitted via the retino-hypothalamic tract to the master clock in the SCN and further via a multisynaptic pathway to the pineal gland (Vollrath, 1981; Falcón, 1999).

A transformation from a directly photosensory pineal organ in "lower vertebrates" to a secretory gland in mammals was suggested early, on the basis of light microscopic observations (see Bargmann, 1943). This suggestion was investigated further in a series of extensive comparative ultrastructural studies of the pineal organ, performed during the 1960s and early 1970s. These studies have formed the basis of the theory that the main parenchymal cell type of all pineal organs, i.e., the pinealocyte, has evolved within the vertebrate radiation through a gradual loss of photoreceptor characters and a gradual increase of neuroendocrine characters. Thus, the non-sensory pinealocyte of mammals was believed to have evolved from a photoreceptor cell, similar to those present in the pineal organ of anamniotes (Collin, 1969; Collin, 1971; Oksche, 1971; Collin and Oksche, 1981).

Recently, we questioned the concept of a gradual regression of photoreceptor structures within specific cell lines and posited that evolution has occurred in the regulatory linkage mechanisms of fate restriction of the cells in the part of the embryonic central nervous system that will give rise to the pineal complex (Ekström and Meissl, 2003). The mechanisms that regulate the possible developmental pathways of the embryonic pineal cells have evolved, and the notion of specific cell lines becomes redundant. All photoreceptors, pinealocytes and neurons are of a neuronal lineage, but have evolved differentially according to evolutionary adaptations of the different vertebrate taxa (Ekström and Meissl, 2003). The different types of pineal photoreceptors and pinealocytes would be the result of specific combinations of regulatory mechanisms in different vertebrates (Fig. 3.5).

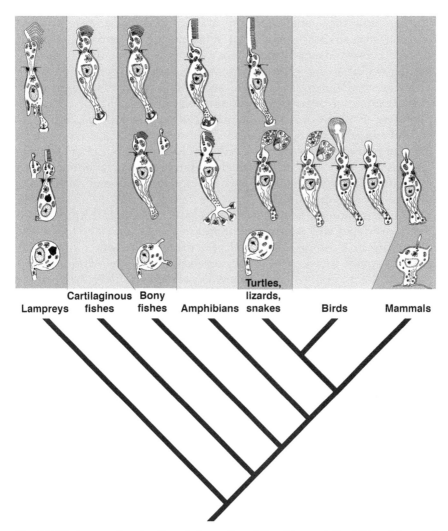

Fig. 3.5 Cladogram showing the phylogenetic relationships and the different types of pinealocytes in vertebrates. Redrawn and adapted from Ekström and Meissl (2003).

Lamb *et al.* (2007) suggested that the pineal photoreceptors of lampreys represent an intermediate stage in the evolution of the ciliary photoreceptors of vertebrates. The notion is interesting in view of the opposing views whether all animal eyes derive from a common ancestor or eyes have evolved independently several times in different animal lineages (for a recent review, see Fernald, 2006). It is beyond the scope of this review to discuss the evidence for or against these differing views, but some aspects of the discussion are highly relevant to the discussion of the possible origins of the pineal photoreceptors.

The molecular genetic machinery that directs location of potential photosensory structures, and their subsequent differentiation, contains strongly conserved core elements (notably *pax6* genes and their homologs). During evolution the gene activation cascade has become modified in different taxa (Zuber *et al.*, 2003), and within taxa with regard to the differentiation of different photosensory structures, like the retina and photosensory pineal organ in vertebrates (Ekström and Meissl, 2003; Mano and Fukada, 2007). Thus, a general mechanism for specification of a location for photosensory structures may have evolved once, but photodetection systems may have appeared many times until selection favoured the two main animal photoreceptor types known today, ciliary and rhabdomeric photoreceptors (Fernald, 2006).

Pineal photoreceptors, and retinal rod and cone photoreceptors, are of the ciliary type, and their differentiation mechanisms have likely been derived from a common ancestral photoreceptor. As other ciliary type photoreceptors, they express ciliary opsins (Arendt and Wittbrodt, 2001; Arendt, 2003), some of which are expressed exclusively in pineal photoreceptors, in rods, or in cones, while other are expressed in both the pineal organ and retina. Rods were probably derived from cones (Kawamura and Tachibanaki, 2008), but the relationship of rods/cones and pineal photoreceptors is still obscure.

In the opsin gene family tree, most pineal-specific opsins like pinopsin, parapinopsin and parietopsin are usually placed basal to the retinal-specific rod and cone opsins (Schichida and Imai, 1998; Lamb *et al.*, 2007; Koyanagi and Terakita, 2008). The pineal-specific exo-rhodopsin (extraocular rhodopsin) of zebrafish, on the other hand, probably arose later in evolution via a duplication of an ancestral gene that gave rise to both retinal rhodopsin and pineal exo-rhodopsin (Asaoka *et al.*, 2002). Thus, on the basis of opsin expression, it could be argued that pineal photoreceptors show more ancestral characteristics. However, opsins evolved before photoreceptor cells (Land and Nilsson, 2002), and opsin genes can be present in the genome without being expressed in photoreceptor cells, offering a substrate for selection of e.g. different visual chromatic sensitivities (Fernald, 2006).

The issue is somewhat confounded by the expression of melanopsin in the pineal organ of chicken (Bailey and Cassone, 2005; Chaurasia *et al.*, 2005; Torii *et al.*, 2007) and *Xenopus* (Chaurasia *et al.*, 2005), and the absence of expression in the pineal organ of zebrafish (Bellingham *et al.*, 2002), Atlantic cod (Drivenes *et al.*, 2003), and ruin lizard (Frigato *et al.*, 2006). However, variable expression of a specific opsin among vertebrates is not so surprising as the expression of a "rhabdomeric opsin", which couples with a Gq type G-protein, in ciliary photoreceptors that otherwise are only known to couple their opsins with transducins, i.e. Gt type G-proteins (Arendt, 2003; Koyanagi and Terakita, 2008). As far as is known today, all pineal and retinal opsins except melanopsin utilize a Gt type phototransduction pathway.

There is some variation regarding the Gt type transduction among pineal photoreceptors. While it has been assumed that pineal photoreceptors generally utilize cone or rod type transducins, photoreceptors in the parietal eye of lizards utilize parietopsin, which relies on gustducin-α and Gα_0 transduction (Su *et al.*, 2006). Taken together, this variability suggests an independent evolution of pineal photoreceptors, where evolution of opsins and transduction mechanisms reflect adaptations to the visual habitat, analogous with the evolution of cone opsins (Fernald, 2006; Bowmaker, 2008).

The notion of pineal photoreceptors of the lamprey type (i.e., the type found also in fish, amphibians, and the parietal eye of lizards) as an "intermediate" chordate type is interesting also from another point of view. According to Lamb *et al.* (2007) pineal photoreceptors would precede the retinal cone and rod photoreceptors in evolutionary ancestry. Indeed, in several respects pineal photoreceptors have "mixed" retinal rod and cone characters. Pineal photoreceptors are "rod-like" in terms of absolute sensitivity, integration times and temporal characteristics of adaptation (Meissl and Ekström, 1988a,b), but "cone-like" in terms of outer segment morphology (Eakin, 1973). The temporal course of their light responses is even slower than that of rods (Pu and Dowling, 1981; Meissl and Ekström, 1988a). In view of the generally accepted view that rods have evolved from cones (Shichida and Imai, 1998; Bowmaker, 2008), and the recent suggestion for a mechanism for the evolution of increased sensitivity in rods (Kawamura and Tachibanaki, 2008), an equally plausible scenario would be that the ancestral vertebrate photoreceptor was "cone-like", and that retinal rods and pineal photoreceptors represent two independent evolutionary adaptations to vision at low intensity levels, and luminance detection, respectively.

A key signature of pineal photoreceptors is their ability to synthesize melatonin. In some vertebrate taxa retinal photoreceptors also produce melatonin, but usually in smaller amounts and predominantly for paracrine functions. Complete biosynthetic machinery for rhythmic production of high levels of melatonin appears to be an exclusive feature of vertebrates. Iyer *et al.* (2004) suggested that the genes for the key enzymes in melatonin biosynthesis, arylalkylamine-N-acetyltransferase (AANAT) and hydroxyindole-O-methyltransferase (HIOMT) were introduced into the genome of an early vertebrate ancestor by horizontal gene transfer. When expressed in photoreceptors and coupled to photosensory functions, AANAT would increase photosensitivity by removal of arylalkylamines (e.g., serotonin) that otherwise react non-enzymatically with retinaldehyde, forming bis-retinyl arylalkylamines and thus decreasing the levels of available retinaldehyde (Klein, 2004). According to Klein's hypothesis (2004), the combined expression of AANAT and HIOMT by photoreceptor cells would (1) further improve arylalkylamine removal by formation of melatonin that readily diffuses over the plasma membrane, (2) provide the animal with a reliable hormonal time signal—melatonin, (3) maintain photoreceptor sensitivity even in the presence of active serotonin synthesis, and (4) set the stage for the evolution of non-sensory pinealocytes that produce melatonin under indirect control of the light-dark cycle. The hypothesis regarding a selective advantage of coupling *aanat* and *hiomt* expression to phototransduction genes (Klein, 2004) has gained support by the findings that AANAT expression is regulated by the transcription factor Otx5 (Appelbaum *et al.*, 2005), which also regulates *exo-rhod* transcription, while *exo-rhod* expression by itself enhances *aanat2* expression in zebrafish (Pierce *et al.*, 2008).

In this scenario (Klein, 2004), an ancestral vertebrate photoreceptor cell of the ciliary type gave rise to pineal photoreceptors that were specialized for melatonin biosynthesis, on the one hand, and retinal photoreceptors that may be capable of melatonin biosynthesis, on the other. Klein's hypothesis is not incompatible with that of Lamb (2007), if (lamprey) pineal photoreceptors and lamprey retinal photoreceptors are considered not as forerunners (see Fig. 2 in Lamb, 2007), but representatives of developmental/ evolutionary lineages that arose in parallel to the gnathostome rod and cone photoreceptors. This latter interpretation conforms well to our hypothesis (Ekström and Meissl, 2003), that different types of pineal (and by inference, retinal) photoreceptors have evolved in parallel through modification of the developmental mechanisms for patterning (location of photosensory structures) and cell differentiation (photosensory cells with different morphological and physiological properties). Available data imply that pineal photoreceptors do not represent ancestral vertebrate photoreceptors. The pineal complex develops from a pineal field at the lateral margins of the neural plate that is competent to give rise to a photosensory organ given an adequate regulatory linkage of fate restriction during development.

ELECTROPHYSIOLOGICAL AND BIOCHEMICAL MEASURES OF DIRECT PHOTOSENSITIVITY AND WAVELENGTH DISCRIMINATION

Early receptor potentials are the earliest sign of photoreceptor activity

The early receptor potential (ERP) of the retina is a rapid biphasic electrical response to intense flashes of light that precedes the a-wave of the electroretinogram (ERG). The ERP shows an extremely short latency and does unlike the ERG not depend on neuronal activity for its formation. It is believed to arise from charge displacements in the visual pigment during bleaching (Brown & Murakami, 1964). In photosensitive pineal organs, early receptor potentials were recorded from the frog pineal organ (Morita and Dodt, 1975), and in fish only in the pike (Falcón and Tanabe, 1983). The shape of the pineal ERP closely resembled that of the retina and also recovery in the dark after bleaching is comparable in pineal and retina. Similarities are also evident in spectral sensitivity measurements. In the pike, the action spectra of pineal and retina showed in the photopic range two peaks with maximal sensitivity in the red (617 nm) and green (533 nm), and in the dark adapted organs only a single peak at 533 nm. In frogs, the action spectra obtained from the pineal organ closely resembled the photopigment absorption of rhodopsin with a λ_{max} at 500 nm (Morita and Dodt, 1975). Thus the structural cone-like pineal photoreceptors appear to contain a photopigment that is different from retinal cones.

ELECTROPINEALOGRAM—AN ELECTRORETINOGRAM-LIKE MASS POTENTIAL

The pineal organ generates a mass slow potential in response to a light stimulus. This response is analogous to the electroretinogram of the eyes, and this potential was consequently termed electropinealogram or EPG (Morita and Dodt, 1973). Measurements of the EPG were performed in lampreys (Morita and Dodt, 1973), as well as in several teleost species like the pike (Falcón and Meissl, 1981), goldfish (Meissl *et al.*, 1986), and the European minnow (Nakamura *et al.*, 1986). The EPG gives detailed information on several physiologically relevant parameters of the photic sensitivity of the pineal like response type, chromatic sensitivity and adaptive properties, without resolving the question which cellular structures are responsible for these potential changes. However, thorough measurements in the frog's pineal make clear that the EPG is probably the representation of summed extracellular currents generated by pineal photoreceptors with only minor contributions of other cellular elements that are postsynaptic to receptors (Donley and Meissl, 1979). Intracellular recordings from photoreceptor cells in several fish species support the idea of a photoreceptor origin because they show potentials with a similar shape and time course to the slow component of the EPG.

Action spectra recorded with the EPG usually correspond closely to the spectra obtained with intracellular recordings or with recordings of the discharge rate of action potentials by ganglion cells. In the lamprey, maximal sensitivity measured with EPG recordings is around 525 nm (ganglion cell activity peaks also at λ_{max} 525 nm). In the goldfish pineal, action spectra showed a single peak at λ_{max} 530 nm, regardless whether they were measured with EPG, with extracellular recordings of action potentials, or with intracellular recordings from individual photoreceptor cells (Meissl *et al.*, 1986). Spectral sensitivity curves in the pineal of the pike reveal photopic and scotopic mechanisms with λ_{max} 530 and 620 nm, regardless of the recording method used (Falcón and Meissl, 1981). In the frog pineal organ, spectral sensitivity measured with the EPG closely resembled the sensitivity spectra obtained with ERP or action potential recordings (Donley and Meissl, 1979).

INTRACELLULAR RECORDINGS CHARACTERIZE THE ELECTROPHYSIOLOGICAL BEHAVIOR OF PINEAL PHOTORECEPTOR CELLS

In the classical phototransduction process in vertebrate rods or cones light induces a cascade of photochemical events that finally terminates in the closure of cGMP-dependent cation channels associated with a hyperpolarization of the membrane potential (Pugh and Lamb, 1990). In the pineal organ of non-mammalian vertebrates this phototransduction cascade appears to be for the most part comparable to retinal phototransduction. Pineal photoreceptors are in darkness partially depolarized and show resting potentials between –20 to –30 mV. In response to brief pulses of light they show hyperpolarizing responses (Pu and Dowling, 1981; Meissl and Ekström, 1988a). The hyperpolarization is graded with light intensity, typically sustained and purely monophasic; only occasionally initial transient responses are observed. Relations between amplitude and light intensity are fundamentally similar to those recorded from retinal photoreceptors with the exception that the sensitivity range of pineal photoreceptors is larger than those of classical cones or rods. Major differences between pineal and retinal photoreceptor responses are also an unusual slow time course of pineal photoresponses and the high sensitivity of pineal photoreceptors which is, despite their relatively poorly developed outer segments, comparable to rods and cones (Meissl and Ekström, 1988a). The extremely slow time course is evident in the rise time of photoreceptor potentials, i.e. the time from response onset to peak potential, which exceeds the time measured in retinal photoreceptors (Baylor and Hodgkin, 1974) five- to six fold, as well as in the response duration and recovery time from peak to the original dark potential (Meissl and Ekström, 1988a).

The light response of retinal photoreceptors is accompanied by a distinct decrease in cGMP levels and as a consequence followed by the closure of cGMP-gated unspecific cation channels (Pugh and Lamb, 1990) which leads then to the hyperpolarization of the cell membrane potential. Recent studies in photoreceptive pineal organs suggest the presence of a similar mechanism in pineal photoreceptors. Light induces a distinct decrease in cGMP levels in pineal photoreceptor cells (Falcón and Gaildrat, 1997; Westermann and

Meissl, 2008) and also a cyclic nucleotide gated (CNG) channel which is activated by cGMP but insensitive to cAMP was recently cloned from trout pineal photoreceptors (Decressac *et al.*, 2002). These molecular biological, biochemical and electrophysiological studies lead to the assumption that light induces a cGMP-dependent signal transduction cascade to close cGMP-activated, non-selective cation channels that produce the membrane hyperpolarization in most pineal photoreceptors (with the possible exception of the chromatic response in the parietal eye of lizards).

THE PINEAL NERVE MESSAGE—AN OUTPUT MESSAGE OF UNKNOWN PHYSIOLOGICAL SIGNIFICANCE

Photoreceptor potentials are converted by second-order neurons into action potentials, a process that is essentially similar in retina and pineal. However, the retinal circuitry comprises numerous interneurons like horizontal cells, bipolar cells and amacrine cells that process the original photoreceptor signal and relay it to ganglion cells before they are transmitted to the target structures in the brain. Only intrinsic photoreceptive retinal ganglion cells (ipRGCs) project directly to their brain targets without the involvement of interneurons, but even these photoreceptive ganglion cells are not completely independent because they receive input from rods and cones *via* amacrine and cone bipolar cells (Belenky *et al.*, 2003; Perez-Leon *et al.*, 2006). In the photoreceptive pineal organ, the complex retinal circuitry is lacking; most photoreceptors are directly connected to pineal ganglion cells which then project to the brain (Ekström and Meissl, 1997). Interneurons were only occasionally detected by electrophysiological recordings followed by intracellular labeling of the neurons and electron microscopy (Ekström and Meissl, 1988). Thus, the basic component of the pineal circuitry simply consists of a bineuronal pathway with a photoreceptive element and an output neuron (Fig. 3.6). Some pineal organs appear to possess an even more reduced circuitry because they contain a novel photoreceptor type that is characterized by direct axonal projections to the brain. Such photoreceptor cells with direct axonal projections to the brain were detected in the pineal of the rainbow trout (Ekström, 1987), the minnow (Ekström *et al.*, 1987) and the lamprey (Samejima *et al.*, 1989), where they comprise a small population of only about 0.3 to 7% of the total number of photoreceptor cells.

The simplicity of the neuronal circuitry is mirrored in the neural message that is transmitted to the brain. Most species show primarily a pure luminance response, originally named achromatic response (Dodt and Heerd, 1962), which characterizes the photosensory pineal organ as a perfect luminance detector. In darkness or under constant illumination, pineal ganglion cells are spontaneously active. Their discharge frequency is

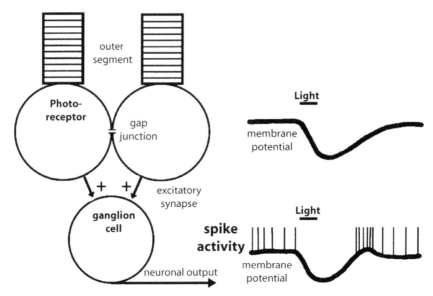

Fig. 3.6 Model of signal transmission from pineal photoreceptor cells to ganglion cells of the luminance (achromatic) type. The neuronal circuit consists of only a bineuronal monosynaptic pathway. The right part of the picture depicts examples of original responses recorded intracellularly from photoreceptors and ganglion cells.

relatively constant and only related to the ambient light level with maximal spike rates in darkness (Fig. 3.7). Upon illumination ganglion cells hyperpolarize (cf. Fig. 3.6), similar to photoreceptors, and the spike rate gradually decreases with increasing brightness and show a nearly linear relationship between discharge frequency and intensity of the ambient light over a range of more than 6 log units (Morita, 1966; Falcón and Meissl, 1981; Meissl *et al.*, 1986). This operating range is slightly broader than that of the presynaptic located pineal photoreceptors (Uchida *et al.*, 1992). From these data the following picture emerges: pineal photoreceptors hyperpolarize in response to light stimulation; this leads to a decrease in the release of an excitatory neurotransmitter (Meissl and George, 1984) followed by a hyperpolarization of second-order ganglion cells accompanied by a decrease of spike discharges (cf. Fig. 3.6). A comparison between the activity of achromatic ganglion cells in the pineal organ with melatonin synthesis shows marked similarities between both output mechanisms of the pineal (Fig. 3.7).

In addition to this basic response pattern of ganglion cells, some species display a special color-coded response type, the so-called chromatic response, especially in the extracranial located parietal eye or frontal organ, but also in several intracranially located pineal organs (see below).

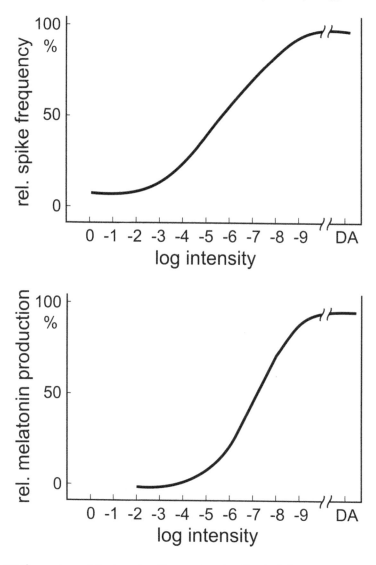

Fig. 3.7 Comparison of the two signalling modes of a directly light sensitive pineal organ of a teleost. The neural mode (luminance response) is reflected by the output (discharge rate) of ganglion cells (upper graph) that shows a sigmoidal response-intensity relation. The organ was exposed to steady white light; log 0 = equal to 2700 µW/cm². Note that the luminance response signals light information over a range of more than 10 log units.

The melatonin response (release of melatonin *in vitro*) shows a similar behavior to steady light (lower graph), only the operating range seems to be restricted to a narrower intensity range, i.e. the slope of the curve is steeper than that of ganglion cells.

MICROSPECTROPHOTOMETRIC EXAMINATION REVEALS THE PRESENCE OF SEVERAL PHOTOPIGMENTS IN THE PINEAL ORGAN

Characterization of pineal photopigments by microspectrophotometric measurements was undertaken only in some rare cases in lamprey, teleosts, and in frogs. In the first microspectrophotometric study, Hartwig and Baumann (1974) reported that the intracranial pineal organ of frogs contains a visual pigment that absorbs maximally at 502 nm. This is comparable to spectral sensitivity data derived from ERP and EPG measurements as well as from spike recordings. The extracranial part, the frontal organ, contained a photolabile substance with λ_{max} between 560 and 580 nm which finds its expression in the so-called achromatic response, a pure luminance response, of the frontal organ (Dodt and Heerd, 1962).

In pineal photoreceptors of fish, only in few cases spectral absorption characteristics of visual pigments have been measured with microspectrophotometry: in trout (Kusmic *et al.*, 1993), in goldfish (Peirson and Bowmaker, 1999), a cavefish (Parry *et al.*, 2003), and in some deep-sea fish (Bowmaker and Wagner, 2004). In the trout, two distinct pigments have been identified with λ_{max} at about 460 and 560 nm; in goldfish the pinealocytes appear to have a mixed pigment pair based on retinal and 3-dehydroretinal with λ_{max} close to 512 nm, and in the cavefish *Astyanax* a pigment with λ_{max} 500 nm or 520 nm. In both goldfish and the cavefish it was assumed that the responsible photopigment is an ERrod-like opsin (Bowmaker and Wagner, 2004). Microspectrophotometric measurements in several mesopelagic fish (hatchet fish, scaly dragon fish, bristlemouths) and also in the deep demersal eel identified one or two visual pigments in the pineal with λ_{max} between 485 and 515 nm. Interestingly, most of these species possess pineal pigments that are displaced to longer wavelengths compared to retinal rods.

From the physiological point of view the data are somewhat confusing, because in some species the pineal photopigments seem to be identical to retinal pigments; in other species they are clearly spectrally distinct from the retina. The differences could lie either in the opsins or in the different composition of the two chromophores 11-cis retinal (vitamin A_1) and 11-cis 3-dehydroretinal (vitamin A_2).

HPLC MEASUREMENTS ESTIMATE THE TYPE OF CHROMOPHORE

The typical chromophores of freshwater teleosts retinaldehyde and 3-dehydroretinaldehyde have been identified in pineal extracts of the rainbow trout by high-performance liquid chromatography (Tabata *et al.*, 1985). In the trout, the chromophores were photoisomerized by bright light illumination to the all-*trans* isomer indicating the presence of functional opsin-based photopigments. In a further study, in the lamprey pineal organ, chromophores consisted mainly of 3-dehydroretinal (92%) with a small amount of retinal (Tamotsu and Morita, 1990). However, the dark-adapted pineal contained surprisingly—contrary to retinal extracts—a high proportion of the all-*trans* isomer. This proportion did not change significantly when the pineal was partially bleached by light of long wavelengths (orange or green), whereas in the retina the amount of the all-*trans* isomer increased and the 11-*cis* isomer was strongly reduced. Bleaching of pineal photoreceptors with blue light (420 nm) caused an increase in the 11-*cis* isomer and a decrease in the all-*trans* isomer, indicating that a blue light photoreceptor is responsible for this isomerization. A recent analysis of the pineal photopigments showed that the lamprey homologue of parapinopsin could be responsible for the high sensitivity in the blue/UV range that had been predicted earlier from careful electrophysiological studies (Morita and Dodt, 1973; Uchida and Morita, 1990). Lamprey parapinopsin was expressed in cultured cells and was successfully reconstituted with 11-*cis* 3-dehydroretinal (retinal$_2$) (Koyanagi *et al.*, 2004). The reconstituted pigment exhibits an absorption maximum in the ultraviolet at λ_{max} 370 nm. Ultraviolet light irradiation converted parapinopsin to the photoproduct with an absorption maximum at 515 nm, but could be reverted by subsequent orange light exposure to the original state which is almost identical to the dark state. These photoreactions can be repetitively achieved by UV and orange light irradiation which shows that the system may consist of two photointerconvertible stable states.

IMMUNOCYTOCHEMISTRY DETECTS MULTIPLE PHOTOPIGMENTS IN THE PINEAL

In early studies, pineal photoreceptors were immunocytochemically labeled by a variety of antibodies raised against retinal opsins from mammals and birds, indicating that both rod- and cone-like opsins may be present (Ekström and Meissl, 1997; Forsell *et al.*, 2002). However, not all pineal photoreceptor outer segments were labeled by these antibodies to known retinal opsins, suggesting the existence of special pineal photopigments. One of these previously unknown photopigments was the chicken pinopsin. Using molecular cloning, a chicken pineal-specific cDNA was identified that encodes a protein (pinopsin) with strong similarities to the known visual opsins with about 48% identities in amino-acid sequence to vertebrate retinal opsins (Okano *et al.*, 1994; Max *et al.*, 1995). Antibodies raised against a 14 amino acids containing part of the chicken pinopsin molecule strongly labeled most bird pinealocytes, but produced only a weak immunoreaction in outer segments of some fish pineal organs (*Raja clavata, Carassius auratus,*

Anguilla anguilla), whereas the pineal of lamprey showed no reaction (Fejér *et al.*, 1987). Also, anti-chicken pinopsin antibody P9 showed no immunoreactivity to pineal cells in the zebrafish, whereas an antiserum raised against bovine rhodopsin strongly labeled pinealocytes (Mano *et al.*, 1999). After the detection of pinopsin the use of molecular biological methods led to the identification of additional opsin based photopigments in pineal and parapineal of fish that are distinct from the known rod and cone classes of opsins: parapinopsin (Soni *et al.*, 1998; Blackshaw and Snyder, 1997), VA opsin (Philp *et al.*, 2000a; Moutsaki *et al.*, 2000), VA-Long (VAL) opsin, a variant of VA opsin (Kojima *et al.*, 2000), and ERrod-like opsin (extra-retinal rod-like opsin) or exo-rhodopsin (Mano *et al.*, 1999; Philp *et al.*, 2000b). However, so far no immunocytochemical studies in the fish pineal are available using specific antibodies raised against these novel opsins (see below).

MOLECULAR SPECIFICATION OF PHOTORECEPTORS IN THE PINEAL ORGAN

The pineal organ develops from "pineal fields" located at the lateral margins of the neural plate, while the retinas develop from a more anteromedially placed "eye field". The specification of the pineal field and the subsequent specification of pineal cell types, emergence of melatonin biosynthesis, and onset of circadian oscillations have been studied in some detail, particularly in the zebrafish embryo (for references, see Ekström and Meissl, 2003; Mano and Fukada, 2007; Pierce *et al.*, 2008). However, while the genetic determination of photoreceptor cells (*otx5* expression) and projection neurons (*pax6* and *onecut* expression) is known, it is less clear what factors direct the expression of phototransduction molecules in the pineal photoreceptors.

In general—although data are fragmentary—it appears that the pineal organ expresses one pineal-specific opsin alongside opsins that are also expressed in the retina. Pineal-specific opsins are pinopsin (Okano *et al.*, 1994; Max *et al.*, 1995), parapinopsin (Blackshaw and Snyder, 1997; Koyanagi *et al.*, 2004), parietopsin (Su *et al.*, 2006) and exo-rhodopsin (Mano *et al.*, 1999; Pierce *et al.*, 2008). It is not known how expression of pinopsin, parapinopsin or parietopsin is directed exclusively to the pineal complex. Pinopsin expression in the chicken pineal organ is controlled by one or more light-responsive elements (LREs), but their function is probably primarily related to enhancement by light and repression by darkness of pinopsin expression (Takanaka *et al.*, 2002; Mano and Fukada, 2007).

Most data on pineal photoreceptor specification come from studies of zebrafish. The zebrafish pineal photoreceptors express exo-rhodopsin and red-sensitive cone opsin, but not other visual opsins, VA (vertebrate ancient opsin) or VAL (VA long) (Ziv *et al.*, 2007). On the other hand, VA opsin is expressed in the pineal organ of the Atlantic salmon (Philp *et al.*, 2000a) as well as the pineal of carp (Moutsaki *et al.*, 2000). In zebrafish, exo-rhodopsin expression in the pineal organ is directed by a *cis*-acting pineal expression-promoting element (PIPE) that acts together with the Crx/Otx binding site that is required for expression of both pineal- and retina-specific genes (Asaoka *et al.*, 2002). The exo-rhodopsin gene is a duplicated form of the rhodopsin gene. The rhodopsin promoter lacks PIPEs and is normally expressed exclusively in retinal rod photoreceptors, but fusing the rhodopsin promoter with PIPE repeats directs GFP-expression in the zebrafish pineal organ (Kojima *et al.*, 2008). Blue cone opsin expression is repressed in zebrafish pineal photoreceptors by an 11bp pineal negative regulatory element (PINE) together with an Otx element (Takechi *et al.*, 2008).

It remains to be clarified in more detail how opsin expression is regulated in the zebrafish pineal—for example why red-sensitive cone opsin is expressed both in the retina and pineal—and how expression of specific opsins is related to the melatonin biosynthetic machinery and the circadian clock. Also, it will be important to determine whether conserved PIPEs and PINEs direct retina- and pineal-specific opsin expressions in other species, or additional regulatory mechanisms have evolved in different vertebrate groups.

SPECTRAL SENSITIVITY—A COMPARISON BETWEEN DIFFERENT EXPERIMENTAL STRATEGIES

Spectral sensitivity measurements of pineal photoreception are restricted to a small number of species and they are partly contradictory with respect to the number of different photopigments or the type of pigments, even in the same species. Most data are derived from intracellular and extracellular recordings from the pineal complex, from microspectrophotometric measurements or from biochemical measurements of the melatonin response. However, electrophysiological recordings from photoreceptors give essentially the same information as recordings from second-order neurons, i.e. recordings of action potentials, or with other methods like recordings of slow mass potentials. This raises the question why carefully collected experimental data from different laboratories are so variable. Let us examine one example. Spectral sensitivity measurements from the trout pineal provide the most complete set of data on the photosensory function of a teleost. Initially extracellular recordings from ganglion cells (Morita, 1966), later intracellular recordings from photoreceptors (Meissl and Ekström, 1988a; Kusmic *et al.*, 1992) and putative interneurons (Ekström and Meissl, 1988) were used to characterize the putative photopigments. High-pressure liquid chromatography provided evidence for the presence of both chromophores, 11-*cis*-retinal and 11-*cis*-3-dehydroretinal, in the pineal (Tabata *et al.*, 1985);

microspectrophotometric measurements showed two distinct types of photopigments (Kusmic *et al.*, 1993) and biochemical estimation of the action spectra for melatonin suppression also identified the responsible photopigment (Max and Menaker, 1992). Analyzing this rich set of data give some inconsistencies in the λ_{max} of the photopigment. All electrophysiological studies are suggestive of the presence of one or two pineal photopigments, a vitamin A_1 and a vitamin A_2-based photopigment with one λ_{max} at around 500 nm and the other between 525 and 533 nm. The action spectra for melatonin suppression peaks also at about 500 nm, but a second peak at longer wavelengths as anticipated from electrophysiology is missing. However, microspectrophotometric analysis of pineal photopigments in the trout indicate also the presence of a vitamin A_1 and a vitamin A_2-based photopigment, but the λ_{max} is shifted to shorter, λ_{max} 463 nm, or longer, λ_{max} 561 nm, wavelengths, respectively (Kusmic and Gualtieri, 2000). The discrepancies between the λ_{max} values obtained from different research groups using varying techniques could mirror the fact that considerable genetic variability exists within farmed rainbow trout strains even when they are derived from the same region (Glover, 2008). Furthermore, it is well known that seasonal and age-related changes in retinal photopigments occur naturally (Bowmaker, 1990), and that light condition itself, i.e. intensity, photoperiod, spectral quality, can control visual pigment ratios (Allen, 1971). It was therefore argued that visual pigments have been selected for maximum sensitivity to the light in their environment (Munz, 1965); others argue that visual contrast, rather than sensitivity, could influence the selection of the opsins (Lythgoe, 1966).

Action spectra for acute light-dependent melatonin suppression in cultured zebrafish pineal organs revealed also the presence of two photopigments with λ_{max} at 500 nm and a second peak at 570 nm (Ziv *et al.*, 2007). Opsin-specific RT-PCR analysis confirmed the expression of exo-rhodopsin and visual red-sensitive opsin in the pineal gland, but not of other retinal visual opsins or one of the novel VA and VAL opsins. The absence of VAL from the pineal is supported by immunohistochemical studies in zebrafish revealing VAL expression in retinal horizontal cells and the diencephalon, but not in the pineal (Kojima *et al.*, 2000).

ULTRAVIOLET RECEPTORS PROVIDE A BASIS FOR CHROMATIC SENSITIVITY

A peculiarity of the pineal complex is the presence of ultraviolet receptors in some species. Pineal photoreceptors containing putative UV-sensitive opsin have been demonstrated by use of immunocytochemistry and *in situ* hybridization (Forsell *et al.*, 2001). It is striking that, in the halibut embryo, transcripts of HPO1 (green-like opsin) and HPO4 (UV-like opsin), as well as immunoreactivity for SW cone/rod opsins become detectable at the same developmental stage. However, during subsequent development putative UV-opsin expressing cells become selectively expressed in the pineal stalk region, while green-opsin ones are ubiquitously distributed (Forsell *et al.* 2002). This suggests that putative UV-opsin expressing photoreceptors are generated only during early development and that the ability to detect UV-light is of greater importance for small, translucent embryos and early larvae.

The UV receptor is part of a special response type of the pineal: the color-coded chromatic response (cf. Fig. 3.8). Chromatic responses were first described and characterized in the frog's frontal organ (Dodt and Heerd, 1962), the extracranial part of the pineal complex. Later this response type was verified in the parietal eye of lizards (Miller and Wolbarsht, 1962), and also in the pineal organ of the pike (Falcón and Meissl, 1981), the trout (Morita, 1966; Meissl and Ekström, unpublished) and the lamprey (Uchida and Morita, 1990). In these species, light stimuli of short wavelengths causes a long-lasting inhibition of spike discharges of pineal neurons, whereas light of longer wavelengths causes activation. The result of this antagonistic flip-flop mechanism depends on the balance of incident ultraviolet light, or blue light, respectively, and the long wavelength part of the visible spectrum (Ekström and Meissl, 1997). The maximal sensitivity of the inhibitory component is at about 360 nm in pike, trout and lamprey and at 430 nm in lizards. The excitatory component is maximally sensitive in the green (530 nm trout, 525 nm lampreys, 496 nm lizards) or in the red (620 nm pike).

It was long disputed whether chromatically antagonistic responses arise through the action of a bistable visual pigment or through chromatic antagonism of two photoreceptor populations with different absorption maxima driving hyperpolarizing or depolarizing mechanisms. This issue is still unresolved, even though considerable progress was made by intracellular recordings from individual photoreceptor cells showing that chromatic antagonism originates primarily from photoreceptors (Uchida and Morita, 1990; Solessio and Engbretson, 1993). Recently it was shown in the lamprey pineal that the putative photopigment is a lamprey homologue of parapinopsin which exhibits an absorption maximum in the ultraviolet range at 370 nm. UV light causes *cis-trans* isomerization of its retinal$_2$ chromophore forming a stable photoproduct with an absorption maximum in the green at 515 nm (Koyanagi *et al.*, 2004). Upon light stimulation with long wavelength the photoproduct reverts to its original pigment. These results demonstrate that the UV pigment is photointerconvertible by UV and light of long wavelength resulting in two stable states. A similar bistable mechanism, consisting of a single photopigment which has two photointerconvertible active states and behaves similarly as a single photopigment (Fig. 3.8), had earlier been hypothesized as the potential mechanism for the color-coded chromatic response in the frog's pineal complex (Dodt, 1963; Eldred and Nolte, 1979).

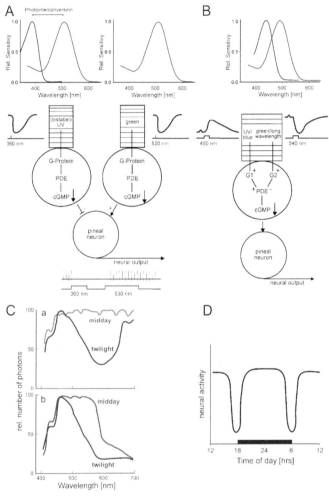

Fig. 3.8 The color-coded chromatic response shows an antagonism between a UV- or a short-wavelengths sensitive mechanism and a mechanism that is sensitive to long wavelengths. Chromatic responses may depend on a photointerconversion of a bistable photopigment or on the presence of two different photoreceptor cells with differing maximal sensitivities (**A**). These photoreceptors make inhibitory and excitatory synapses with a second-order neuron. The neural output would thus depend on the interaction between both inhibitory and excitatory mechanisms. An alternative mechanism was proposed for the chromatic response in the parietal eye where chromatic antagonism seems to depend on two opsins found in the same photoreceptor cell that drive different phototransduction components (**B**). These components cause depolarizing (λ_{max} = 450 nm) or hyperpolarizing (λ_{max} = 540 nm) responses of the photoreceptor cell. (**C**) Changes of the spectral composition of light in air (a) and in 3 m depth (b) of a sea at different times of the day. Note the dramatic changes in the light spectrum at twilight. The chromatic response is ideally suited to detect these changes at twilight (redrawn and adapted from McFarland, 1986). (**D**) Model for the activity of chromatic neurons throughout the day.

However, the mechanism that leads to chromatic antagonism in the pineal of lampreys, fish or frogs remains to be determined. If specific UV receptors and a bistable photopigment are responsible for the inhibitory component, which receptor type or which mechanism generates then the green/long wavelength dependent excitation? One scenario could lie in the existence of a separate green/long wavelengths sensitive photoreceptor that makes synaptic contacts with a common chromatic ganglion cell. Because of the differences in the maximal sensitivities, this photoreceptor type would be separate from the luminosity photoreceptor type. Another scenario is conceivable that includes the antagonistic control of the cGMP phosphodiesterase by two different G-proteins (cf. Fig. 3.8) as it was proposed for the first time for the lizard (Xiong *et al.*, 1998). In the parietal eye, chromatic antagonism seems to depend on two light signaling pathways within the same cell—a hyperpolarizing mechanism maximally sensitive to blue light and a depolarizing mechanism maximally sensitive to green light (Solessio and Engbretson, 1993). However, unlike the chromatic response in the lamprey pineal it seems to be based on two opsins found in the same photoreceptor cell: a blue-sensitive pinopsin and a green-sensitive parietopsin which both drive different phototransduction components (Su *et al.*, 2006). The hyperpolarizing response is produced, similarly to retinal rods and cones, by the activation of a cGMP-phosphodiesterase that lowers cGMP content and closes cyclic nucleotide gated (CNG) channels, whereas the depolarizing component is produced by the inhibition of the same cGMP-phosphodiesterase, elevated cGMP concentrations and opening of CNG channels (cf. Fig. 3.8) (Finn *et al.*, 1997). Recent cloning of the molecular components underlying this antagonism revealed the following putative mechanisms: blue-sensitive pinopsin activates a cGMP-phosphodiesterase by means of gustducin to lower cGMP concentrations and to close CNG channels, while green-sensitive parietopsin inhibits the phosphodiesterase by means of $G\alpha_0$ which results in opening of CNG channels through elevation of cGMP (Su *et al.*, 2006).

However, we have presently no information about the photo-transduction components that drive the chromatic response in the pineal organ of fish; no information whether it is composed of two photopigments residing in the same photoreceptor, as possibly in the parietal eye, or of photopigments in separate photoreceptor cells as in the lamprey pineal organ.

THE PINEAL CLOCK

Pineal clock mechanisms will be described in detail elsewhere in this book and will be only briefly outlined here.

Unlike the mammalian pineal organ, melatonin synthesis in the fish pineal is due to the presence of intrapineal photoreceptors directly controlled by the ambient light-dark cycle. Melatonin levels are high during the dark period and low during the light period. The melatonin synthesis profile which follows a diurnal rhythm in all vertebrates studied so far is reflected either in plasma levels *in situ* or in the release of melatonin from explanted pineal organs *in vitro* (Ekström and Meissl, 1997; Falcón, 1999). In fish, two different forms of circadian organization of melatonin synthesis have been suggested: (a) clock controlled melatonin synthesis and (b) light-darkness controlled melatonin synthesis. In the first case (a) it appears that the pineal organ combines all features of a clock system: the endogenous oscillator, an input mechanism (e.g. photoreception) that is able to entrain the oscillator and an output mechanism (melatonin synthesis and release) that controls the effector sides of the clock. Surprisingly, all these three basic clock mechanisms seem to be localized in a single cell type, the pineal photoreceptor cell (Bolliet *et al.*, 1997). Thus, the direct action of light is to adjust rhythmicity in such a way that a light pulse during the dark phase immediately suppresses melatonin production as a short-term effect and phase shifts the subsequent rhythms as a persistent effect (Falcón *et al.*, 1987; Cahill, 1996). In the second case (b), represented by all salmonids studied so far, melatonin synthesis appears to be directly controlled by the prevailing light-dark cycle. There is no evidence for an endogenous rhythm in melatonin synthesis in the pineal organ in salmonids *in vitro* as well as *in vivo*; the pineal organ acts more or less as a kind of light dosimeter and reflects the length of photoperiod and light intensity (Gern and Greenhouse, 1988; Iigo *et al.*, 1998). It is assumed that salmonids do not contain either an intrapineal oscillator that regulates melatonin production or, as an alternative, that the coupling between clock mechanism and melatonin synthesis is relatively weak (Iigo *et al.*, 2007). However, it should be emphasized that only melatonin synthesis was used as a marker for the presence or absence of an endogenous circadian oscillator. So far no attempts were undertaken in any species to investigate the possible presence of a circadian component in the second output mechanism, the neuronal output of the pineal organ.

TEMPORAL PHYSIOLOGY

The role of the pineal organ in the temporal physiology of fish is clear-cut in the sense that the pineal transduces and transmits information about the photic environment, and thus about the duration of day and night during the 24 h cycle. The information transmitted from the pineal has two signalling modes, a neuroendocrine mode (melatonin) that is systemic, and a neural mode that reaches selected brain centers. The signals from the pineal organ may be used to synchronize daily cycles, but also to convey information about the season (e.g., long duration of melatonin release

= "winter"). Figure 3.9 outlines the possible signalling modes of chromatic and achromatic (luminance) responses of the pineal organ during long summer days and short winter days. Whereas the luminance response displays a perfect picture of the length of the day sending action potentials with a high discharge rate during the night and low rate during the day to selected brain centers, the chromatic response exhibits a more sophisticated signalling mode which depends not only on intensity of the ambient illumination, but also on wavelength (Fig. 3.9). In a distinct spectral domain, the ultraviolet range and the long wavelength range, chromatic cells signal information about the relative contribution of both parts of the spectrum. Because both spectral domains drive antagonistic inhibitory or excitatory responses (cf. Fig. 3.8A, B), this color-coded response is ideally suited to monitor relative changes in the spectral composition of the ambient illumination which are pronounced during beginning and the end of the daily photophase (cf. Fig. 3.8C). Thus the chromatic mechanism could potentially serve as a dawn and dusk detector (Fig. 3.8D), although, and this should be brought up in this context, nothing is known concerning the physiological significance of neuronal signals from the pineal organ. Although the brain areas that are the targets of the pineal ganglion cells are known (Ekström *et al.*, 1994), the neurons that receive the innervation have not yet been identified and characterized morphologically or physiologically.

The impact of the pineal signals on the temporal physiology of the organism is less clear-cut. It apparently depends on the time of day, season, stage of the animal's life-cycle, the activity pattern of the animal (nocturnal, diurnal, or crepuscular), and environmental adaptations (cf. Fig. 3.9). Although the potential targets of melatonin are fairly well known, very little is known about how these signals are transduced into physiological and behavioral responses by these targets (Ekström and Meissl, 1997; Falcón *et al.*, 2007). If we focus on melatonin as a signal that synchronizes physiology in fish, it is of course crucial to understand the regulation of the cyclic production of melatonin. Melatonin biosynthesis is controlled by the ambient light (light blocks the synthesis) and by a circadian clock mechanism inherent to the pineal photoreceptor cells. In recent years, important advances in the analysis of the pineal clock have been made through molecular biological studies of, primarily, the zebrafish *Danio rerio* (e.g. Gamse *et al.*, 2002; Appelbaum *et al.*, 2004, 2005; Vuilleumier *et al.*, 2006; Pierce *et al.*, 2008). The pineal circadian clock mechanism is reviewed elsewhere in this volume and will not be dealt with further here.

In the present context, it is important that melatonin acts at different levels. First, within the photosensory organs themselves, i.e. in the retina and the pineal organ. Second, in visual centers in the brain. Third, in brain centers that regulate endocrine functions, as well as in the adenohypophysis. Finally, melatonin may act directly in peripheral organs. Cassone and

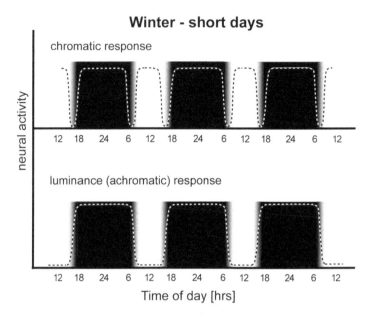

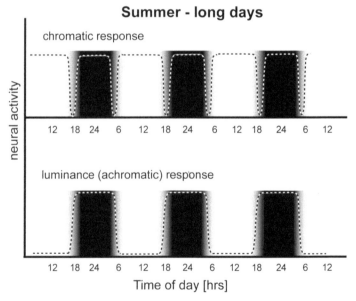

Fig. 3.9 Model of activity changes (dashed lines) of pineal neurons during short winter days and during long summer days. Both response types, the color-coded chromatic response, as well as the luminance (achromatic) response type signal the end and the beginning of the photophase. Whereas the luminance response reflects the prevailing light intensities with a high activity at night and low activity at daytime, is the chromatic response a sensor for twilight detection.

Natesan (1997) suggested an evolutionary trend, where direct actions in the visual system is the ancestral function, and loss of melatonin receptors in the visual centers of the brain and appearance in the pituitary are derived features of mammals. In this scenario fish have retained the ancestral functions of melatonin as a signaling molecule and melatonin has been shown to affect a large number of physiological parameters (Ekström and Meissl, 1997)—consistent with its different levels of action. It has not been possible to pin-point any particular function of melatonin; when experimentally manipulated its effects apparently depend on the time of day, season, stage of the animal's life-cycle, the activity pattern of the animal (nocturnal, diurnal, or crepuscular), and environmental adaptations (Cassone and Natesan, 1997; Ekström and Meissl, 1997; Falcón, 1999). This is, however, consistent with the "simple message" of melatonin: darkness, or subjective night during free-running conditions. This reliable message has allowed multiple adaptations among vertebrates to optimize the organism's physiology over the 24-h light-dark cycle, and over the seasons.

References

Allen, D.W. 1971. Photic control of the visual proportions of two visual pigments in fish. *Vision Research* 11: 1077–1112.

Appelbaum, L., R. Toyama, I.B. Dawid, D.C. Klein, R. Baler and Y. Gothilf. 2004. Zebrafish serotonin-N-acetyltransferase-2 gene regulation: pineal-restrictive downstream module contains a functional E-box and three photoreceptor conserved elements. *Molecular Endocrinology* 18: 1210–1221.

Appelbaum, L., A. Anzulovich, R. Baler and Y. Gothilf. 2005. Homeobox-clock protein interaction in zebrafish. A shared mechanism for pineal-specific and circadian gene expression. *Journal of Biological Chemistry* 280: 11544–11551.

Arendt, D. 2003. Evolution of eyes and photoreceptor cell types. *International Journal of Developmental Biology* 47: 563–571.

Arendt, D. and J. Wittbrodt. 2001. Reconstructing the eyes of Urbilateria. *Philosophical Transactions of the Royal Society B: Biological Sciences* 356: 1545–1563.

Asaoka, Y., H. Mano, D. Kojima, and Y.A. Fukada. 2002. Pineal expression-promoting element (PIPE), a cis-acting element, directs pineal-specific gene expression in zebrafish. *Proceedings of the National Academy of Sciences of the United States of America* 99: 15456–15461.

Bailey, M.J. and V.M. Cassone. 2005. Melanopsin expression in the chick retina and pineal gland. *Molecular Brain Research* 134: 345–348.

Bargmann, W. 1943. Die Epiphysis cerebri. In: *Handbuch der mikroskopischen Anatomie des Menschen* VI/4. W. von Möllendorff (ed). Springer-Verlag, Berlin, pp. 309.

Baylor, D.A. and A.L. Hodgkin. 1974. Changes in the time scale and sensitivity in turtle photoreceptors. *Journal of Physiology* 242: 729–758.

Belenky, M.A., C.A. Smeraski, I. Provencio, P.J. Sollars and G.E. Pickard. 2003. Melanopsin retinal ganglion cells receive bipolar and amacrine cell synapses. *Journal of Comparative Neurology* 460: 380–393.

Bellingham, J., D. Whitmore, A.R. Philp, D.J. Wells and R.G. Foster. 2002. Zebrafish melanopsin: isolation, tissue localisation and phylogenetic position. *Molecular Brain Research* 107: 128–136.

Blackshaw, S. and S.H. Snyder. 1997. Parapinopsin, a novel catfish opsin localized to the parapineal organ, defines a new gene family. *Journal of Neuroscience* 17: 8083–8092.

Bolliet, V., V. Bégay, C. Taragnat, J.P. Ravault, J.P. Collin and J. Falcón. 1997. Photoreceptor cells of the pike pineal as cellular circadian oscillators. *European Journal of Neuroscience* 9: 643–653.

Borg, B., P. Ekström and Th. van Veen. 1983. The parapineal organ of teleosts. *Acta Zoologica* (Stockholm) 64: 211–218.

Bowmaker, J.K. 1990. Visual pigments of fishes. In: *The Visual System of Fish*, R.H. Douglas and M.B.A. Djamgoz. (eds.). Chapman and Hall, London, pp. 81–107.

Bowmaker, J.K. 2008. Evolution of vertebrate visual pigments. *Vision Research* 48: 2022–2041.

Bowmaker, J.K. and H.J. Wagner. 2004. Pineal organs of deep-sea fish: photopigments and structure. *Journal of Experimental Biology* 207: 2379–2387.

Brown, K.T. and M. Murakami. 1964. A new receptor potential of the monkey retina with no detectable latency. *Nature (London)* 201: 626–628.

Cahill, G.M. 1996. Circadian regulation of melatonin production in cultured zebrafish pineal and retina. *Brain Research* 708: 177–181.

Cassone, V.M. and A.K. Natesan. 1997. Time and time again: the phylogeny of melatonin as a transducer of biological time. *Journal of Biological Rhythms* 12: 489–497.

Chaurasia, S.S., M.D. Rollag, G. Jiang, W.P. Hayes, R. Haque, A. Natesan, M. Zatz, G. Tosini, C. Liu, H.W. Korf, P.M. Iuvone and I. Provencio. 2005. Molecular cloning, localization and circadian expression of chicken melanopsin (*Opn4*): differential regulation of expression in pineal and retinal cell types. *Journal of Neurochemistry* 92: 158–170.

Cole, W.C. and J.H. Youson. 1981. The effect of pinealectomy, continuous light, and continuous darkness on metamorphosis of anadromous sea lamprey, *Petromyzon marinus* L. *Journal of Experimental Zoology* 218: 397–404.

Collin J.-P. 1969. Contribution à l'étude de l'organe pinéal. De l'épiphyse sensorielle à la glande pinéale: modalités de transformation et implications fonctionelles. *Annales de la Station Biologique de Besse-en-Chandesse* Suppl. 1: 1–359.

Collin J.-P. 1971. Differentiation and regression of the cells of the sensory line in the epiphysis cerebri. In: *The Pineal Gland*, G.E.W. Wolstenholme and J. Knight (eds.). JA Churchill, London, pp. 79–125.

Collin, J.-P. and A. Meiniel. 1971. L'organe pinéal. Études combinées ultrastructurales, cytochimiques (monoamines) et expérimentales, chez *Testudo mauritanica*. Grains denses des cellules de la lignée "sensorielle" chez les vertébrés. *Archives d'Anatomie Microscopique et de Morphologie Experimentale* 60: 269–304.

Collin J.-P. and A. Oksche. 1981. Structural and functional relationships in the nonmammalian pineal gland. In: *The Pineal Gland*, Vol. I: *Anatomy and Biochemistry*. R.J. Reiter (ed.). CRC Press, Boca Raton, pp. 27–67.

Decressac, S., A. Grechez-Cassiau, J. Lenfant, J. Falcon and P. Bois. 2002. Cloning, localization and functional properties of a cGMP-gated channel in photoreceptor cells from fish pineal gland. *Journal of Pineal Research* 33: 225–233.

Dodt, E. 1963. Reversible Umsteuerung lichtempfindlicher Systeme bei Pflanzen und Tieren. *Experientia* 19: 53–59.

Dodt, E. and E. Heerd. 1962. Mode of action of pineal nerve fibers in frogs. *Journal of Neurophysiology* 25: 405–429.

Donley, C.S. and H. Meissl. 1979. Characteristics of slow potentials from the frog epiphysis (*Rana esculenta*); possible mass photoreceptor potentials. *Vision Research* 19: 1343–1349.

Drivenes, O., A.M. Soviknes, L.O.E. Ebbesson, A. Fjose, H.C. Seo and J.V. Helvik. 2003. Isolation and characterization of two teleost melanopsin genes and their differential expression within the inner retina and brain. *Journal of Comparative Neurology* 456: 84–93.

Eakin, R.M. 1973. *The Third Eye.* University California Press, Berkeley.

Ekström, P. 1987. Photoreceptors and CSF-contacting neurons in the pineal organ of a teleost fish have direct axonal connections with the brain: an HRP-electron-microscopic study. *Journal of Neuroscience* 7: 987–995.

Ekström, P. and H.W. Korf. 1985. Pineal neurons projecting to the brain of the rainbow trout, *Salmo gairdneri* Richardson (Teleostei). *In-vitro* retrograde filling with horseradish peroxidase. *Cell and Tissue Research* 240: 693–700.

Ekström, P. and H.W. Korf. 1986. Putative cholinergic elements in the photosensory pineal organ and retina of a teleost, *Phoxinus phoxinus* L. (Cyprinidae). Distribution of choline acetyltransferase immunoreactivity, acetylcholinesterase-positive elements, and pinealofugally projecting neurons. *Cell and Tissue Research* 246: 321–329.

Ekström, P. and H. Meissl. 1988. Intracellular staining of physiologically identified photoreceptor cells and hyperpolarizing interneurons in the teleost pineal organ. *Neuroscience* 25: 1061–1070.

Ekström, P. and H. Meissl. 1997. The pineal organ of teleost fish. *Reviews in Fish Biology and Fisheries* 7: 199–284.

Ekström, P. and H. Meissl. 2003. Evolution of pineal photosensory organs in new light. The fate of neuroendocrine photoreceptors. *Philosophical Transactions of the Royal Society B: Biological Sciences* 358: 1679–1700.

Ekström, P., R.G. Foster, H.W. Korf, and J.J. Schalken. 1987. Antibodies against retinal photoreceptor-specific proteins reveal axonal projections from the photosensory pineal organ in teleosts. *Journal of Comparative Neurology* 265: 25–33.

Ekström, P., T. Östholm and B.I. Holmqvist. 1994. Primary visual projections and pineal neural connections in fishes, amphibians and reptiles. *Advances in Pineal Research* 8: 1–18.

Eldred, W.D. and J. Nolte. 1978. Pineal photoreceptors: Evidence for a vertebrate visual pigment with two physiologically active states. *Vision Research* 18: 29–32.

Falcón, J. 1979. L'organe pinéal du Brochet (*Esox lucius* L.). II. Étude en microscopie électronique de la différenciation et de la rudimentation partielle des photorécepteurs; conséquences possibles sur l'élaboration des messages photosensoriels. *Annales de Biologie Animale Biochimie Biophysique* 19: 661–688.

Falcón, J. 1999. Cellular circadian clocks in the pineal. *Progress in Neurobiology* 58: 121–162.

Falcón, J. and H. Meissl. 1981. The photosensory function of the pineal organ of the pike (*Esox lucius* L.). Correlation between structure and function. *Journal of Comparative Physiology* A 144: 127–137.

Falcón, J. and J. Tanabe. 1983. Early receptor potential of pineal organ and lateral eye of the pike. *Naturwissenschaften* 70: 149–150.

Falcón, J. and P. Gaildrat. 1997. Variations in cyclic adenosine 3′,5′-monophosphate and cyclic guanosine 3′,5′-monophosphate content and efflux from the photosensitive pineal organ of the pike in culture. *Pflügers Archiv European Journal of Physiology* 433: 336–342.

Falcón, J., J. Guerlotté, P. Voisin and J.P. Collin. 1987. Rhythmic melatonin biosynthesis in a photoreceptive pineal organ: a study in the pike. *Neuroendocrinology* 45: 479–486.

Falcón, J., L. Besseau, S. Sauzet, G. Boeuf. 2007. Melatonin effects on the hypothalamo-pituitary axis in fish. *Trends in Endocrinology and Metabolism* 18: 81–88.

Fejér, Z. S., A. Szél, P. Röhlich, T. Görcs, M.J. Manzano E Silva and B. Vigh. 1997. Immunoreactive pinopsin in pineal and retinal photoreceptors of various vertebrates. *Acta Biologica Hungarica* 48: 463–471.

Fernald, R.D. 2006. Casting a genetic light on the evolution of eyes. *Science* 313: 1914–1918.

Finn, J.T., E.C. Solessio and K.-W. Yau. 1997. A cGMP-gated cation channel in depolarizing photoreceptors of the lizard parietal eye. *Nature (London)* 385: 815–819.

Forsell, J., P. Ekström, I.N. Flamarique and B. Holmqvist. 2001. Expression of pineal ultraviolet- and green-like opsins in the pineal organ and retina of teleosts. *Journal of Experimental Biology* 204: 2517–2525.

Forsell, J., B. Holmqvist and P. Ekström. 2002. Molecular identification and developmental expression of UV and green mRNAs in the pineal organ of the Atlantic halibut. *Developmental Brain Research* 136: 51–62.

Frigato, E., D. Vallone, C. Bertolucci and N.S. Foulkes. 2006. Isolation and characterization of melanopsin and pinopsin expression within photoreceptive sites of reptiles. *Naturwissenschaften* 93: 379–385.

Gamse, J.T., Y.C. Shen, C. Thisse, B. Thisse, P.A. Raymond, M.E. Halpern and J.O. Liang. 2002. Otx5 regulates genes that show circadian expression in the zebrafish pineal complex. *Nature Genetics* 30: 117–121.

Gern, W.A. and S.S. Greenhouse. 1988. Examination of in vitro melatonin secretion from superfused trout (*Salmo gairdneri*) pineal organs maintained under diel illumination or continuous darkness. *General and Comparative Endocrinology* 71: 163–174.

Glover, K.A. 2008. Genetic characterisation of farmed rainbow trout in Norway: intra- and inter-strain variation reveals potential for identification of escapees. *BMC Genetics* 9: 87.

Hafeez, M.A. 1971. Light microscopic studies on the pineal organ in teleost fishes with special regard to its function. *Journal of Morphology* 134: 281–314.

Hamasaki, D.I. and P. Streck. 1971. Properties of the epiphysis cerebri of the small-spotted dogfish shark, *Scyliorhinus caniculus* L. *Vision Research* 11: 189–198.

Hartwig, H.-G. and Ch. Baumann. 1974. Evidence for photosensitive pigments in the pineal complex of the frog. *Vision Research* 14: 597–598.

Iigo, M., S. Kitamura, K. Ikuta, F.J. Sánchez-Vázquez, R. Ohtani-Kaneko, M. Hara, K. Hirata, M. Tabata and K. Aida. 1998. Regulation by light and darkness of melatonin secretion from the superfused masu salmon (*Oncorhynchus masou*) pineal organ. *Biological Rhythm Research* 29: 86–97.

Iigo, M., T. Abe, S. Kambayashi, K. Oikawa, T. Masuda, K. Mizusawa, S. Kitamura, T. Azuma, Y. Takag, K. Aida and T. Yanagisawa. 2007. Lack of circadian regulation of *in vitro* melatonin release from the pineal organ of salmonid teleosts. *General and Comparative Endocrinology* 154: 91–97.

Iyer, L.M., L. Aravind, S.L. Coon, D.C. Klein and E.V. Koonin. 2004. Evolution of cell-cell signaling in animals: did late horizontal gene transfer from bacteria have a role. *Trends in Genetics* 20: 292–299.

Kawamura, S. and S. Tachibanaki. 2008. Rod and cone photoreceptors: molecular basis of the difference in their physiology. *Comparative Biochemistry and Physiology* A150: 369–377.

Klein, D.C. 2004. The 2004 Aschoff-Pittendrigh lecture: Theory of the origin of the pineal gland—a tale of conflict and resolution. *Journal of Biological Rhythms* 19: 264–279.

Kojima, D., H. Mano and Y. Fukada. 2000. Vertebrate ancient-long opsin: a green-sensitive photoreceptive molecule present in zebrafish deep brain and retinal horizontal cells. *Journal of Neuroscience* 20: 2845–2851.

Kojima, D., J.E. Dowling and Y. Fukada. 2008. Probing pineal-specific gene expression with transgenic zebrafish. *Photochemistry and Photobiology* 84: 1011–1015.

Korf, H.W. 1974. Acetylcholinesterase-positive neurons in the pineal and parapineal organs of the rainbow trout, *Salmo gairdneri* (with special reference to the pineal tract). *Cell and Tissue Research* 155: 475–489.

Koyanagi, M and A. Terakita. 2008. Gq-coupled rhodopsin subfamily composed of invertebrate visual pigment and melanopsin. *Photochemistry and Photobiology* 84: 1024–1030.

Koyanagi, M., E. Kawano, Y. Kinugawa, T. Oishi, Y. Shichida, S. Tamotsu and A. Terakita. 2004. Bistable UV pigment in the lamprey pineal. *Proceedings of the National Academy of Sciences of the United States of America* 101: 6687–6691.

Kusmic, C. and P. Gualtieri. 2000. Morphology and spectral sensitivities of retinal and extraretinal photoreceptors in freshwater teleosts. *Micron* 31: 183–200.

Kusmic, C., P.L. Marchiafava and E. Strettoi. 1992. Photoresponses and light adaptation of pineal photoreceptors in the trout. *Proceedings of the Royal Society* (London) B248: 149–157.

Kusmic, C., L. Barsanti, V. Passarelli and P. Gualtieri. 1993. Photoreceptor morphology and visual pigment content in the pineal organ and in the retina of juvenile and adult trout, *Salmo irideus*. *Micron* 24: 279–286.

Lamb, T.D., S.P. Collin and E.N. Pugh, Jr. 2007. Evolution of the vertebrate eye: opsins, photoreceptors, retina and eye cup. *Nature Reviews Neuroscience* 8: 960–975.

Land, M.F. and D.-E. Nilsson. 2002. *Animal Eyes*. Oxford University Press, Oxford.

Lythgoe, J.N. 1966. Visual pigments and underwater vision. In: *Light as an Ecological Factor*, R. Bainbridge, G.C. Evans and O. Rackham (eds.). Blackwell, Oxford, pp. 375–391.

Mano, H. and Y. Fukada. 2007. A median third eye: pineal gland retraces evolution of vertebrate photoreceptive organs. *Photochemistry and Photobiology* 83: 11–18.

Mano, H., D. Kojima and Y. Fukada. 1999. Exo-rhodopsin: a novel rhodopsin expressed in the zebrafish pineal gland. *Molecular Brain Research* 73: 110–118.

Max, M. and M. Menaker. 1992. Regulation of melatonin production by light, darkness, and temperature in the trout pineal. *Journal of Comparative Physiology* A170: 479–489.

Max, M., P.J. McKinnon, K.J. Seidenman, R.K. Barrett, M.L. Applebury, J.S. Takahashi and R.F. Margolskee. 1995. Pineal opsin: a nonvisual opsin expressed in chick pineal. *Science* 267: 1502–1506.

McFarland, W.N. 1986. Light in the sea—Correlations with behaviors of fishes and invertebrates. *American Zoologist* 26: 389–401.

McNulty, J.A. 1984. Responses of synaptic ribbons in pineal photoreceptors under normal and experimental lighting conditions. *Journal of Pineal Research* 1: 139–147.

Meiniel, A. 1980. Ultrastructure of serotonin-containing cells in the pineal organ of *Lampetra planeri* (Petromyzontidae). A second sensory cell line from photoreceptor cell to pinealocyte. *Cell and Tissue Research* 207: 407–427.

Meissl, H. and P. Ekström. 1988a. Photoreceptor responses to light in the isolated pineal organ of the trout, *Salmo gairdneri*. *Neuroscience* 25: 1071–1076.

Meissl, H. and P. Ekström. 1988b. Dark and light adaptation of pineal photoreceptors. *Vision Research* 28: 49–56.

Meissl, H. and P. Ekström. 1991. Action of γ-aminobutyric acid (GABA) in the isolated photosensory pineal organ. *Brain Research* 562: 71–78.

Meissl, H. and S.R. George. 1984. Electrophysiological studies on neuronal transmission in the frog's photosensory pineal organ. The effect of amino acids and biogenic amines. *Vision Research* 24: 1727–1734.

Meissl, H., T. Nakamura and G. Thiele. 1986. Neural response mechanisms in the photoreceptive pineal organ of goldfish. *Comparative Biochemistry and Physiology* A84: 467–475.

Miller, W.H. and M.L. Wolbarsht. 1962. Neural activity in the parietal eye of a lizard. *Science* 135: 316–317.

Morita, Y. 1966. Entladungsmuster pinealer Neurone der Regenbogenforelle (*Salmo irideus*) bei Belichtung des Zwischenhirns. *Pflügers Archiv European Journal of Physiology* 289: 155–167.

Morita, Y. and E. Dodt. 1973. Slow photic responses of the isolated pineal organ of lampreys. *Nova Acta Leopoldina* 38: 331–339.

Morita, Y. and E. Dodt. 1975. Early receptor potential from the pineal photoreceptor. *Pflügers Archiv European Journal of Physiology* 354: 273–280.

Moutsaki, P., J. Bellingham, B.G. Soni, Z.K. David-Gray and R.G. Foster. 2000. Sequence, genomic structure and tissue expression of carp (*Cyprinus carpio* L.) vertebrate ancient (VA) opsin. *FEBS Letters* 473: 316–322.

Munz, F.W. 1965. Adaptation of visual pigments to the photic environment. *Ciba Foundation Symposium. Colour Vision, Physiology and Experimental Psychology.* J. & A. Churchill, London, pp. 27–45.

Nakamura, T., G. Thiele and H. Meissl. 1986. Intracellular responses from the photosensitive pineal organ of the teleost, *Phoxinus phoxinus*. *Journal of Comparative Physiology* 159A: 325–330.

Okano, T., T. Yoshizawa and Y. Fukada. 1994. Pinopsin is a chicken pineal photoreceptive molecule. *Nature* 327: 94–97.

Oksche, A. 1965. Survey of the development and comparative morphology of the pineal organ. *Progress in Brain Research* 10: 3–29.

Oksche, A. 1971. Sensory and glandular elements of the pineal organ. In: *The Pineal Gland* G.E.W. Wolstenholme and J. Knight (eds). JA Churchill, London, pp. 127–146

Oksche, A. and H. Kirchstein. 1971. Weitere elektronenmikroskopische Untersuchungen am Pinealorgan von *Phoxinus laevis* (Teleostei, Cyprinidae). *Zeitschrift für Zellforschung* 112: 572–588.

Omura, Y. 1984. Pattern of synaptic connections in the pineal organ of the ayu, *Plecoglossus altivelis* (Teleostei). *Cell and Tissue Research* 236: 611–617.

Omura, Y. and M.A. Ali. 1981. Ultrastructure of the pineal organ of the killifish, *Fundulus heteroclitus*, with special reference to the secretory function. *Cell and Tissue Research* 219: 355–369.

Omura, Y. and M. Oguri. 1969. Histological studies on the pineal organ of 15 species of teleosts. *Bulletin of the Japanese Society of Scientific Fisheries* 35: 991–1000.

Parry, J.W.L., S.N. Peirson, H. Wilkens and J.K. Bowmaker. 2003. Multiple photopigments from the Mexican blind cavefish, *Astyanax fasciatus*: a microspectrophotometric study. *Vision Research* 43: 31–41.

Perez-Leon, J.A., E.J. Warren, C.N. Allen, D.W. Robinson and R.L. Brown. 2006. Synaptic inputs to retinal ganglion cells that set the circadian clock. *European Journal of Neuroscience* 24: 1117–1123.

Peirson, S.N. and J.K. Bowmaker. 1999. The photopigment content of the goldfish pineal organ. *Investigative Ophthalmology & Visual Science* 40(4): S158.

Philp, A.R., J. Garcia-Fernandez, B.G. Soni, R.J. Lucas, J. Bellingham and R.G. Foster. 2000a. Vertebrate ancient (VA) opsin and extraretinal photoreception in the atlantic salmon (*Salmo salar*). *Journal of Experimental Biology* 203: 1925–1936.

Philp, A.R., J. Bellingham, J.M. Garcia-Fernandez and R.G. Foster. 2000b. A novel rod-like opsin from the extra-retinal photoreceptors of teleost fish. *FEBS Letters* 468: 181–188.

Pierce, L.X., R.R. Noche, O. Ponomareva, C. Chang and J.O. Liang. 2008. Novel functions for Period 3 and Exo-rhodopsin in rhythmic transcription and melatonin biosynthesis within the zebrafish pineal organ. *Brain Research* 1223: 11–24.

Pombal. M.A., J. Yáñez, O. Marín, A. González and R. Anadón. 1999. Cholinergic and GABAergic neuronal elements in the pineal organ of lampreys, and tract-tracing observations of differential connections of pinealofugal neurons. *Cell and Tissue Research* 295: 215–233.

Pu, G.A. and J.E. Dowling. 1981. Anatomical and physiological characteristics of pineal photoreceptor cell in the larval lamprey, *Petromyzon marinus*. *Journal of Neurophysiology* 46: 1018–1038.

Pugh, E.N. Jr. and T.D. Lamb. 1990. Cyclic GMP and calcium: The internal messengers of excitation and adaptation in vertebrate photoreceptors. *Vision Research* 30: 1923–1948.

Rüdeberg, C. 1968. Structure of the pineal organ of the sardine, *Sardina pilchardus sardine* (Risso) and some further remarks on the pineal organ of *Mugil* spp. *Zeitschrift für Zellforschung* 84: 219–237.

Rüdeberg, C. 1969. Structure of the parapineal organ of the adult rainbow trout, *Salmo gairdneri* Richardson. *Zeitschrift für Zellforschung* 93: 282–304.

Samejima, M., S. Tamotsu, K. Watanabe and Y. Morita. 1989. Photoreceptor cells and neural elements with long axonal processes in the pineal organ of the lamprey, *Lampetra japonica*, identified by use of the horseradish peroxidase method. *Cell and Tissue Research* 258: 219–224.

Shichida, Y. and H. Imai. 1998. Visual pigment: G-protein-coupled receptor for light signals. *Cellular and Molecular Life Sciences* 54: 1299–1315.

Solessio, E. and G.A. Engbretson. 1993. Antagonistic chromatic mechanisms in photoreceptors of the parietal eye of lizards. *Nature* 364: 442–445.

Soni; B.G., A.R. Philp and R.G. Foster. 1998. Novel retinal photoreceptors. *Nature* 394: 27–28.

Su, C.Y., D.G. Luo, A. Terakita, Y. Shichida, H.W. Liao, M.A. Kazmi, T.P. Sakmar and K.W. Yau. 2006. Parietal-eye phototransduction components and their potential evolutionary implications. *Science* 311: 1617–1621.

Tabata, M., T. Suzuki and H. Niwa. 1985. Chromophores in the extraretinal photoreceptor (pineal organ) of teleosts. *Brain Research* 338: 173–176.

Takanaka, Y., T. Okano, K. Yamamoto and Y. Fukada. 2002. A negative regulatory element required for light-dependent pinopsin gene expression. *Journal of Neuroscience* 22: 4357–4363.

Takechi, M., S. Seno and S. Kawamura. 2008. Identification of cis-acting elements repressing blue opsin expression in zebrafish UV cones and pineal cells. *Journal of Biological Chemistry* 14: 31625–31632.

Tamotsu, S. and Y. Morita. 1990. Blue sensitive visual pigment and photoregeneration in pineal photoreceptors measured by high-performance liquid chromatography. *Comparative Biochemistry and Physiology* B96: 487–490.

Torii, M., D. Kojima, T. Okano, A. Nakamura, A. Terakita, Y. Shichida, A. Wada and Y. Fukada. 2007. Two isoforms of chicken melanopsins show blue light sensitivity. *FEBS Letters* 581: 5327–5331.

Uchida, K. and Y. Morita. 1990. Intracellular responses from UV-sensitive cells in the photosensory pineal organ. *Brain Research* 534: 237–242.

Uchida, K., T. Nakamura and Y. Morita. 1992. Signal transmission from pineal photoreceptors to luminosity-type ganglion cells in the lamprey, *Lampetra japonica*. *Neuroscience* 47: 241–247.

van Veen, T. 1982. The parapineal and pineal organs of the elver (glass eel), *Anguilla anguilla* L. *Cell and Tissue Research* 222: 433–444.

Vigh, B., I. Vigh-Teichmann and B. Aros. 1975. Comparative ultrastructure of cerebrospinal fluid-contacting neurons and pinealocytes. *Cell and Tissue Research* 158: 409–424.

Vigh-Teichmann, I. and B. Vigh. 1983. The system of cerebrospinal fluid-contacting neurons. *Archives of Histology and Cytology* 46: 427–468.

Vollrath, L. 1981. The pineal organ. In: *Handbuch der mikroskopischen Anatomie des Menschen* VI/7, A. Oksche and L. Vollrath (eds), Springer-Verlag, Heidelberg.

Vuilleumier, R., L. Besseau, G. Boeuf, A. Piparelli, Y. Gothilf, W.G. Gehring, D.C. Klein and J. Falcón. 2006. Starting the zebrafish pineal circadian clock with a single photic transition. *Endocrinolology* 147: 2273–2279.

Wake, K. 1973. Acetylcholinesterase-containing nerve cells and their distribution in the pineal organ of the goldfish, *Carassius auratus*. *Zeitschrift für Zellforschung* 145: 287–298.

Westermann, B.A. and H. Meissl. 2008. Nitric oxide in photoreceptive pineal organs of fish. *Comparative Biochemistry and Physiology* A151: 198–204.

Xiong, W.H., E.C. Solessio and K.W. Yau. 1998. An unusual cGMP pathway underlying depolarizing light response of the vertebrate parietal-eye photoreceptor. *Nature Neuroscience* 1: 359–365.

Yáñez, J., M.A. Pombal and R. Anadón. 1999. Afferent and efferent connections of the parapineal organ in lampreys: a tract tracing and immunocytochemical study. *Journal of Comparative Neurology* 403: 171–189.

Ziv, L., A. Tovin, D. Strasser and Y. Gothilf. 2007. Spectral sensitivity of melatonin suppression in the zebrafish pineal gland. *Experimental Eye Research* 84: 92–99.

Zuber, M.E., G. Gestri, A.S. Viczian, G. Barsacchi and W.A. Harris. 2003. Specification of the vertebrate eye by a network of eye field transcription factors. *Development* 130: 5155–5167.

4

MELATONIN IN FISH: CIRCADIAN RHYTHM AND FUNCTIONS

Russel J. Reiter,[CA] *Dan-Xian Tan* and *Lucien C. Manchester*

INTRODUCTION

Melatonin (*N*-acetyl-5-methoxytryptamine) was isolated from bovine pineal tissue and chemically characterized by a group of dermatologists (Lerner *et al.*, 1958, 1959) five decades ago who intended to use it in the treatment of pigmentary skin disorders. These studies were initiated because of a discovery 50 years earlier that something of pineal origin had a powerful blanching effect in the skin of tadpoles (McCord and Allen, 1917). At the time of melatonin's discovery, there was widespread agreement among scientists that the pineal gland was a non-functional neural vestige. Since the characterization of melatonin, its synthesis in the pineal of all species has been documented and, because of its almost exclusive production at night, it is known as the chemical expression of darkness (Reiter, 1991). The light:dark dependency of the melatonin rhythm also has been characterized in a variety of fish species as in other classes of vertebrates (Bromage *et al.*, 2001; Nikaido *et al.*, 2009).

Research on the physiology of melatonin in fish has not matured to the level of that in other vertebrates, in particular mammals. There are, however, excellent scientists working in this area and new data are accumulating regarding the multiplicity of functions of melatonin in fish. Many of these findings are summarized in this review.

Department of Cellular and Structural Biology, University of Texas Health Science Center, San Antonio, Texas, USA.
[CA]Corresponding author: reiter@uthscsa.edu

THE CIRCADIAN MELATONIN RHYTHM: MULTIPLE REGULATORY FACTORS

Melatonin is rhythmically produced in both the pineal gland and the retinas of teleosts (Bromage *et al.*, 2001) with the former considered to be the major contributor to blood levels of melatonin (Kezuka *et al.*, 1989). Among teleosts, both solar and lunar light may influence pineal melatonin production. The rabbitfish (*Siganus guttatus*) is a restricted lunar-synchronized spawner which lays its eggs during a specific lunar phase. One obvious change during the lunar cycle is the widely divergent degree of brightness between the periods of the full moon and new moon. Using this information, Takemura *et al.* (2006) tested whether the cultured pineal gland of the rabbitfish is able to perceive and respond to moonlight intensities. Rabbitfish fry, collected in the wild, were used as a source of pineal glands. Glands were cultured and exposed to either natural or artificial light conditions with the release of melatonin into the medium being the important endpoint. Under artificial light:dark conditions, maximal melatonin release was always associated with the dark period while under continual lighting, melatonin release was suppressed. When organ cultured pineal glands were maintained under natural lighting conditions at the time of the full and new moon, only small amounts of melatonin were secreted at night. The findings show that, as in virtually all fish species, pineal gland function changes over a light:dark cycle and that the gland of the rabbitfish does respond to the light intensity of the bright moon.

The same group followed this study with the examination of the effects of light, including moonlight, on *period 2* (*Per 2*), a light-inducible clock gene in lower vertebrates, in the pineal gland of the rabbitfish (Sugama *et al.*, 2008). In intact fish, *Per 2* expression in the pineal gland exhibited a daily rhythm with an elevated number of transcripts occurring during the day. Likewise, exposure of fish to light at night rapidly increased *Per 2* levels in the pineal. Finally, exposure of fish to two hours of full moonlight intensity upregulated *Per 2* expression in the pineal. This documents that the pineal of the intact rabbitfish also is capable of detecting and responding to light of the full moon. This may relate to the lunar cycle-dependent spawning cycle in this, and other, species.

The pineal gland of different teleosts are differentially sensitive to environmental light. The photoneuroendocrine cells of two teleosts, the Atlantic salmon (*Salmo salar*) and the European sea bass (*Dicentrarchus labrax* L.) in terms of their light sensitivity was compared by Migaud *et al.* (2006). In both species, the exposure of *ex vivo* glands or intact fish to light inhibited melatonin synthesis and release; however, the cultured gland studies documented that the pineal of the sea bass was 10 times more sensitive to light than was the gland of the salmon. Migaud and co-workers (2006) also examined the degree of light penetration through the cartilaginous pineal window in the skull. These studies established that more light penetrates the window of the sea bass skull; this penetration was especially noticeable toward the red end of the visible spectrum and suggests that these specific wavelengths may be important in regulating melatonin production in this species. The differences in light sensitivity of the pineal of these species presumably relate to adaptation to their particular niches.

Plasma melatonin in the sea bass is most sensitive to suppression by blue light (half-band width, 434–477nm) with the minimal light intensity required for inhibition being $6.0\mu W/cm^2$ (Bayarri *et al.*, 2002). As expected, downwards directed light had a greater inhibitory effect in suppressing plasma melatonin than did upwards directed light.

In fish in their natural habitat, seasonal changes in daylength are accompanied by fluctuations in water temperature as well. Using the Senegal sole (*Solea senegalensis* L.), the interaction of seasonal fluctuations in daylength and water temperature in determining plasma melatonin levels was investigated (Vera *et al.*, 2007). Higher mid-night plasma melatonin levels were measured in the summer than in the winter with the highest values being reached when water temperature was 25°C. While water temperature influenced nighttime melatonin concentrations, it was without effect on low daytime concentrations. Since melatonin levels in this study were only measured at mid-dark and mid-light, it is difficult to determine whether more or higher melatonin levels actually were produced during the summer than during the winter. Such infrequent blood sampling times may have jeopardized identifying the highest or lowest melatonin levels.

The Arctic charr (*Salvelinus alpinus*) is the northern most freshwater fish. It lives in habitats with extreme variations in both daylength and temperature. Strand *et al.* (2008) compared plasma melatonin levels in Arctic charr at different seasons in Lake Storvatnet (70° N) in northern Norway. From September to April a daily melatonin was present with highest levels at night. During these months the lake was covered by thick ice and snow. This fluctuation reflected the prevailing above-surface photoperiod even in the months when there were very slight changes in sub-surface irradiance measured between day and night. In June, plasma melatonin levels exhibited no daily variation and were uniformly low. In this month, it was presumed that the light irradiance at night exceeded the minimal brightness required to suppress melatonin production. These studies demonstrate that the Arctic charr, living at high latitudes, still keeps track of time under extreme conditions even when lakes are covered with thick ice and snow. These findings also suggested that other seasonal rhythms in this species, e.g., reproduction, may be mediated by a seasonally changing melatonin cycle.

Some euryhaline fish can tolerate water with very marked degrees of salinity. Two such species are the gilthead sea bream (*Sparus aurata*) and the

European sea bass (*Dicentrarchus labrax*). These species easily move from fresh water to sea water with high salinity. Kleszczynska *et al.* (2006) and Lopez-Olmeda and co-workers (2009) tested whether the salinity of the water influenced melatonin levels in euryhaline fish. In both species mentioned above (sea bream and sea bass), greater salinity was associated with higher melatonin levels. In the sea bass, water salinity also influenced melatonin receptor density. One implication of this finding is that melatonin may be involved in the migration of fish between waters of different salinities.

MELATONIN RECEPTORS IN FISH

In fish, as in the other vertebrates, many of the effects of melatonin involve its interaction with specific high- and low-affinity membrane receptors. Cloning and pharmacokinetic studies using the radioligand 2-[^{125}I]-iodomelatonin (^{125}Mel) aided in the precise identification and classification of melatonin receptor subtypes (Vanecek, 1998; Dubocovich and Markowska, 2005). The high-affinity receptors are members of the G-protein-coupled, seven transmembrance domain family. While three membrane melatonin receptors have been identified, the Mel1c subtype is relegated to non-mammalian vertebrates whereas MT1 (formerly known as Mel1a) and MT2 (formerly Mel1b) subtypes have been identified in all vertebrate species investigated. The MT2 receptor, however, may be non-functional in some species (Barrett *et al.*, 2003).

The membrane melatonin receptors are coupled to several intracellular messenger systems including the cyclic AMP (cAMP) pathway which is inhibited through members of the G_i-protein family which suppress adenylate cyclase activity when melatonin interacts with its receptor (Barrett *et al.*, 2003; Rimler *et al.*, 2006). In addition to inhibition of cAMP, melatonin may also activate the phospholipase C pathway through G_q proteins (Steffens *et al.*, 2003; Rimler *et al.*, 2006). Activation of phospholipase C leads to the formation of two messengers, diacylglycerol and inositol triphosphate; the former activates protein kinase C (PKC). Both the phospholipase C pathway as well as the cAMP pathway modulate intracellular calcium [Ca^{2+}]$_i$ via PKC and protein kinase A (PKA). The kinases control plasma membrane voltage-gated calcium channels while inositol triphosphate regulates intracellular calcium stores (Vanecek, 1998; Balik *et al.*, 2004). Finally, the coupling of cyclic GMP to melatonin receptors has been documented (Saenz *et al.*, 2002; Barrett *et al.*, 2003; Huang *et al.*, 2005).

Attempts to clone the fish melatonin receptor have only yielded partial sequences. The species in which these attempt have been made include the zebrafish (*Danio rerio*), the pike (*Esox lucius*) and the rainbow trout (*Oncorhynchus mykiss*). In the zebrafish, MT1, MT2 and Mel1c receptors have been identified and there is evidence that several copies of each exist in this species (Gaildrat and Falcon, 1999, 2000; Mazurias *et al.*, 1999). In fish, the MT2 receptor appears to be negatively coupled to the cAMP pathway (Gaildrat *et al.*, 2002).

The distribution of melatonin receptors in fish is widespread; this is particularly true of the brain with many neural structures containing the receptors (Vernadakis *et al.*, 1998; Gaildrat and Falcon, 1999; Bayarri *et al.*, 2004a; Park *et al.*, 2007a; Oliveira *et al.*, 2008). The percentage of each subtype of receptor present, however, varies widely (Falcon *et al.*, 2007). Gene expression and ^{125}Mel binding are found in neural areas of lampreys, chondrichthyans and teleosts that are receptive sites for sensory information; some of the major areas that are the recipients of the sensory information include the cerebellum, telecephalon, diencephalon, olfactory bulbs and optic tectum. A major input to these areas is derived from the retinas and the pineal gland itself. In the deep sea teleost (*Coryphaenoides armatus*), a species that survives at a depth where solar light does not penetrate, melatonin binding sites are not present in the optic tectum, diencephalon or cerebellum (Smith *et al.*, 1996). When the number of binding sites for melatonin are compared in the brain of surface-dwelling fish compared to those that are deep dwellers, the latter generally contain fewer receptors (Falcon *et al.*, 2007).

Melatonin binding sites/receptors were also identified in tissues involved in osmoregulation in three species of fish, the flounder (*Platichthys flesus*), rainbow trout (*Oncorhynchus mykiss*) and sea bream (*Sparus aurata*) (Kulczykowska *et al.*, 2006). The tissues of interest, i.e., gills, small intestine and kidney, were studied during the light and during darkness. The receptors were found to be of high affinity and G-protein coupled but they exhibited no day/night differences. The Bmax of the receptors varied by tissue and species.

TWENTY-FOUR HOUR VARIATIONS IN MELATONIN RECEPTORS

Measurable melatonin receptors (binding sites) commonly vary over a 24 hour period in mammals. This seems to be common for fish as well. In the catfish (*Silurus asotus*) brain, where melatonin receptors are well characterized (Iigo *et al.*, 1997), the Bmax values for the receptors did not exhibit daily variations maintained under a light:dark cycle of 12:12, but the Kd did fluctuate in a rhythmic manner. In the brain of the pike (*Esox lucius*), the Bmax of ^{125}Mel binding was greatest at the light-to-dark transition while the Kd exhibited a biphasic rhythm with depressed levels at mid-day and mid-night relative to other time points (Gaildrat *et al.*, 1998). In the retina of the sea bass (*Dicentrarchus labrax*), both the receptor affinity (Kd) and binding capacity (Bmax) exhibit clear 24 hour rhythms; these changes were in contrast to the brain of this species where no rhythms in Kd or Bmax for melatonin receptors were detected (Bayarri *et al.*, 2004b). In only two brain areas (optic tectum and cerebellum) of the Senegal sole (*Solea senegalensis*) were daily fluctuations in the Bmax of melatonin binding sites uncovered; in both neural regions, binding capacity was greatest at mid-dark. According to Amano *et al.* (2006), the Kd of melatonin binding sites in the brain of the masu salmon (*Oncorhynchus masou*) varied slightly over the course of a light:dark cycle with the fluctuation disappearing when the fish were kept under constant light or constant dark.

The most complete study of 24 hour changes in a melatonin receptor in the brain of a fish is the report of Park and co-workers (2007b). By measuring MT1 mRNA, they documented maximal levels at night in whole brain, retina, liver and kidney. These rhythms, initially measured under a light:dark cycle, persisted under conditions of continuous light or dark indicating that the rhythms had a circadian basis. This is also one of the few studies where melatonin receptors have been identified in non-neural structures, i.e., liver and kidney.

The golden rabbitfish (*Siganus guttatus*) exhibits restricted lunar-related rhythms and characteristically spawns synchronously at the first quarter moon. Park *et al.* (2006) cloned the melatonin receptor [Mel (1b)] cDNA and then examined mRNA expression in several tissues in this species. Highest expression was observed during the daytime in the liver and kidney, while in the brain a biphasic rhythm was apparent with peaks during the day and night. In the retina only a single peak occurred during the nighttime.

MELATONIN RECEPTOR FUNCTION IN BRAIN AND PITUITARY

While the receptor-mediated actions of melatonin in the mammalian brain are reasonably well known, in the fish brain, this information remains rudimentary. In mammals, the suprachiasmatic nuclei (SCN), the biological clock, and the pars tuberalis of the adenohypophysis are rich in melatonin receptors and are likely involved in regulation and synchronization of circadian and circannual rhythms. While the SCN of some fish have melatonin receptors, whether these nuclei function as a circadian oscillator remains unknown (Ekström and Vanecek, 1992; Vernadakis *et al.*, 1998). Other diencephalic nuclei also bind melatonin and, moreover, some of these structures express molecules normally involved with photoreception, for example, opsins, transducin and arrestin (Philip *et al.*, 2000a,b). Unraveling the association of the retinas, the SCN and the photopigments in different brain areas of teleosts will be a challenge for future researchers.

The preoptic area (POA) of fish brain is of particular interest given that it seems to integrate photoperiodic information derived from neurons originating in the retinas and in the photoreceptive pineal gland as well as the melatonin message from the pineal (Ekström and Vanecek, 1992; Ekström and Meissl, 1997; Vernadakis *et al.*, 1998). Neural inputs from the pineal gland provide information on changes in illumination arising from above while the eyes provide photic information in the horizontal plane. Additionally, the POA receives the daily and seasonally changing melatonin signal from the pineal gland. Axons originating from neurons in the POA then project to the pituitary gland; the neurotransmitters released from these axons are both monoamines and peptides. These data strongly suggest that photic information, with the POA as an intermediate, influences the release of hormones from the pituitary gland in fish. Certainly, this seems to be the case in the masu salmon (*Oncorhynchus masou*) where melatonin administration reportedly reduces pituitary levels of gonadotropin releasing hormone (GnRH) and luteinizing hormone (LH) while it stimulates follicle stimulating hormone (FSH) (Amano *et al.*, 2004). Somewhat more direct evidence was provided by findings in the Atlantic croaker (*Micropogonias undulatus*) where the intraventricular injection of melatonin depressed the release of LH from the pituitary gland (Khan and Thomas, 1996). Clearly, how photic information and the melatonin message are integrated in the POA (or other neural structures) and how the resulting message is passed to the pituitary gland is poorly understood in fish.

Besides an apparent action in the POA, melatonin may also modulate neuroendocrine physiology directly at the level of the pituitary gland. MT1 and MT2 receptor subtypes have been detected in the pike and trout pituitary gland using PCR; moreover, this structure binds [125]Mel (Gaildrat and Falcon 1999; Falcon *et al.*, 2003). Thus, some of the molecular machinery is present to mediate the effects of melatonin directly on pituitary hormone secretion. The outcome of these studies may be confounded by the sex of the animal being studied as well as its reproductive status (Falcon *et al.*, 2007).

Melatonin seems also to influence pituitary growth hormone (GH) release in fish, but the mechanisms are seemingly complex and require identification. Melatonin both inhibits and stimulates GH release with the former requiring the preactivation of the cAMP pathway. Since *in vivo*, GH is elevated at night and reduced during the day, it is possible that these changes are mediated by the diurnal melatonin fluctuation (Boeuf and Falcon, 2001).

MELATONIN AND CIRCADIAN/CIRCANNUAL RHYTHMS

Time keeping is a major function of melatonin in most species where it is capable of providing both clock and calendar information (Reiter, 1993). Melatonin, in many mammals, is produced in a circadian manner in two major sites, the pineal gland and the retinas. Despite many evolutionary changes, the day:night variation of melatonin production in these two organs has been maintained intact from fish to mammals (Falcon, 1999). In virtually all these species the melatonin peak occurs at night. However, in the case of the teleost retina, peak melatonin levels may occur at any time during the light:dark cycle or there may be no peak at all; this variation or lack thereof may relate to any number of factors including the species being investigated, the time of year, the reproductive state of the animal being studied, etc. (Beaseau *et al.*, 2006). The plasma melatonin rhythm persists in all fish studied to date with greatest concentrations always being associated with the daily dark period. In fish, the nocturnal duration of melatonin is short, but high, in the summer and long with a low amplitude in the winter (Falcon *et al.*, 2007). As in other classes of animals, melatonin serves to signal circadian and annual time (Reiter, 1987; Underwood, 1989).

There are data linking the daily melatonin cycle to various circadian fluctuations including those of locomotor activity, thermal preference, rest, food intake, vertical migration, shoaling, skin pigmentation, osmoregulation, metabolism, and sleep (Ekström and Meissl, 1997; Zhdanova *et al.*, 2001). Depending on the species, in fish the feeding pattern can be either diurnal or nocturnal and this can change depending on the physiological state of the species (Sanchez-Vazquez *et al.*, 2000). The regulation of locomotor activity in fish as influenced by melatonin differs among species with pinealectomy causing either a free-running rhythm, inducing splitting into two components or resulting in arrhythmicity. According to Falcon and colleagues (2007), this is consistent with the idea that the fish pineal organ functions as a circadian pacemaker with a multioscillatory circadian system; thus, differences between species reflect variations in coupling between the various oscillators (Underwood, 1989).

Under normal light:dark conditions, a demand feeding rhythm has been documented in some fish species; under constant light and temperature conditions this rhythm can be synchronized by food availability (Bolliet *et al.*, 2001; Pinillos *et al.*, 2001; Herrero *et al.*, 2005; Lopez-Olmeda *et al.*, 2006a,b). In terms of locomotor activity, the influence of melatonin seems to relate to the time of day it is administered.

In fish, circannual cycles include those of reproduction, growth and smoltification (in migrating salmonids). The changing seasonal photoperiod, as in some mammals (Reiter, 1980), plays a central role in annual changes in reproductive competence. In most, but not all, fish species breeding is associated with increasing day lengths. In these long day breeders, removal of the pineal gland during the winter months typically stimulates reproduction while this procedure in the summer inhibits sexual function. The opposite occurs in fish that are short-day breeders (Ekström and Meissl, 1997; Mayer *et al.*, 1997). While the observations summarized above have been reported, there are numerous expectations to these findings. The results differ with gender, reproductive status, and the prevailing light:dark cycle. Also, melatonin supplementation does not always compensate for the surgical removal of the pineal gland. As a result of the markedly diverse findings in this area, Ekström and Meissl (1997) concluded that the association of the melatonin rhythm with the annual reproductive cycle in fish is unproven.

In an attempt to document a potential relationship between seasonal changes in melatonin and the annual cycle of reproduction in the three spined stickleback (*Gasterosteus aculeatus*), the concentration of melatonin was measured in the brain of fish collected during different seasons and reproductive states (Sokolowska *et al.*, 2004). Very high levels of melatonin were identified in the brain in early spring (March) and in autumn (October) compared to values measured in other seasons. On the basis of the results, the authors surmised that melatonin may function as a calendar to determine annual changes in reproductive competence in the stickleback.

While the photoperiod has been shown to determine growth in salmonids (Boeuf and La Bail, 1999; Taylor *et al.*, 2005), in other species the findings are contradictory. For example, in goldfish (*Carassius auratus*) the effect of pinealectomy on growth was dependent on the photoperiod and melatonin reversed the change under short days but not under long days (EkstrÖm and Meissl, 1997). In the Atlantic salmon par (*Salmo salar*), pineal removal reduced growth rates at a time of year when day lengths were increasing but after the summer solstice pinealectomy accelerated the growth of the fish (Mayer *et al.*, 1997). The published information on the subject of photoperiod, the pineal gland, melatonin and growth provides no basis for any definitive conclusions concerning these associations.

MELATONIN AND SELECTION OF A SEXUAL MATE

A newly-discovered function of melatonin in fish may relate to the ability of the indoleamine to increase the sexual attraction of females to males by enhancing nuptial coloration and the associated behavior. This potential function of melatonin in regulating temporary changes in nuptial coloration was investigated in the two-spotted goby (*Gobiusculus flavescens*) (Sköld *et al.*, 2008). While ornamental or nuptial coloration in teleost fish is generally more conspicuous in males, in several fish including the goby, females develop regionally specific coloration patterns when they become sexually receptive (McLennan, 1995; Takahashi, 2000). In the case of the goby, females develop an orange belly which serves as a sexual attractant for the male. The coloration is a consequence of the color of the internal gonads along with the chromatophore-based pigmentation of the skin which also becomes transparent to display the underlying brightly-colored gonads (Fujii and Oshima, 1994).

When the belly skin of female gobies was treated with a combination of melatonin and melanocyte stimulating hormone (MSH), Sköld and co-workers (2008) noted that the belly skin exhibited increased orange coloration while melatonin combined with prolactin rendered the skin samples more transparent. These treatments exaggerated what was referred to as the chromatic "glow" effect of the skin, a change that typically accompanies courtship behavior. The visible change in the color of the skin was due to the aggregation of dark pigment granules in the melanophores of the epidermis as a result of melatonin (this is a well known action of melatonin in amphibian skin) while prolactin increased the orange coloration by inducing erythrophore (red/orange) and xanthophore (yellow) pigment dispersal.

The overall result was an increased "glowing" appearance of the belly skin which would serve to attach the male for courtship and mating. While Sköld *et al.* (2008) did not examine the effect of the hormone treatments on the *in vivo* behavioral changes that accompany revealing the belly skin to the male, i.e., what is referred to as sigmoid display, it seems likely that melatonin, alone or in combination with the other hormones used in their study would also influence these responses *in vivo*.

The findings of Sköld and colleagues (2008) may have far-reaching implications. Since enhanced nuptial coloration increases an individual's chances of mating and delivering offspring, the fact that melatonin is involved in improving attractive skin pigmentation suggests this indoleamine influences the selection of a sexual mate (Anderson, 1994). Moreover, given that the brighter coloration is usually associated with healthier individuals of a species, the implication is that melatonin aids in determining which individuals of a species will pass their genes to the next generation (Amundsen and Forsgren, 2001; Cotton *et al.*, 2006) and ensure survival of the fittest.

The actions of melatonin in contracting the pigment granules to an area around the nucleus as well as the effects on prolactin are likely receptor-mediated although that was not tested in this study. Preservation of the erythrophore and xanthophore pigments by melatonin may have been receptor-independent effects given that melatonin, as well as several of its metabolites, are free radical scavengers thereby functioning to preserve the other pigments (Tan *et al.*, 1993, 2007; Reiter, 2000; Peyrot and Ducrocq, 2008).

MELATONIN: EMBRYO DEVELOPMENT AND HATCHING

Hatching of the Atlantic halibut (*Hippoglossus hippoglossus*) has long been thought to be mediated by the photoperiod but the photosensory organ involved in this process has never been identified. In an attempt to resolve this issue, Forsell *et al.* (1997) examined the neurochemical differentiation of both the pineal gland and the retina of Atlantic salmon during embryonic development. Using immunocytochemical methods, they compared the appearance of the integral components of the phototransduction process in the developing pineal and retinas. The measures included opsins, arrestin (S-antigen) and alpha-transducin. Although melatonin per se was not measured, these workers did measure the levels of serotonin, a necessary precursor of melatonin.

The results revealed that the pineal anlage immunocytochemically contains each of the proteins required for phototransduction and serotonin between 11 and 15 days. Hatching normally takes place on day 15. Conversely, no immunoreactive cells were found in the embryonic retinal tissues during this time or in the newly hatched larva. The implication of the findings is that, if in fact hatching is a photoperiod-dependent process it is likely the developing pineal gland that is responsible for light detection; the most likely pineal secretory product would presumably be melatonin.

In salmonids, melatonin may influence estrogen receptor (ER) expression in various tissues. Considering this assumption, the effects of melatonin on ER and/or vitellogenin expression in the rainbow trout (*Onorhynchus mykiss*) were examined in a combination of *in vitro* and *in vivo* studies (Mazurais *et al.*, 2000). The findings showed, however, that melatonin had no influence on either basal or estradiol-beta stimulated ER expression in cells expressing the ER. Moreover, incubation of aggregates of hepatocytes with melatonin (10^{-8} to 10^{-4} M) did not change estrogen-stimulated ER or vitellogenin RNA expression. Similarly, neither pinealectomy nor melatonin treatment altered hepatic ER expression or vitellogenin levels. Seemingly, melatonin does not have a major effect or ER expression in the liver of the rainbow trout.

In the catfish (*Clarias batrachus L.*), the effect of melatonin on plasma vitellogenin levels varied with the season and the reproductive state of the fish (Ghosh and Nath, 2005). Melatonin, administered for 15 days, during the preparatory, prespawning, spawning and postspawning periods in both intact and pinealectomized catfish had either an inhibitory or no effect on blood levels of vitellogenin concentrations; the predominant effect was no response to melatonin treatment with administration of the indole depressing plasma vitellogenin levels only during the early prespawning period. This response seemed to be independent of gonadotrophin levels since, in all seasons, these were inhibited by melatonin.

MELATONIN, CALCIUM METABOLISM AND BONE STRUCTURE

A proposed association between melatonin and calcium metabolism has a long history (Kiss *et al.*, 1969; Csaba and Barath, 1974; Csaba and Bokay,

1977). In the subsequent years, the results of many studies uncovered a relationship between melatonin and bone formation in a variety of vertebrates (Cardinali *et al.*, 2003; Cutando *et al.*, 2007). One morphological feature that has become a characteristic of surgical removal of the pineal gland is the development of spinal scoliosis in birds (Machida *et al.*, 1995; Kanemura *et al.*, 1997; Wang *et al.*, 1998; Turgut *et al.*, 2003, 2005). In this case, it was presumed that the loss of the pineal gland resulted in altered calcium metabolism which weakened the osseous structure of the vertebrae causing, in bipedal birds, the development of spinal curvature. Initial studies in rodents failed to document an effect of pinealectomy on spinal scoliosis, however; the lack of effect was presumably due to their quadrupedal ambulatory gait which relieved the vertebral column of a great deal of stress. This was borne out when several groups (Tanaka *et al.*, 1982; Machida *et al.*, 1999; Oyama *et al.*, 2006) showed that pinealectomy, in fact, caused the same deformation of the vertebral column in rodents rendered bipedal due to early surgical removal of the forelimbs. These findings are consistent with the observations of Roth *et al.* (1999) and Ladizesky and colleagues (2001, 2003) who documented the positive effects of melatonin on osteoblastic activity and on bone calcification, respectively. Similar observations have been reported regarding the actions of melatonin on human osteoblasts (Satomura *et al.*, 2007).

In at least one species of fish (*Salmo salar L.*), as well, long term pinealectomy results in a number of negative changes in skeletal structure (Fjelldal *et al.*, 2004). This study is of particular interest inasmuch as the number of fish used was large and numerous parameters of bone structure were measured. In this case, 86 salmon were surgically pinealectomized while 30 were sham operated; the average weight of the fish was 3.22 kg at the time of surgical removal of the pineal. At 11 months after surgery, the 30 sham operated fish exhibited normal straight vertebral columns. Of the 86 pinealectomized fish, 80% (69 fish) had clear radiographic deformities of the spinal column. Abnormal curvature of the spine was apparent in both lateral and dorsal radiographs (Fjelldal *et al.*, 2004). A measurement of stiffness, yield load and resilience also revealed that the vertebral bodies of the pinealectomized fish were all worse than these parameters in fish with an intact pineal gland. Moreover, the vertebrae from pinealectomized salmon had a decreased calcium and phosphorus content and a reduced calcium to phosphorus molar ratio relative to these indices in sham-operated fish. These changes were also reflected in a lower total mineral content of the bones of the pinealectomized fish.

The outcome of the study of Fjelldal and co-workers (2004) confirmed the marked effect of the loss of the pineal gland, and the associated reduction of melatonin, in contributing to abnormal spinal curvature and reduced mechanical strength of the vertebrae. Given the locomotor pattern of fish, the observed changes appear to be independent of gravity or posture. The authors conclude by noting that the skeletal effects of pinealectomy in a salmonid are similar to those in birds and mammals and, therefore, the actions of melatonin on bone have been conserved for an estimated 400 million years. The authors also feel that the Atlantic salmon may be a useful model in which to study the relationship between the pineal gland and skeletal development in addition to the pathogenesis of idiopathic scoliosis in humans. Also, given that the observed changes occurred in melatonin deficient fish, it may be a model in which to examine osteoporosis in humans which typically becomes manifested in individuals after middle age when endogenous levels of melatonin are normally diminished (Reiter, 1995).

As in bone, osteoclasts (bone resorptive cells) and osteoblasts (bone formative cells) are involved in scale formation and turnover. Considering the effects of the pineal gland/melatonin on calcium homeostasis and bone structure, Suzuki and Hattori (2002) examined the effects of melatonin on cultured goldfish (*Carassius auratus*) scales. In this study, tartrate-resistant acid phosphatase (TRACP) activity was used as a marker of osteoclasts while alkaline phosphatase (ALP) activity was used to monitor osteoblasts. The addition of melatonin (10^{-9} to 10^{-5}M) to the incubation medium significantly suppressed both TRACP and ALP for 6 hours. Conversely, estradiol-17β enhanced the activities of both osteoblasts and osteoclasts; these responses were likewise inhibited by melatonin. Suzuki and Hattori (2002) also showed that melatonin reduced insulin-like growth factor -1 (IGF-1) mediated growth and differentiation of osteoblasts in goldfish scales. This was the first study to identify the inhibitory action of melatonin on osteoblastic and osteoclastic activities in scales. These actions on osteoblasts and osteoclasts in scales are reminiscent of those that occur during intramembraneous bone formation and suggest that melatonin may also influence this process as well.

Suzuki *et al.* (2008) also used the cultured scale assay to assess the actions of novel bromomelatonin derivatives as potentially useful drugs to treat bone diseases. Of particular interest was 1-benzyl-2, 4, 6-tribromomelatonin (10^{-9} to 10^{-6} M) which significantly activated scale osteoblasts (which would enhance scale formation) while inhibiting osteoclastic activity (10^{-9} to 10^{-6}M) (which degrades bone). The combined actions of this drug would generally enhance scale formation and mineralization. When tested, 1-benzyl-2, 4, 6-tribromomelatonin in fact augmented total bone mineral density of the femoral metaphysis in ovariectomized rats and enhanced the stress-strain index of the diaphysis.

MELATONIN AND SLEEP

It is generally believed and supported by considerable experimental data that melatonin is a circadian sleep promoting agent (Claustrat *et al.*, 2005;

Arendt, 2006; Gorfine *et al.*, 2006; Jan *et al.*, 2007) although there are dissenting opinions as to its efficacy in this regard (Buscemi *et al.*, 2006). The zebrafish (*Danio rerio*) is a diurnal vertebrate that is commonly used by researchers in developmental biological and genetic studies. This species has also been used, although sparingly, as a model to examine the effects of melatonin on sleep-like states in fish. In the initial study, Zhdanova and co-workers (2001) used indirect observations to evaluate sleep; these included prolonged quietness associated with specific postures, elevated arousal threshold and a more prolonged sleep-like state and heightened arousal threshold after sleep deprivation (referred to as sleep rebound). In this study, the animals were larval zebrafish 7–14 days of age.

The larval zebrafish were individually maintained in 350 µl of water and melatonin (concentrations in the water ranging from 10nM to 100µM) was added at Zeitgeber Time (ZT) 5. At all doses melatonin promoted the behavioral features considered to represent sleep. Similar behavioral changes were induced by other hypnotics including diazepam and pentobarbital. The actions of melatonin on the sleep state were prevented when luzindole, an MT1 and MT2 melatonin receptor blocker, was added in advance of melatonin treatment. Zhdanova *et al.* (2001) emphasize that these observations support the conclusion that melatonin is evolutionarily conserved as a sleep promoting agent in diurnal fish species. In mammals, electrophysiological criteria are commonly used to evaluate sleep; those were not possible to apply in these larval fish studies.

As in many other species, sleep efficiency deteriorates with increasing age in the zebrafish. This species typically survives to six years of age and exhibits age-associated physiological changes at two years of age (Tsai *et al.*, 2007). Using the same behavioral criteria/responses described above, Zhdanova and colleagues (2007) evaluated the effects of melatonin on the sleep-like state in aged (3–5 years) zebrafish. Prior to melatonin treatment, the aged fish exhibited an overall reduction in total sleep time, accompanied by reduced activity levels and a higher arousal threshold. The addition of melatonin to the water improved the criteria used to evaluate the sleep-state; these findings were consistent with the preserved levels of mRNA expression for melatonin receptors.

MELATONIN/IMMUNE SYSTEM INTERACTION

In mammals, the melatonin rhythm has a major impact on the immune system; these responses involve an interaction of melatonin with both membrane receptors and nuclear binding sites (Carrillo-Vico *et al.*, 2005). Information regarding a melatonin/immune system association in fish is meager and the potential interactions must be inferred from indirect evidence. Esteban *et al.* (2006) used two species, the hermaphrodite protandrous seawater gilthead seabream (*Sparus aurata* L.) and the gonocoric seawater sea bass (*Dicentrarchus labrax* L.) to examine possible associations of the photoperiod and the innate immune system. Both species of teleosts were maintained in a 12:12 light:dark cycle (light on daily from 08:00–20:00 h) with blood samples collected both during the dark and light periods. The humoral innate immune indices that were monitored included complement, and lysozyme and peroxidase activities. In both fish species, complement levels were elevated in the day compared to nighttime values. In the seabream, lysozymal activity was elevated early in the dark period, a response not measured in the plasma of the sea bass. Likewise, in the seabream the peroxidase activity was higher at the dark to light transition (08:00 h) than at any other time. Again, in the sea bass peroxidase activity exhibited no substantial changes over a light:dark cycle.

The authors (Estaban *et al.*, 2006) presumed that these observations are consistent with an effect of the pineal gland and the associated plasma melatonin rhythm on the immune system. In the absence of additional data, however, this presumption remains tenuous.

MELATONIN AS AN ANTIOXIDANT

Melatonin as well as several of its metabolites, e.g., cyclic 3-hydroxymelatonin, N^1-acetyl-N^2-formyl-5-methoxykynuramine (AFMK) and N^1-acetyl-5-methoxykynuramine (AMK), are highly effective in reducing molecular damage resulting from oxidative and nitrosative agents that are persistently generated in aerobic organisms (Tan *et al.*, 2007; Peyrot and Ducrocq, 2008; Reiter *et al.*, 2008). While the role of melatonin as an antioxidant is an active area of research in many organisms, there is a dearth of information in this investigative arena from fish with only a single incomplete and inconclusive report being uncovered (Sreejith *et al.*, 2007). While the minimal evidence presented in this report is consistent with melatonin's ability to reduce the peroxidation of fatty acids and to modulate antioxidative enzymes, the experimental design of these *in vitro* studies was questionably appropriate. This is an area that would benefit from more intensive investigation.

CONCLUDING REMARKS

Identification of the physiological actions of melatonin in fish is gradually yielding to an onslaught of studies in a variety of disciplines. The actions of melatonin are extremely complex in all vertebrates and that is no less so for fish, as shown in this review. Major discoveries have been made in clarifying the function of the pineal and melatonin in the last decade and it is anticipated that this research is on the threshold of significant advances.

Currently, numerous fish species are under investigation in a variety of different laboratories. There are no commonly-used fish models for the investigation of melatonin's actions on specific functions, e.g., circadian rhythm regulation. More uniformity among the different research laboratories in terms of the fish species investigated may assist in identifying precise actions of melatonin. Thereafter, the well established findings could be investigated in other fish species.

References

Amano, M., M. Iigo, K. Ikuta, S. Kitamura, K. Okuzawa, H. Yamadu and K. Yamumori. 2004. Disturbance of plasma melatonin profile by high dose melatonin administration inhibits testicular maturation of precocious male masu salmon. *Zoological Science* 21: 79–85.

Amano, N., M. Iigo, S. Kitamura, N. Amiya and K. Yamamori. 2006. Changes in melatonin binding sites under artificial light:dark, constant light and constant dark conditions in the masu salmon brain. *Comparative Biochemistry and Physiology* A 144: 509–513.

Amundsen, T. and E. Forsgren. 2001. Male mate choice selects for female coloration in a fish. *Proceedings of the National Academy of Science of the United States of America* 98: 13155–13160.

Anderson, M. 1994. *Sexual Selection*. Princeton University Press, Princeton.

Arendt, J. 2006. Melatonin and human rhythms. *Chronobiology International* 23: 21–37.

Balik, A., K. Kretschmannova, P. Mazna, I. Svobodova and H. Zemkova. 2004. Melatonin actions in neonatal gonadotrophs. *Physiological Research* 53: 153–166.

Barrett, P., S. Conway and P.J. Morgan. 2003. Digging deep—structure-function relationships in the melatonin receptor family. *Journal of Pineal Research* 35: 221–230.

Bayarri, M.J., J.A. Madrid and F.J. Sanchez-Vazquez. 2002. Influence of light intensity, spectrum and orientation on sea bass plasma and ocular melatonin. *Journal of Pineal Research* 32: 34–40.

Bayarri, M.J., R. Garcia-Allegue, J.A. Munoz-Cueto, J.A. Madrid, M. Tabota, F.J. Sanchez-Vazquez and M. Iigo. 2004a. Melatonin binding sites in the brain of European sea bass (*Dicentrarchus labrax*). *Zoological Science* 21: 427–434.

Bayarri, M.J., M. Iigo, J.A. Munoz-Custo, E. Isorna, M.J. Delgado, J.A. Madrid, F.J. Sanchez-Vazquez and A.L. Alonso-Gomez. 2004b. Binding characteristics and daily rhythms are distinct in the retina and the brain areas of the European sea bass retina (*Dicentrarchus labrax*). *Brain Research* 1029: 241–250.

Besseau, L., A. Benyassi, M. Moller, S.C. Coon, J.L. Weller, G. Boeuf and D.C. Klein. 2006. Melatonin pathway: breaking the "high-at-night" role in the trout retina. *Experimental Eye Research* 82: 620–627.

Boeuf, G. and P.Y. La Bail. 1999. Does light have an influence on fish growth? *Aquaculture* 177: 129–152.

Boeuf, G. and J. Falcon. 2001. Photoperiod and growth in fish. *Vic et Milieu* 51: 237–346.

Bolliet, V., A. Aranda and T. Boujard. 2001. Demand-feeding rhythm in rainbow trout and European catfish: synchronization by photoperiod and food availability. *Physiology and Behavior* 73: 625–633.

Bromage, N., M. Porter and C. Randall. 2001. The environmental regulation in farmed finfish with special reference to the role of photoperiod and melatonin. *Aquaculture* 197: 63–98.

Buscemi, N., B. Vandemeer, N. Houton, R. Pandya, L. Tjosvold, L. Hartling, S. Vobra, T.P. Klassen and G. Baker. 2006. Efficacy and safety of exogenous melatonin for secondary sleep disorders and sleep disorders accompanying sleep restriction: meta-analysis. *British Medical Journal* 332: 385–393.

Cardinali, D.P., M.G. Ladizesky, V. Boggio, R.A. Cutrera and C. Mautalen. 2003. Melatonin effects on bone: experimental facts and clinical perspectives. *Journal of Pineal Research* 34: 81–87.

Carrillo-Vico, A., J.M. Guerrero, P.J. Lardone and R.J. Reiter. 2005. A review of the multiple actions of melatonin on the immune system. *Endocrine* 27: 189–200.

Csaba, G. and P. Barath 1974. The effect of pinealectomy on the parathyroid follicular cells of the rat thyroid gland. *Acta Anatomica* 88: 137–146.

Csaba, G. and J. Bokay. 1977. The effect of melatonin and corpus pineale extract on serum electrolytes in the rat. *Acta Biologica Academia Science Hungaria* 28: 143–144.

Claustrat, B., J. Brun and G. Chazot. 2005. The basic physiology and pathophysiology of melatonin. *Sleep Medicine Reviews* 9: 11–24.

Cotton, S., J. Small and A. Pomiankowski. 2006. Sexual selection and condition-dependent made preferences. *Current Biology* 16: R755–R765.

Cutando, A., G. Gomez-Moreno, G. Arana, D. Acuna-Castroviejo and R.J. Reiter. 2007. Melatonin: potential functions in the oral cavity. *Journal of Periodontology* 78: 1094–1102.

Dubocovich, M.L. and M. Markowska. 2005. Functional MT1 and MT2 melatonin receptors in mammals. *Endocrine* 27: 101–110.

Ekström, P. and J. Vanecek. 1992. Localization of 2-[^{125}I] iodomelatonin binding sites in the brain of the Atlantic salmon, *Salmo salar* L. *Neuroendocrinology* 55: 529–537.

Ekström, P. and H. Meissl. 1997. The pineal organ of teleost fishes. *Review of Fish Biology and Fisheries* 7: 199–284.

Estaban, M.A., A. Cuesta, A. Rodriguez and J. Meseguer. 2006. Effect of photoperiod of the fish innate immune system: a link between fish pineal gland and the immune system. *Journal of Pineal Research* 41: 261–266.

Falcon, J. 1999. Cellular circadian clocks in the pineal. *Progress in Neurobiology* 58: 121–162.

Falcon, J., L. Besseau, S. Sauzet and G. Boeuf. 2007. Melatonin effects on the hypothalamo-pituitary axis of fish. *Trends in Endocrinology and Metabolism* 18: 81–88.

Falcon, J., L. Besseau, D. Fazzari, J. Attia, P. Gaildrat, M. Beauchaud and G. Boeuf. 2003. Melatonin modulates secretion of growth hormone and prolactin by trout pituitary glands and cells in culture. *Endocrinology* 144: 4648–4658.

Fjelldal, P.G., S. Grotmol, H. Kryvi, N.B. Gjerdet, G.L. Taranger, T. Hansen, M.J.R. Porter and G.K. Totland. 2004. Pinealectomy induces malformation of the spine and reduces the mechanical strength of the vertebrae in Atlantic salmon, *Salmo salar*. *Journal of Pineal Research* 36: 132–139.

Forsell, J., B. Holmqvist, J.V. Helvik and P. Ekström. 1997. Role of the pineal organ in photoregulated hatching of the Atlantic salmon. *International Journal of Developmental Biology* 41: 591–595.

Fujii, R. and N. Oshima. 1994. Factors influencing motile activities of fish chromatophores. In: *Advances in Comparative and Environmental Physiology*, Vol. 2, R. Gilles (ed.). Springer-Verlag, Berlin, pp. 1–54.

Gaildrat, P. and J. Falcon. 1999. Expression of melatonin receptors and 2-[^{125}I] iodomelatonin binding sites in the pituitary of a teleost fish. *Advances in Experimental Biology and Medicine* 46: 61–72.

Gaildrat, P. and J. Falcon. 2000. Melatonin receptors in the pituitary of a teleost fish: mRNA expression, 2-[^{125}I] iodomelatonin binding and cyclic AMP response. *Neuroendocrinology* 72: 57–66.

Gaildrat, P., B. Ron and J. Falcon. 1998. Daily and circadian variations in 2-[^{125}I]-iodomelatonin binding sites in pike brain (*Esox lucius*). *Journal of Neuroendocrinology* 10: 511–517.

Gaildrat , P., F. Becq and J. Falcon. 2002. First cloning and functional characterization of a melatonin receptor in fish brain: a novel one? *Journal of Pineal Research* 32: 74–84.

Gorfine, T., Y. Assaf, Y. Goshen-Gottstein, Y. Yeshurun and N. Zisapel. 2006. Sleep-anticipating effects of melatonin in the human brain. *Neuroimage* 31: 410–418.

Ghosh, J. and P. Nath. 2005. Seasonal effects of melatonin on ovary and plasma gonadotropin and vitellogenin levels in intact and pinealectomized catfish, *Clarias batrachus* (Linn.). *Indian Journal of Experimental Biology* 43: 224–232.

Herrero, M.J., M. Pascual, J.A. Madrid and F.J. Sanchez-Vazquez. 2005. Demand-feeding rhythms and feeding entrainment of locomotor activity rhythms in tench (*Tinca tinca*). *Physiology and Behavior* 84: 595–605.

Huang, H., S.C. Lee and X.L. Yang. 2005. Modulation by melatonin of glutamateric synaptic transmission in the carp retina. *Journal of Physiology* 569: 857–871.

Iigo, M., F.J. Sanchez-Vazquez, M. Hara, R. Ohtani-Kaneko, K. Herata, H. Shinohara, M. Tabata and K. Aida. 1997. Characterization, guanosine 5'-O-(3-thiotriphosphate) modulation, daily variation, and localization of melatonin-binding sites in the catfish (*Silurus asotus*) brain. *General and Comparative Endocrinology* 108: 45–55.

Jan, J.E., M.B. Wasdell, R.J. Reiter, M.D. Weiss, K.P. Johnson, A. Ivanenko and R.P. Freeman. 2007. Melatonin therapy in sleep disorders: recent advances, why it works, who are the candidates and how to treat. *Current Pediatric Reviews* 3: 214–224.

Kanemura, T., N. Kawakami, M. Deguchi, K. Mimatsu and H. Iwata. 1997. Natural course of experimental scoliosis in pinealectomized chickens. *Spine* 22: 1563–1567.

Kezuka, H., K. Aida and I. Hanyu. 1989. Melatonin secretion from goldfish pineal gland in organ culture. *General and Comparative Endocrinology* 75: 217–221.

Khan, I.A. and P. Thomas. 1996. Melatonin influences gonadotropin II secretion in the Atlantic croaker (*Micropogonics undulates*). *General and Comparative Endocrinology* 104: 231–242.

Kiss, J., D. Bankegyi and G. Csaba. 1969. The effect of pinealectomy on blood calcium level. II. Relationship between the pineal body and the parathyroid glands. *Acta Medica Academia Science Hungarica* 26: 363–370.

Kleszczynska, A., L. Varga-Chacoff, M. Gozdowska, H. Kalamarz, G. Martinez-Rodriquez, J.M. Mancera and E. Kulczykowska. 2006. Arginine vasotocin, isotocin and melatonin responses following acclimation of gilthead sea bream (*Sparus aurata*) to different environmental salinities. *Comparative Biochemistry and Physiology* A 145: 268–273.

Kulczykowska, E., H. Kalamarz, J.M. Warne and R.J. Balment. 2006. Day-night specific binding of 2-[^{125}I]iodomelatonin and melatonin content of the gills, small intestine and kidney of three fish species. *Journal of Comparative Physiology* B176: 277–285.

Ladizesky, M.G., R.A. Cutrera, V. Boggio, J. Somoza, J.M. Centrella, C. Mautalen and D.P. Cardinali. 2001. Effect of melatonin on bone metabolism in ovariectomized rats. *Life Sciences* 70: 557–565.

Ladizesky, M.G., V. Boggio, L.E. Albornaz, P.O. Castrillon, C. Mautalen and D.P. Cardinali. 2003. Melatonin increases oestradiol-induced bone formation in ovariectomized rats. *Journal of Pineal Research* 34: 143–151.

Lerner, A.B., J.D. Case and R.V. Heinzlmann. 1959. Structure of melatonin. *Journal of the American Chemical Society* 81: 6084–6085.

Lerner, A.B., J.D. Case, Y. Takahashi, Y. Lee and W. Mori. 1958. Isolation of melatonin, the pineal gland factor that lightens melanocytes. *Journal of the American Chemical Society* 80: 2587.

Lopez-Olmeda, J.F., M.J. Bayarri, M.A. Rol de Lama, J.A. Madrid and F.J. Sanchez-Vazquez. 2006a. Effects of melatonin administration on oxidative stress and daily locomotor activity patterns in goldfish. *Journal of Physiology and Biochemistry* 62: 17–25.

Lopez-Olmeda, J.F., J.A. Madrid and F.J. Sanchez-Vazquez. 2006b. Melatonin effects on food intake and activity rhythms in two fish species with different activity patterns: diurnal (goldfish) and nocturnal (tench). *Comparative Biochemistry and Physiology* A 144: 180–187.

Lopez-Olmeda, J.F., G. Oliveira, H. Kalamarz, E. Kulczykowska, M.J. Delgado and E.J. Sanchez-Vazquez. 2009. Effects of melatonin salinity on melatonin levels in plasma and peripheral tissues and on melatonin binding sites in European sea bass (*Dicentrarchus labrax*). *Comparative Biochemistry and Physiology* A 152: 486–490.

Machida, M., J. Dubousset, Y. Imamura, T. Iwaya, T. Yamada and J. Kimura. 1995. Role of melatonin deficiency in the development of scoliosis in pinealectomized chickens. *British Journal of Bone and Joint Surgery* 77: 134–138.

Machida, M., I. Murai, Y. Miyashita, J. Dubousset, T. Yamada and J. Kimura. 1999. Pathogenesis of idiopathic scoliosis: experimental study in rats. *Spine* 24: 1985–1989.

Mayer, I., C. Bornestaf, L. Wettenberg and B. Borg. 1997. Melatonin in non-mammalian vertebrates: physiological role in reproduction? *Comparative Biochemistry and Physiology* A118: 515–531.

Mazurais, D., I. Brierley, I. Anglade, J. Drew, C. Randall, M. Bromage, D. Michel, O. Kah and L.M. Williams. 1999. Central melatonin receptors in the rainbow trout: comparative distribution of ligand binding and gene expression. *Journal of Comparative Neurology* 409: 313–324.

Mazurais, D., M. Porter, G. Lethimonier, G. LeDrean, P. Le Goff, C. Randall, F. Pakdel, N. Bromage and O. Kah. 2000. Effects of melatonin on liver estrogen receptor and vitellogenin expression in rainbow trout: an *in vitro* and *in vivo* study. *General and Comparative Endocrinology* 118: 344–353.

McCord, C.P. and F.B. Allen. 1917. Evidence associating pineal function with alterations in pigmentation. *Journal of Experimental Zoology* 23: 207–224.

McLennan, D.A. 1995. Male mate choice based upon female nuptial coloration in the brook stickleback, *Culaea inconstans* (Kirtland). *Animal Behavior* 50: 213–221.

Migaud, H., J.F. Taylor, G.L. Taranger, A. Davie, J.M. Cerda-Reverter, M. Carrillo, T. Hansen and N.R. Bromage. 2006. A comparative *ex vivo* and *in vivo* study of day and night perception in teleosts species using the melatonin rhythm. *Journal of Pineal Research* 41: 42–53.

Nikaido, Y., S. Ueda and A. Takemura. 2009. Photic and circadian regulation of melatonin production in the Mozambique tilapia *Oreochromis mossambicus*. *Comparative Biochemistry and Physiology* A 152: 77–82.

Oliveira, C., J.F. Lopez-Olmeda, M.J. Delagado, A.L. Alonso-Gomez and F.J. Sanchez-Vazquez. 2008. Melatonin binding sites in Senegal sole: day/night changes in density and location in different regions of the brain. *Chronobiology International* 25: 645–652.

Oyama, J., I. Murai, K. Kanazawa and M. Michida. 2006. Bipedal ambulation induces experimental scoliosis in C57BL/6J mice with reduced plasma and pineal melatonin levels. *Journal of Pineal Research* 40: 219–224.

Park, Y.J., J.G. Park, S.J. Kim, Y.D. Lee, M. Saydur Rahmon and A. Takamura. 2006. Melatonin receptor of a reef fish with lunar-related rhythmicity: cloning and daily variations. *Journal of Pineal Research* 41: 166–174.

Park, Y.J., J.G. Park, H.B. Jeong, Y. Takeuchi, S.J. Kim, Y.D. Lee and A. Takamura. 2007a. Expression of the melatonin receptor Mel (1c) in neural tissues of the reef fish *Siganus guttatus*. *Comparative Biochemistry and Physiology* A 147: 103–111.

Park, Y.J., J.G. Park, N. Hiyakawa, Y.D. Lee, S.J. Kim and A. Takamura. 2007b. Diurnal and circadian regulation of a melatonin receptor, MT1, in the golden rabbitfish, *Siganus guttatus*. *General and Comparative Endocrinology* 150: 253–262.

Peyrot, F. and C. Ducrocq. 2008. Potential role of tryptophan derivatives in stress responses characterized by the generation of reactive oxygen and reactive nitrogen species. *Journal of Pineal Research* 45: 235–246.

Philip, A.R., J. Bellingham, J. Garcia-Fernandez and R.G. Foster. (2000a). A novel rod-like opsin isolated from the extra-retinal photoreceptors of teleost fish. *FEBS Letters* 468: 181–188.

Philip, A.R., J.M. Garica-Fernandez, B.G. Soni, R.J. Lucas, J. Bellingham and R.G. Foster. (2000b). Vertebrate ancient (VA) opsin and extraretinal photoreception in the Atlantic salmon (*Salmo salar*). *Journal of Experimental Biology* 203: 1925–1936.

Pinillos, M.L., N. De Pedro, A.L. Alonso-Gomez, M. Alonso-Bedate and M. Delgado. 2001. Food intake inhibition by melatonin in goldfish (*Carassius auratus*). *Physiology and Behavior* 72: 629–634.

Reiter, R.J. 1980. The pineal gland and its hormones in the control of reproduction in mammals. *Endocrine Reviews* 1: 109–131.

Reiter, R.J. 1987. The melatonin message: duration versus coincidence hypothesis. *Life Sciences* 46: 2119–2131.

Reiter, R.J. 1991. Melatonin: the chemical expression of darkness. *Molecular and Cellular Endocrinology* 79: C153–C158.

Reiter, R.J. 1993. The melatonin rhythm: both a clock and a calendar. *Experientia* 49: 654–664.

Reiter, R.J. 1995. The pineal gland and melatonin in relation to aging: a summary of the theories and of the data. *Experimental Gerontology* 30: 199–212.

Reiter, R.J. 2000. Melatonin: lowering the high price of free radicals. *News in Physiological Sciences* 15: 246–250.

Reiter, R.J., S.D. Paredes, A. Korkmay, M.J. Jou and D.X. Tan. 2008. Melatonin combats molecular terrorism at the mitochondrial level. *Interdisciplinary Toxicology* 1: 137–146.

Rimler, A., R. Jockers, Z. Lupowitz, S.R. Sampson and N. Zisapel. 2006. Differential effects of melatonin and its downstream effector PKC $\propto$ on subcellular localization of RGS proteins. *Journal of Pineal Research* 40: 144–152.

Roth, J.A., B.G. Kim, W.L. Lin and M.I. Cho. 1999. Melatonin promotes osteoblast differentiation and bone formation. *Journal of Biological Chemistry* 274: 22041–22047.

Saenz, D.A., A.G. Turjanski, G.B. Sacca, M. Marti, F. Doctorovich, M.I. Sarmiento, D.A. Estrin and R.E. Rosenstein. 2002. Physiological concentrations of melatonin inhibit the nitridergic pathway in the Syrian hamster retina. *Journal of Pineal Research* 33: 31–36.

Sanchez-Vazquez, F.J., M. Iigo, J.A. Madrid and M. Tabata. 2000. Pinealectomy does not effect the entrainment of light nor the generation of the circadian demand-feeding rhythms of rainbow trout. *Physiology and Behavior* 69: 455–461.

Satomura, K., S. Tobiuma, R. Tokuyama, Y. Yamasaki, K. Kudoh, E. Maeda and M. Nagayama. 2007. Melatonin at pharmacological doses enhances human osteoblastic differentiation *in vitro* and promotes mouse cortical bone formation *in vivo*. *Journal of Pineal Research* 42: 231–239.

Sköld, H.N., T. Amundsen, P.A. Svensson, I. Mayer, J. Bjelvenmark and E. Forsgren. 2008. Hormonal regulation of female nuptial coloration in a fish. *Hormones and Behavior* 54: 549–556.

Smith, A., V.L. Trudeau, L.M. Williams, M.G. Martinoli and I.G. Priede. 1996. Melatonin receptors are present in some non-optic regions of the brain of a deep-sea fish living in the absence of solar light. *Journal of Neuroendocrinology* 8: 655–658.

Sokolowska, E., N. Kalamarz and E. Kulczykowska. 2004. Seasonal changes in brain melatonin concentration in the three-spined stickleback (*Gasterosteus aculeatus*): towards an endocrine calendar. *Comparative Biochemistry and Physiology* A 139: 365–369.

Sreejith, P., R.S. Beyo, L. Divya, A.S. Vijayasree, M. Manju and C.V. Oommen. 2007. Triiodothyronine and melatonin influence antioxidant defense mechanism in a teleost *Anabas testudineus* (Block): *in vitro* study. *Indian Journal of Biochemistry and Biophysics* 44: 154–158.

Steffens, F., X.B. Zhou, U. Sausbier, C. Sailer, K. Motejlek, P. Ruth, J. Olcese, M. Korth and T. Weiland. 2003. Melatonin receptor signaling in pregnant and nonpregnant rat uterine myocytes as probed by large conductance Ca^{2+}-activated K^+ channel activity. *Molecular Endocrinology* 17: 2103–2115.

Strand, J.E.T., J.J. Aarseth, T.L. Hanebrekke and E.H. Jorgensen. 2008. Keeping track of time under ice and snow in a sub-arctic lake: plasma melatonin rhythms in Arctic charr overwintering under natural conditions. *Journal of Pineal Research* 44: 227–233.

Sugama, N., J.G. Park, Y.J. Park, Y. Takenchi, S.J. Kim and A. Takemura. 2008. Moonlight affects nocturnal Period 2 transcript levels in the pineal gland of the reef fish, *Siganus guttatus*. *Journal of Pineal Research* 45: 133–141.

Suzuki, N. and A. Hattori. 2002. Melatonin suppresses osteoclastic and osteoblastic activities in the scales of goldfish. *Journal of Pineal Research* 33: 253–258.

Suzuki, N., M. Somei, A. Seki, R.J. Reiter and A. Hattori. 2008. Novel bromomelatonin derivatives as potentially effective drugs to treat bone diseases. *Journal of Pineal Research* 45: 229–234.

Takahashi, D. 2000. Conventional sex roles in an amphidromous *Rhinogobius* goby in which females exhibit nuptial coloration. *Ichthyological Research* 47: 303–306.

Takamura, A., S. Ueda, N. Hiyakawa and Y. Nikaido. 2006. A direct influence of moonlight intensity on changes in melatonin production by cultured pineal glands on the golden rabbitfish, *Siganus guttatus*. *Journal of Pineal Research* 40: 236–241.

Tan, D.X., L.D. Chen, B. Poeggeler, L.C. Manchester and Reiter, R.J. 1993. Melatonin: a potent, endogenous hydroxyl radical scavenger. *Endocrine Journal* 1: 57-60.

Tan, D.X., L.C. Manchester, M.P. Terron, L.J. Flores and R.J. Reiter. 2007. One molecule, many derivatives: a never-ending interaction of melatonin with reactive oxygen and nitrogen species? *Journal of Pineal Research* 42: 28–42.

Tanaka, H., Y. Kimura and Y. Ujino. 1982. The experimental study of scoliosis in bipedal rat in the lathyrism. *Archives of Orthopedic and Traumatic Surgery* 101: 1–27.

Taylor, J.F., H. Migaud, M.J. Porter and N.R. Bromage. 2005. Photoperiod influences growth rate and plasma insulin-like growth factor-1 levels in juvenile rainbow trout, *Oncorhynchus mykiss*. *General and Comparative Endocrinology* 142: 169–185.

Tsai, S.B., J. Tucci, J. Uchiyama, N.J. Fabian, M.C. Lin, I.V. Zhdanova and S. Kishi. 2007. Differential effects of genotoxic effects on both concurrent body growth and gradual senescence in adult zebrafish. *Aging Cell* 6: 209–224.

Turgut, M., C. Yenisey, A. Uysal, M. Bozkurt and M. Yurtseven. 2003. The effects of pineal gland transplantation on the production on spinal deformity and serum melatonin level following pinealectomy in the chicken. *European Spine Journal* 12: 487–494.

Turgut, M., S. Kaplan, A.T. Turgut, H. Aslan, T. Güvenc, E. Cullu and S. Erdogan. 2005. Morphological, stereological and radiological changes in pinealectomized chicken cervical vertebrae. *Journal of Pineal Research* 39: 392–399.

Underwood, H. 1989. The pineal and melatonin: regulators of circadian function in lower vertebrates. *Experientia* 45: 914–922.

Vanecek, J. 1998. Cellular mechanisms of melatonin action. *Physiological Reviews* 78: 687–721.

Vera, L.M., C. De Oliveira, J.F. Lopez-Olmeda, J. Ramos, E. Mananos, J.A. Madrid and F.J. Sanchez-Vazquez. 2007. Seasonal and daily plasma melatonin rhythms and reproduction in Senegal sole kept under natural photoperiod and natural or controlled water temperature. *Journal of Pineal Research* 43: 50–55.

Vernadakis, A.J., W.E. Bemis and E.L. Bittman. 1998. Localization and partial characterization of melatonin receptors in amphioxus, hagfish, lamprey and skate. *General and Comparative Endocrinology* 110: 67–78.

Wang, X., M. Moreau, J. Raso, J. Zhao, H. Jiang, J. Mahood and K. Bagnall. 1998. Changes in serum melatonin levels in response to pinealectomy in the chicken and its correlations with development of scoliosis. *Spine* 23: 2377–2381.

Zhdanova, I.V., S.Y. Wang, O.U. Leclair and N.P. Danilova. 2001. Melatonin promotes sleep-like state in zebrafish. *Brain Research* 903: 263–268.

Zhdanova, I.V., L. Yu, M. Lopez-Patino, E. Shang, S. Kishi and E. Guelin. 2007. Aging of the circadian system in zebrafish and the effects of melatonin on sleep and cognitive performance. *Brain Research Bulletin* 75: 433–441.

5

FISHING FOR LINKS BETWEEN THE CIRCADIAN CLOCK AND CELL CYCLE

Kajori Lahiri and *Nicholas S. Foulkes*

INTRODUCTION

Depending on where you are on our planet, the environment can change considerably during the course of the day night cycle. For this reason adaptations to time-of-day dependent changes in for example sunlight, temperature, food availability and predation are critical for the survival of organisms. The circadian clock plays an essential role in these adaptations by enabling organisms to anticipate environmental changes. Most aspects of physiology are controlled by this clock and consequently show circadian rhythms. Since it is not a precise 24-hour timer, it requires daily resetting by environmental cues such as light and temperature to ensure it remains synchronized with the day-night cycle. However, the clock also continues to function ("free-run") under artificial, constant conditions (Pittendrigh, 1993).

As a result of considerable work over the past two decades, involving genetic and biochemical studies in a wide range of model organisms ranging from cyanobacteria and plants to higher vertebrates, we have now gained a detailed understanding of the molecular and cellular nature of the circadian clock (Pittendrigh, 1993; Brown *et al.*, 1999; Vitaterna *et al.*, 2001). The clock is a highly conserved, cell autonomous mechanism where transcription-translation feedback loops constitute the core regulatory

Institute of Toxicology and Genetics, Karlsruhe Institute of Technology, Hermann-von-Helmholtz Platz 1, Eggenstein-Leopoldshafen 76344, Germany.
e-mail: nicholas.foulkes@itg.fzk.de

elements. Both positive and negative transcriptional regulator factors as well as kinases and phosphatases that modulate their activity constitute core clock components. Strict temporal control of the turnover, subcellular localization and interaction of the core clock elements results in the characteristic slow progression, but impressive accuracy of this feedback loop, that generates the circadian cycle. In vertebrates, the principal transcriptional activators are Clock and Bmal, two bHLH transcriptional activators that bind as a heterodimer to E-box promoter elements (Fig. 5.1 (Reppert *et al.*, 2002; Bell-Pedersen *et al.*, 2005)). They in turn upregulate expression of genes encoding the Period (Per) and Cryptochrome (Cry) proteins, which serve as negatively acting transcriptional cofactors. Per and Cry interact with Clock-Bmal and thereby down regulate their own expression (Reppert *et al.*, 2002; Bell-Pedersen *et al.*, 2005). The nuclear receptors Rev-erb α and Rora are involved in an additional regulatory feedback loop, which confers stability and robustness on the circadian clock (Preitner *et al.*, 2002; Ueda *et al.*, 2002; Sato *et al.*, 2004). In this loop, the transcription of Bmal itself is regulated by Rev-erb α and Rora *via* RORE sequences in the promoter region. Rev-erb α exerts a repressive effect, while Rora has an activating effect on Bmal transcription (Preitner *et al.*, 2002; Ueda *et al.*, 2002; Sato *et al.*, 2004).

Via both direct and indirect regulation, the clock mechanism in turn imposes a circadian rhythm on many other gene expression regulatory networks, the so-called clock output pathways (Panda *et al.*, 2004). In this regard, of considerable interest are recent observations linking the circadian clock with another key cellular cyclic mechanism: the cell cycle. The cell cycle is a process of fundamental importance that involves the process of cell division and the associated replication and subsequent segregation of genetic material into the new daughter cells. The cell cycle entails the following well defined stages: G0 (resting cell state), G1 (cells increase in size), S-phase (DNA replication), G2 (the cells continue to grow) and finally M-phase (mitosis) (Fig. 5.1 (Murray, 1993; Sanchez *et al.*, 2005).) Fundamental aspects of the mechanism that controls this process are highly conserved from yeast to vertebrates (Murray, 1993). In order to ensure that cells divide only at the appropriate time and place, and also to protect against errors in the whole process, the cell cycle involves passage through critical "check points" at key transition points within the cycle (at the transition from G2 to M, G1 to S and the so-called metaphase checkpoint). These represent points where cells survey various key parameters such as the presence of DNA damage or the availability of essential extracellular or intracellular signals. If necessary, at these check points the cell cycle can be paused until repair mechanisms have been implemented or until essential cellular signals are available. Progression through the cell cycle is regulated by complex signaling cascades, which involve cyclins and cyclin-dependent kinases (cdks) (Fig. 5.1 (Murray, 1993; Hochegger *et al.*, 2008)).

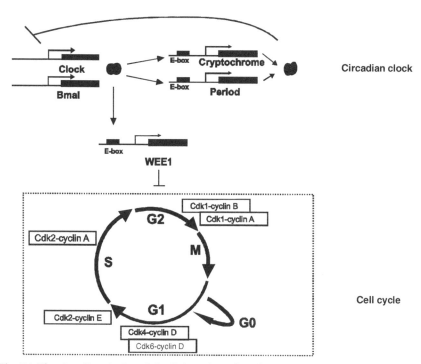

Fig. 5.1 The circadian clock and the link to the cell cycle.
Clock and Bmal heterodimers promote the transcription of Period and Cryptochrome genes through E-box enhancer elements. Periods and Cryptochromes inhibit their own transcription by repressing Clock and Bmal.
The WEE1 kinase is regulated by the core clock components Clock and Bmal. WEE1 kinase inhibits the CDC2 kinase and thus represses the G2/M transition. The progression through the cell cycle phases (M, G1, S and G2) is strictly controlled by cell-cycle-dependent kinases (cdk) and different cyclins.

It is now apparent that cells are frequently restricted to pass through certain key phases of the cell cycle only at certain times of the day-night cycle. Thus, DNA replication and mitosis (S and M phase respectively) often seem to be "gated" to occur predominantly during the night period (Klevecz *et al.*, 1987). Evidence points to the circadian clock playing a key role in this striking temporal regulation. The discovery of this property in species ranging from cyanobacteria and algae to mammals strongly points to clock regulation of the cell cycle representing a fundamental strategy for cells to survive the day - night cycle (Burns *et al.*, 1976; Lakatua *et al.*, 1983; Scheving *et al.*, 1992; Mori *et al.*, 1996).

Many basic questions remain concerning the links between the cell cycle and the circadian clock: What are the molecular mechanisms that link the two systems? What is the biological relevance of restricting cells to divide only at certain times of day? Does the clock also time associated processes such as DNA damage repair? Fish, and more specifically the zebrafish (*Danio rerio*), have become increasingly important models for studying links between the circadian clock and the cell cycle in vertebrates. The aim of this chapter is to overview our general understanding of how the clock regulates the cell cycle, focusing in particular detail on the contribution of fish-based studies to this knowledge.

CIRCADIAN CELL BIOLOGY

Circadian rhythms of cell cycle were first extensively documented in Cyanobacteria (Mori *et al.*, 2000). In rapidly growing *Synechococcus elongatus* cultures, cells grow with doubling times that are considerably faster than once per 24 h but nevertheless express robust circadian rhythms of cell division and gene expression (Mori *et al.*, 1996). It was shown that the circadian clock controls the timing of cell division; however the clock itself is not affected by passage through the cell cycle (Mori *et al.*, 2001). Control of cell cycle by the circadian clock has also been encountered in eukaryotic unicellular organisms such as the photoautotrophic *Euglena gracilis* (Hagiwara *et al.*, 2002). Encountering circadian clock regulation of the cell cycle both in prokaryotes and unicellular eukaryotes implies that this cellular property constitutes a fundamental strategy for surviving day night changes in the environment that has been highly conserved during evolution.

Consistent with this notion, unicellular organisms are not alone in restricting key steps of their cell cycle to occur at certain times of day. Studies in mammalian cell cultures have shown that the circadian clock gates the timing of mitosis and S-phase in different tissues and cell types. In rodents, cells of the hematopoietic and immune systems as well as liver, gastro-intestinal tract, skin, cornea, rectal mucosa cells and healthy bone marrow have all been shown to display circadian rhythms of DNA synthesis (Burns *et al.*, 1976; Lakatua *et al.*, 1983; Scheving *et al.*, 1992). Furthermore, single cell imaging experiments have confirmed that as in cyanobacteria, passage through the cell cycle does not appear to affect the circadian clock in mammalian cells (Nagoshi *et al.*, 2004).

ZEBRAFISH: NEW INSIGHT INTO CIRCADIAN CELL CYCLE RHYTHMS

A detailed understanding of how the circadian clock influences the cell cycle *in vivo* has been obtained as a result of studies in the zebrafish. This species is now firmly established as a valuable model system for studying many aspects of vertebrate biology including the circadian clock. The advantages it offers include its utility for large-scale genetic screens and live *in vivo* imaging as well as the accessibility of the earliest developmental stages. Interestingly, as a result of genome duplication events that occurred early during the evolution of the teleost lineage (Postlethwait *et al.*, 1998; Taylor *et al.*, 2001) zebrafish have extra copies of clock genes when compared with mammals. Thus, four period genes (Delaunay *et al.*, 2000, 2003; Pando *et al.*, 2001; Vallone *et al.*, 2004), six cryptochrome genes (Kobayashi *et al.*, 2000), three Bmal genes (Cermakian *et al.*, 2000; Ishikawa *et al.*, 2002) as well as three clock genes (Whitmore *et al.*, 1998; Ishikawa *et al.*, 2002) have been cloned and characterized in zebrafish. A key advantage of using zebrafish to study the circadian timing system originates from the entrainment of its peripheral clocks by direct exposure to light (Whitmore *et al.*, 2000). This property is encountered *in vivo* as well as *in vitro* and even in zebrafish cell lines (Whitmore *et al.*, 2000; Vallone *et al.*, 2004). This contrasts with the situation of mammalian peripheral clocks, where entrainment by LD cycles *in vivo* occurs indirectly via the retina and the hypothalamic suprachiasmatic nucleus. Furthermore, cultured mammalian cells require non-physiological treatments such as serum shocks to transiently synchronize their individual cell clocks (Balsalobre *et al.*, 2000b).

The zebrafish is particularly attractive for studying links between the circadian clock and the cell cycle during early development. The embryonic and larval stages show high levels of cell proliferation and also a light-entrainable circadian clock matures very early during embryonic development (by 24 hours post fertilization (Dekens *et al.*, 2008). Furthermore, since zebrafish larvae normally only start to feed after 5 days post fertilization, by studying earlier developmental stages, the potentially indirect effects of feeding time on the timing of cell cycle progression are avoided.

The timing of the cell cycle of zebrafish has been studied extensively by using the incorporation of BrdU (a thymidine analog) into replicating DNA as an assay for S-phase (Dekens *et al.*, 2003)). BrdU can be delivered non-invasively either by adding it directly to fish water or to the medium of cultured zebrafish cells. The compound diffuses rapidly into eggs, embryos, larvae or even adult tissues so that treatments as short as 15 minutes are sufficient to label cell nuclei progressing through S phase. BrdU incorporation can then be visualized by fixation with paraformaldehyde, acid treatment and subsequent staining with a monoclonal antibody directed against the BrdU epitope or in the case of cell cultures, by the use of an ELISA assay (Dekens *et al.*, 2003). In addition, the frequency of mitotic cells has been documented by staining fixed cells or tissues with a phospho-specific Histone H3 antibody. This antibody specifically labels Histone H3 when it is phosphorylated at serine 10 in the context of the compacted chromatin that characterizes mitotic nuclei. By using these assays, studies have revealed that proliferating cells within tissues of larval and adult zebrafish raised in light-dark (LD) cycle conditions, all show robust day-night rhythms in S phase as well as mitosis (Dekens *et al.*, 2003). Importantly, exposure of various zebrafish cell lines to LD cycles also establishes daily rhythms in S phase pointing to, at least in part, a cell autonomous mechanism driving S phase rhythms (Dekens *et al.*, 2003). The persistence of the rhythms following transfer from LD to constant darkness (DD) conditions supports the hypothesis that the circadian clock is part of the driving mechanism. However, given the direct light sensing ability of peripheral tissues in zebrafish and the possibility that the cell cycle length may approximate to the length of the circadian clock cycle, a direct light-driven effect on cell cycle control still cannot be excluded. Interestingly, the amplitude of these S-phase rhythms differs considerably between tissues *in vivo*. The highest amplitude rhythms are encountered in the epidermis and the gut epithelium, both sites of sustained, high levels of cell proliferation (Dekens *et al.*, 2003). During the first 3 days of embryonic development, the period involving the most rapid and extensive cell proliferation and tissue morphogenesis, no circadian rhythms of S phase are evident, thus circadian rhythmicity in cell cycle progression is clearly developmentally regulated. Exposure to a light dark cycle is essential for the appearance of these rhythms around 4 days post fertilization. Thus, larvae or adults raised in DD do not show circadian cell cycle rhythms. However, the ability of larvae to develop perfectly normally when maintained in DD conditions points to these rhythms being dispensable for normal growth and development in the laboratory environment. (Fig. 5.2)

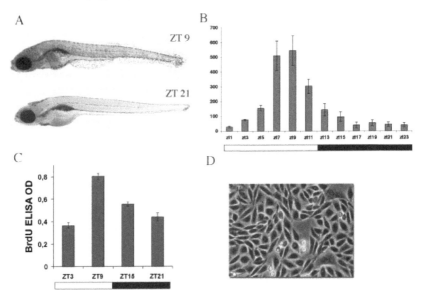

Fig. 5.2 A. Whole mount staining for BrdU incorporation in 6-day old zebrafish larvae raised under light dark cycles. Peak point at ZT 9 and trough point at ZT 21 are shown. B. Quantification of BrdU positive nuclei from larvae raised in LD conditions and labelled each 2 h during a 24 h time course are shown. C. ELISA results of BrdU incorporation in PAC-2 cells under light dark cycles are shown. D. Image of the PAC-2 cell line of zebrafish.

Zebrafish are certainly not the first teleost species where cell cycle rhythms have been studied. For example, one early report already described the link between the cell division and clock in retinal cells of an African cichlid fish *Haplochromis burtomi.* (Chiu *et al.*, 1995). By using PCNA expression as a marker, cell proliferation was monitored in the outer nuclear layer of the retina where rod precursor cells reside. Comparing results from fish maintained under LD and DD conditions the authors hypothesized a circadian regulation of S-phase with a peak of cell division occurring during the night. Thus as would expected for such a highly conserved mechanism, circadian clock control of the cell cycle is likely to be generally encountered in teleost species.

These observations raise some basic questions concerning the nature of the cell-autonomous mechanisms that link the cell cycle and the circadian clock. Furthermore, are these cell autonomous mechanisms sufficient to explain circadian cell cycle rhythms at the whole animal level? These issues are addressed in the next section of this chapter.

MOLECULAR LINKS BETWEEN THE CLOCK AND THE CELL CYCLE

The direct links that exist between core components of the circadian clock and cell cycle control in vertebrates have been mostly explored in the mouse. Systematic surveys have demonstrated that more than 10% of the transcripts in various vertebrate tissues show a circadian rhythm of expression (Akhtar *et al.*, 2002; Panda *et al.*, 2002). Amongst the different classes of cycling transcripts are genes involved in cellular metabolism and cell cycle control. More mechanistic insight into links between the circadian clock and the cell cycle has been obtained by detailed studies addressing hepatocyte proliferation that follows partial hepatectomy (surgical removal of sections of the liver) and leads to liver regeneration (Matsuo *et al.*, 2003). Hepatocytes are restricted by the circadian clock to pass through the G2/M checkpoint of the cell cycle preferentially only at certain times of day. Specifically, the key checkpoint control kinase, WEE1 appears to be regulated by core circadian clock components. WEE1 in turn phosphorylates CDC2 kinase on tyr15 thereby inactivating this key cell cycle regulator. *Wee1* mRNA expression shows a characteristic circadian rhythm of abundance. Circadian regulation is achieved directly via transcriptional activation by the CLOCK/ BMAL heterodimer, which can bind to E-box elements located in the promoter regions of the *wee1* kinase gene. In Cryptochrome deficient mice (which are arrhythmic following transfer from LD cycles to constant darkness (DD) conditions) there is an impairment of hepatocyte proliferation. Interestingly, in an independent study addressing the zebrafish homologue of the *wee1* kinase gene, besides E-box elements, AP-1 binding sites were also identified and were implicated in mediating light driven expression (Hirayama *et al.*, 2005). This result would tend to support the notion that direct light exposure may also play a role in regulating the timing of cell cycle in the zebrafish.

Another direct link between circadian clock genes and cell cycle regulation was demonstrated in a study of mice carrying a mutation in the core clock gene *mper2* (Fu *et al.*, 2002). Interestingly, *mper2* knockout mice are more prone to cancer than wild type siblings. Gamma irradiation leads to an increase in tumor development and reduced apoptosis in thymocytes. Furthermore, genes involved in cell cycle regulation such as *cyclinD1, cyclinA, c-myc, mdm-2 and gadd45α* all show a deregulated expression pattern in these mutant mice. In particular, *c-myc* appears to be regulated by the circadian clock via E-box elements in its promoter. In *mper2* mutant mice, basal levels of *c-myc* are elevated implying that *mper2* might act as a tumor suppressor via its regulation of DNA-damage response pathways. This hypothesis was supported by an independent study where the expression for Per1, Per2 and Per3 were shown to be deregulated in breast cancer tissues (Chen *et al.*, 2005). Furthermore, it has been shown that Per1 expression is absent in endometrial carcinomas (Yeh *et al.*, 2005). It remains unclear as to whether other clock genes might also function as tumor suppressors. In *clock* mutant mice several cell cycle inhibitory genes are up-regulated (such as *p21, p27, chk1, chk2* and *atr*), while other genes involved in cell proliferation are down-regulated (*jak2, akt, cdk2, ER-α, TGFβ,* and *EGF*) (Miller *et al.*, 2007). Also embryonic mouse fibroblast lines derived from these animals exhibited a reduction in cell proliferation and DNA synthesis as compared with wild type (WT) controls (Miller *et al.*, 2007).

ZEBRAFISH: THE SYSTEMIC CONTROL OF CELL CYCLE RHYTHMS

Given that a cell autonomous mechanism has been shown to link the circadian clock and the cell cycle in zebrafish, the question arises whether this is sufficient to drive cell cycle rhythms at the whole animal level. This question has been addressed by studying panels of zebrafish mutants, that exhibit abnormal development of a wide range of tissues and structures. These animals represent powerful tools to dissect the contribution of specific organs and tissues to the generation of clock outputs at the whole animal level.

In this study the contribution of the visual system as well as the hypothalamic-pituitary axis to the control of circadian cell cycle rhythms was explored (Dickmeis *et al.*, 2007). In *lakritz* mutants, that lack the retinal ganglion cell layer and are functionally blind, normal circadian cell cycle rhythms were still evident. This result suggests that in zebrafish, unlike the situation in mammals, the retina is not required to relay lighting information to time cell cycle progression (Dickmeis *et al.*, 2007). This conclusion is consistent with peripheral clocks in zebrafish being entrained by direct exposure to light. However, a study of functionally blind, eyeless mutants that result from mutant alleles of the *chokh/rx3* gene revealed a more complex situation. While "weak" mutations still exhibited normal cell cycle rhythms, a strong mutant allele resulted in a severe attenuation of these rhythms (Dickmeis *et al.*, 2007). Detailed characterization of these strong mutants revealed an absence of *pomc* expressing cells of the corticotrope lineage in the anterior pituitary and arcuate nucleus. That this corticotrope cell deficiency might underlie the cell cycle phenotype was supported by studies of a panel of pituitary mutants that were deficient in various cell types within the developing pituitary gland. In all cases, loss of corticotrope cells was associated with a loss of cell cycle rhythms. Subsequently it was confirmed that deficiency of the corticotrope lineage in these mutants was associated with a considerable reduction in circulating cortisol levels (Dickmeis *et al.*, 2007). Furthermore, normal cell cycle rhythms could be "rescued" by raising these mutant larvae in the presence of the glucocorticoid receptor agonist dexamethasone (Dickmeis *et al.*, 2007). Circulating levels of glucocorticoids show a characteristic circadian rhythm that has previously been implicated in the entrainment of mammalian peripheral clocks *in vivo* (Balsalobre *et al.*, 2000a). Furthermore, in Rat1 fibroblasts transient dexamethasone treatment synchronizes rhythms of clock gene expression (Balsalobre *et al.*, 2000b). Based on these observations, glucocorticoids have been implicated as signals conveying timing information from the SCN pacemaker to the peripheral oscillators. (Balsalobre *et al.*, 2000a,b).

The results in the zebrafish point to the glucocorticoid signaling pathway being important for the regulation of circadian cell cycle rhythms. However, the fact that tonic levels of dexamethasone rescue cell cycle rhythms in corticotrope deficient mutants would tend to argue that the natural rhythmic profile of cortisol secretion is not necessary for its support of cell cycle rhythms. (Fig. 5.3)

RELEVANCE OF CIRCADIAN CELL CYCLE RHYTHMS

DNA damage and the circadian clock

One obvious next question is what is the adaptive significance of restricting certain steps in the cell cycle to occur only at certain times of day? The current favored hypotheses point to this mechanism playing a role in the avoidance of the damaging effects of UV radiation in sunlight (Nikaido *et al.*, 2000; Tamai *et al.*, 2004). UV light irradiation of cells induces cytotoxic, mutagenic and carcinogenic lesions within DNA (Pfeifer *et al.*, 2005). Stages in the cell cycle when cells are particularly sensitive to the damaging effects of UV light (e.g. S phase when DNA is being replicated) are thereby restricted to stages of the day night cycle when there is little or no UV light in the environment. The UV wavelengths of sunlight are certainly able to penetrate

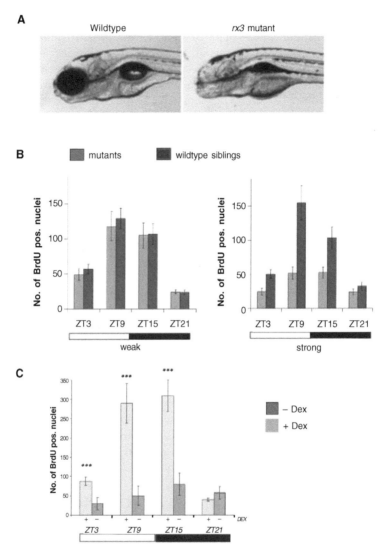

Fig. 5.3 A. Image of zebrafish wildtype larvae and the *rx3* mutant larvae. B. Quantification of BrdU labelled nuclei of the *rx3* weak and strong mutant allele from larvae raised under LD cycle. Wildtype siblings are shown in comparaison to the mutant larvae. C. Effects of dexamethasone treatment on BrdU incorporation in *rx3* strong mutants. Means of BrdU-positve nuclei are shown for Dex-treated versus untreated mutants.

shallow bodies of water and so potentially have harmful effects on aquatic species such as fish (Tamai *et al.*, 2004; Dong *et al.*, 2007, 2008). Thus, the circadian clock control of cell cycle as a strategy to avoid the damaging effects of UV light would certainly also be relevant for fish.

The transparency of the larval stages of zebrafish together with the fact that they typically inhabit shallow bodies of water might place them at particular risk from experiencing UV damage. One study has tested whether the survival rate of embryos raised under LD cycles differs from those raised under DD conditions (Tamai *et al.*, 2004) following exposure to UV light. The results clearly showed that LD cycle raised embryos had a significantly better survival rate and so that early light detection of the embryos had a beneficial effect on their survival. Light exposure not only is required for establishing cell cycle rhythms in the larval stages but also leads to upregulation of the expression of DNA repair enzymes such as 6-4 DNA Photolyase. This enzyme plays a vital role in the repair of UV damaged DNA and furthermore, the reaction it catalyzes, requires exposure to light (Zhao *et al.*, 1998). It is tempting to speculate therefore that zebrafish exploit light in a two-part strategy to avoid UV damaged DNA. Firstly, light is used to entrain the clock and thereby, via its links with the cell cycle machinery, to restrict S phase and mitosis to occur during times in the day night cycle when there is minimal exposure to UV light. The second part of this strategy is to respond to light exposure by acutely inducing the expression of genes that are important to repair DNA damage. In this regard, a fascinating issue is the absence of 6-4 photolyase from placental mammals. This observation raises many fundamental issues concerning how the strategy to avoid and repair UV damaged DNA has changed during evolution and the selective pressures that have lead to these changes. 6-4 DNA photolyase is a close relative of the cryptochrome family of genes that are involved in circadian clock photoreception in many species as well as being core clock components in vertebrates (Ozturk *et al.*, 2007). Together, these findings point to a fundamental link between the molecular elements of the clock mechanism, its response to light and the UV damage repair system.

Consistent with the evolutionary conservation of links between the circadian clock and UV damage avoidance, several additional studies have documented a tight link between components of the circadian clock and the cellular response to DNA damage.

In Neurospora, the clock gene *period-4* (*prd-4*) was originally identified as a single allele displaying shortened circadian period and altered temperature compensation. PRD-4 is an ortholog of the mammalian checkpoint kinase 2 (Chk2). While the expression of *prd-4* is regulated by the circadian clock the protein also physically interacts with the core clock protein FRQ. Resetting of the clock has been reported by exposure to DNA-damaging agents. The effect of these "zeitgebers" is dependent on the time of the day and is also mediated by *prd-4* (Pregueiro *et al.*, 2006).

Similar to the situation in *Neurospora crassa*, DNA damage induced by ultraviolet light and tert-butyl hydroperoxide in the mammalian system also acts as a zeitgeber. By using Rat-1 fibroblasts expressing an *mper2* promoter-driven luciferase reporter it has been shown that ionizing radiation induces a phase shift in a dose dependent manner. This shift interestingly also translates into an effect on the behavioral rhythms in mice (Oklejewicz *et al.*, 2008).

DNA damage response pathways have been highly conserved through evolution. The two kinases ATM and ATR play a major role in this response by phosphorylating their downstream kinases such as Chk1 and Chk2, which in turn activate targets that lead to a pause in cell cycle progression and the initiation of DNA repair (Hurley *et al.*, 2007). In mammals, recent findings have indicated that the circadian clock has a direct link with these DNA repair pathways. Per1, one of the major core clock proteins interacts directly with Atm and Chk1 in human colon cancer cells (Shiloh, 2003).The Atm kinase and its downstream effector are upregulated by DNA double strand breaks and subsequently initiates DNA repair by phosphorylation of target proteins. Additional studies have shown that ectopic expression of Per1 can induce Chk2 phosphorylation, while it's silencing impairs this phosphorylation (Gery *et al.*, 2006). The homolog of the Timeless protein in humans has been shown to be involved in the DNA damage response pathway by directly interacting with the Chk1 and ATR-Atrip complex (Unsal-Kacmaz *et al.*, 2005).

CIRCADIAN CELL CYCLE RHYTHMS: IMPLICATIONS FOR HEALTH AND MEDICINE

The regulation of the cell cycle by the circadian clock has many important implications for understanding and treating major diseases such as cancer. Growing evidence points to disruptions in the sleep wake cycle in shift workers leading to an increased risk of developing tumors and also accelerated tumor development (Rafnsson *et al.*, 2003; Schernhammer *et al.*, 2003; Haus *et al.*, 2006; Kubo *et al.*, 2006). Clinical studies have shown that the survival rate is higher in cancer patients who exhibit normal circadian physiology outputs compared with patients who have dampened or abnormal circadian rhythms (Mormont *et al.*, 2000; Sephton *et al.*, 2000). Consistently, chronically jet-lagged mice and SCN ablated mice show an accelerated tumor growth which translates into a significant reduction in survival rate when compared with control animals (Filipski *et al.*, 2004a,b, 2006). Thus, ablation of the central circadian clock pacemaker as well as disruption of the sleep—wake cycle can both impact on cancer development and tumor growth.

Furthermore, there have been many attempts to develop strategies for treating cancer that involve the delivery of chemotherapeutic agents only at certain times of day. These strategies aim to synchronize chemotherapy with windows of time when healthy cells occupy stages of the cell cycle where they are more resistant to its toxic effects. Disruption of the circadian clock regulation of the cell cycle in cancer cells leaves them more susceptible to this timed therapy (Levi *et al.*, 2008). This strategy, so-called "Chronotherapeutics" also relies on the fact that the circadian clock not only regulates the cell cycle but also the metabolic functions of a cell (Levi *et al.*, 2008) The toxicity and efficacy of 30 anticancer agents has been tested and shown to vary by more than 50% depending on the dosing time in experimental models (Mormont *et al.*, 2003). Chronotherapeutics requires the careful assessment of the circadian rhythms of each individual patient as well as a clear understanding of the dynamic cross talk between the circadian clock cell cycle regulation and the course of the cancer processes.

UNRESOLVED QUESTIONS AND FUTURE DIRECTIONS

Our understanding of the biological significance of circadian cell cycle rhythms is still far from complete. Furthermore, there is much work still to be done in order to document precisely the mechanisms that link the circadian clock with the cell cycle. Studies in zebrafish have specifically given us a unique insight into the origin, regulation and function of these rhythms during early development. However, it remains unclear why during the first 3 days of development, a period of sustained high levels of cell proliferation, the embryos do not show robust circadian cell cycle rhythms, even though a functional clock is established by the first and second day post fertilization (Ziv *et al.*, 2006; Dekens *et al.*, 2008). It is tempting to speculate that this developmental timing might reflect fundamental constraints of embryonic development. Specifically, by restricting the timing of key steps in the cell cycle to occur only at certain times of day, it would delay the whole process of embryonic development and so be disadvantageous for survival. Other mechanisms might protect the early embryos from the damaging effects of sunlight. These might involve precisely where eggs are laid in the aquatic environment as well as the maternal inheritance of high levels of transcripts encoding 6-4 DNA Photolyase. These predictions suggest that the precise time of emergence of these cell cycle rhythms in teleosts may vary from species to species, depending on differences in their rate of early development and also other factors such as their egg laying behavior. Alternatively, late emergence of circadian cell cycle rhythms might simply reflect the delay before the establishment of a fully functioning endocrine system. This hypothesis is based on the observations that corticotropes and cortisol are essential for cell cycle rhythms in the context of the whole animal (Dickmeis *et al.*, 2007).

Another issue that remains unresolved is why only certain cell types exhibit high amplitude cell cycle rhythms. Notably actively proliferating cells in the skin and the gut show the highest amplitude circadian rhythms in cell cycle. The skin and gut epithelium are sites of sustained high levels of cell proliferation. It is tempting to speculate that these cells are more susceptible to the damaging effects of sunlight. Alternatively, these high amplitude rhythms might reflect underlying fundamental differences in skin or gut epithelium biology between the day and the night. Zebrafish are a diurnal species, and therefore it might be advantageous to time cell proliferation to coincide with the inactive phase of the animal. Again, comparisons with other teleost species may help to shed light on the biological relevance of these cell-type specific properties.

Thus fish are placed in an ideal position to explore the control and relevance of this fascinating clock output. The availability of model species such as zebrafish and medaka and their respective molecular and genetic tools together with the great diversity of teleost species will surely be exploited in future studies and help to advance our understanding.

Acknowledgements

We thank the Karlsruhe Institute of Technology and the Max Planck Gesellschaft (MPI for Developmental Biology, Tübingen) for funding. K.L was funded by the EU project zf-Tools.

References

Akhtar, R.A., A.B. Reddy, E.S. Maywood, J.D. Clayton, V.M. King, A.G. Smith, T.W. Gant, M.H. Hastings and C.P. Kyriacou. 2002. Circadian cycling of the mouse liver transcriptome, as revealed by cDNA microarray, is driven by the suprachiasmatic nucleus. *Current Biology* 12: 540–550.

Balsalobre, A., S.A. Brown, L. Marcacci, F. Tronche, C. Kellendonk, H.M. Reichardt, G. Schutz and U. Schibler. 2000a. Resetting of circadian time in peripheral tissues by glucocorticoid signaling. *Science* 289: 2344–2347.

Balsalobre, A., L. Marcacci and U. Schibler. 2000b. Multiple signaling pathways elicit circadian gene expression in cultured Rat-1 fibroblasts. *Current Biology* 10: 1291–1294.

Bell-Pedersen, D., V.M. Cassone, D.J. Earnest, S.S. Golden, P.E. Hardin, T.L. Thomas and M.J. Zoran. 2005. Circadian rhythms from multiple oscillators: lessons from diverse organisms. *Nature Reviews Genetics* 6: 544–556.

Brown, S.A. and U. Schibler. 1999. The ins and outs of circadian timekeeping. *Current Opinion in Genetics and Development* 9: 588–594.

Burns, E.R., L.E. Scheving, J.E. Pauly and T. Tsai. 1976. Effect of altered lighting regimens, time-limited feeding, and presence of Ehrlich ascites carcinoma on the circadian rhythm in DNA synthesis of mouse spleen. *Cancer Research* 36: 1538–1544.

Cermakian, N., D. Whitmore, N.S. Foulkes and P. Sassone-Corsi. 2000. Asynchronous oscillations of two zebrafish CLOCK partners reveal differential clock control and function. *Proceedings of the National Academy of Sciences of the United States of America* 97: 4339–4344.

Chen, S.T., K.B. Choo, M.F. Hou, K.T. Yeh, S.J. Kuo and J.G. Chang. 2005. Deregulated expression of the PER1, PER2 and PER3 genes in breast cancers. *Carcinogenesis* 26: 1241–1246.

Chiu, J.F., A.F. Mack and R.D. Fernald. 1995. Daily rhythm of cell proliferation in the teleost retina. *Brain Research* 673: 119–125.

Dekens, M.P. and D. Whitmore. 2008. Autonomous onset of the circadian clock in the zebrafish embryo. *EMBO Journal* 27: 2757–2765.

Dekens, M.P., C. Santoriello, D. Vallone, G. Grassi, D. Whitmore and N.S. Foulkes. 2003. Light regulates the cell cycle in zebrafish. *Current Biology* 13: 2051–2057.

Delaunay, F., C. Thisse, B. Thisse and V. Laudet. 2003. Differential regulation of Period 2 and Period 3 expression during development of the zebrafish circadian clock. *Gene Expression Patterns* 3: 319–324.

Delaunay, F., C. Thisse, O. Marchand, V. Laudet and B. Thisse. 2000. An inherited functional circadian clock in zebrafish embryos. *Science* 289: 297–300.

Dickmeis, T., K. Lahiri, G. Nica, D. Vallone, C. Santoriello, C.J. Neumann, M. Hammerschmidt and N.S. Foulkes. 2007. Glucocorticoids play a key role in circadian cell cycle rhythms. *PLos Biology* 5: e78.

Dong, Q., K. Svoboda, T.R. Tiersch and W.T. Monroe. 2007. Photobiological effects of UVA and UVB light in zebrafish embryos: evidence for a competent photorepair system. *Journal Photochemistry Photobiology* B88: 137–146.

Dong, Q., W. Todd Monroe, T.R. Tiersch and K.R. Svoboda. 2008. UVA-induced photo recovery during early zebrafish embryogenesis. *Journal Photochemistry Photobiology* B93: 162–171.

Filipski, E., X.M. Li and F. Levi. 2006. Disruption of circadian coordination and malignant growth. *Cancer Causes Control* 17: 509–514.

Filipski, E., F. Delaunay, V.M. King, M.W. Wu, B. Claustrat, A. Grechez-Cassiau, C. Guettier, M.H. Hastings and L. Francis. 2004a. Effects of chronic jet lag on tumor progression in mice. *Cancer Research* 64: 7879–7885.

Filipski, E., V.M. King, M.C. Etienne, X. Li, B. Claustrat, T.G. Granda, G. Milano, M.H. Hastings and F. Levi. 2004b. Persistent twenty-four hour changes in liver and bone marrow despite suprachiasmatic nuclei ablation in mice. *American Journal of Physiology—Regulatory, Integrative and Comparative Physiology* 287: R844–851.

Fu, L., H. Pelicano, J. Liu, P. Huang and C. Lee. 2002. The circadian gene Period 2 plays an important role in tumor suppression and DNA damage response *in vivo*. *Cell* 111: 41–50.

Gery, S., N. Komatsu, L. Baldjyan, A. Yu, D. Koo and H.P. Koeffler. 2006. The circadian gene per1 plays an important role in cell growth and DNA damage control in human cancer cells. *Molecular Cell* 22: 375–382.

Hagiwara, S.Y., A. Bolige, Y. Zhang, M. Takahashi, A. Yamagishi and K. Goto. 2002. Circadian gating of photoinduction of commitment to cell-cycle transitions in relation to photoperiodic control of cell reproduction in *Euglena*. *Photochemistry Photobiology* 76: 105–115.

Haus, E. and M. Smolensky. 2006. Biological clocks and shift work: circadian dysregulation and potential long-term effects. *Cancer Causes Control* 17: 489–500.

Hirayama, J., L. Cardone, M. Doi and P. Sassone-Corsi. 2005. Common pathways in circadian and cell cycle clocks: light-dependent activation of Fos/AP-1 in zebrafish controls CRY-1a and WEE-1. *Proceedings of the National Academy of Sciences of the United States of America* 102: 10194–10199.

Hochegger, H., S. Takeda and T. Hunt. 2008. Cyclin-dependent kinases and cell-cycle transitions: does one fit all? *Nature Reviews Molecular Cellular Biology* 9: 910–916.

Hurley, P.J. and F. Bunz. 2007. ATM and ATR: components of an integrated circuit. *Cell Cycle* 6: 414–417.

Ishikawa, T., J. Hirayama, Y. Kobayashi and T. Todo. 2002. Zebrafish CRY represses transcription mediated by CLOCK-BMAL heterodimer without inhibiting its binding to DNA. *Genes Cells* 7: 1073–1086.

Klevecz, R.R., R.M. Shymko, D. Blumenfeld and P.S. Braly. 1987. Circadian gating of S phase in human ovarian cancer. *Cancer Research* 47: 6267–6271.

Kobayashi, Y., T. Ishikawa, J. Hirayama, H. Daiyasu, S. Kanai, H. Toh, I. Fukuda, T. Tsujimura, N. Terada, Y. Kamei, S. Yuba, S. Iwai and T. Todo. 2000. Molecular analysis of zebrafish photolyase/cryptochrome family: two types of cryptochromes present in zebrafish. *Genes Cells* 5: 725–738.

Kubo, T., K. Ozasa, K. Mikami, K. Wakai, Y. Fujino, Y. Watanabe, T. Miki, M. Nakao, K. Hayashi, K. Suzuki, M. Mori, M. Washio, F. Sakauchi, Y. Ito, T. Yoshimura and A. Tamakoshi. 2006. Prospective cohort study of the risk of prostate cancer among rotating-shift workers: findings from the Japan collaborative cohort study. *American Journal of Epidemiology* 164: 549–555.

Lakatua, D.J., M. White, L.L. Sackett-Lundeen and E. Haus. 1983. Change in phase relations of circadian rhythms in cell proliferation induced by time-limited feeding in BALB/c X DBA/2F1 mice bearing a transplantable Harding-Passey tumor. *Cancer Research* 43: 4068–4072.

Levi, F., A. Altinok, J. Clairambault and A. Goldbeter. 2008. Implications of circadian clocks for the rhythmic delivery of cancer therapeutics. *Philosophical Transactions. Series A Mathematical, Physical, and Engineering Science* 366: 3575–3598.

Matsuo, T., S. Yamaguchi, S. Mitsui, A. Emi, F. Shimoda and H. Okamura. 2003. Control mechanism of the circadian clock for timing of cell division in vivo. *Science* 302: 255–259.

Miller, B.H., E.L. McDearmon, S. Panda, K.R. Hayes, J. Zhang, J.L. Andrews, M.P. Antoch, J.R. Walker, K.A. Esser, J.B. Hogenesch and J.S. Takahashi. 2007. Circadian and CLOCK-controlled regulation of the mouse transcriptome and cell proliferation. *Proceedings of the National Academy of Sciences of the United States of America* 104: 3342–3347.

Mori, T. and C.H. Johnson. 2000. Circadian control of cell division in unicellular organisms. *Progress in Cell Cycle Research* 4: 185–192.

Mori, T. and C.H. Johnson. 2001. Independence of circadian timing from cell division in cyanobacteria. *The Journal of Bacteriology* 183: 2439–2444.

Mori, T., B. Binder and C.H. Johnson. 1996. Circadian gating of cell division in cyanobacteria growing with average doubling times of less than 24 hours. *Proceedings of the National Academy of Sciences of the United States of America* 93: 10183–10188.

Mormont, M.C. and F. Levi. 2003. Cancer chronotherapy: principles, applications, and perspectives. *Cancer* 97: 155–169.

Mormont, M.C., J. Waterhouse, P. Bleuzen, S. Giacchetti, A. Jami, A. Bogdan, J. Lellouch, J.L. Misset, Y. Touitou and F. Levi. 2000. Marked 24-h rest/activity rhythms are associated with better quality of life, better response, and longer survival in patients with metastatic colorectal cancer and good performance status. *Clinical Cancer Research* 6: 3038–3045.

Murray, A. and T. Hunt. 1993. *The Cell Cycle: An introduction*. Oxford University Press Inc., New York.

Nagoshi, E., C. Saini, C. Bauer, T. Laroche, F. Naef and U. Schibler. 2004. Circadian gene expression in individual fibroblasts: cell-autonomous and self-sustained oscillators pass time to daughter cells. *Cell* 119: 693–705.

Nikaido, S.S. and C.H. Johnson. 2000. Daily and circadian variation in survival from ultraviolet radiation in *Chlamydomonas reinhardtii*. *Photochemistry Photobiology* 71: 758–765.

Oklejewicz, M., E. Destici, F. Tamanini, R.A. Hut, R. Janssens and G.T. van der Horst. 2008. Phase resetting of the mammalian circadian clock by DNA damage. *Current Biology* 18: 286–291.

Ozturk, N., S.H. Song, S. Ozgur, C.P. Selby, L. Morrison, C. Partch, D. Zhong and A. Sancar. 2007. Structure and function of animal cryptochromes. *Cold Spring Harbor Symposia on Quantitative Biology* 72: 119–131.

Panda, S. and J.B. Hogenesch. 2004. It's all in the timing: many clocks, many outputs. *Journal of Biological Rhythms* 19: 374–387.

Panda, S., M.P. Antoch, B.H. Miller, A.I. Su, A.B. Schook, M. Straume, P.G. Schultz, S.A. Kay, J.S. Takahashi and J.B. Hogenesch. 2002. Coordinated transcription of key pathways in the mouse by the circadian clock. *Cell* 109: 307–320.

Pando, M.P., A.B. Pinchak, N. Cermakian and P. Sassone-Corsi. 2001. A cell-based system that recapitulates the dynamic light-dependent regulation of the vertebrate clock. *Proceedings of the National Academy of Sciences of the United States of America* 98: 10178–10183.

Pfeifer, G.P., Y.H. You and A. Besaratinia. 2005. Mutations induced by ultraviolet light. *Mutation Research* 571: 19–31.

Pittendrigh, C.S. 1993. Temporal organization: reflections of a Darwinian clock-watcher. *Annual Review of Physiology* 55: 16–54.

Postlethwait, J.H., Y.L. Yan, M.A. Gates, S. Horne, A. Amores, A. Brownlie, A. Donovan, E.S. Egan, A. Force, Z. Gong, C. Goutel, A. Fritz, R. Kelsh, E. Knapik, E. Liao, B. Paw, D. Ransom, A. Singer, M. Thomson, T.S. Abduljabbar, P. Yelick, D. Beier, J.S. Joly, D. Larhammar, F. Rosa, M. Westerfield, L.I. Zon, S.L. Johnson and W.S. Talbot. 1998. Vertebrate genome evolution and the zebrafish gene map. *Nature Genetics* 18: 345–349.

Pregueiro, A.M., Q. Liu, C.L. Baker, J.C. Dunlap and J.J. Loros. 2006. The Neurospora checkpoint kinase 2: a regulatory link between the circadian and cell cycles. *Science* 313: 644–649.

Preitner, N., F. Damiola, L. Lopez-Molina, J. Zakany, D. Duboule, U. Albrecht and U. Schibler. 2002. The orphan nuclear receptor REV-ERBalpha controls circadian transcription within the positive limb of the mammalian circadian oscillator. *Cell* 110: 251–260.

Rafnsson, V., J. Hrafnkelsson, H. Tulinius, B. Sigurgeirsson and J.H. Olafsson. 2003. Risk factors for cutaneous malignant melanoma among aircrews and a random sample of the population. *Journal of Occupational and Environmental Medicine* 60: 815–820.

Reppert, S.M. and D.R. Weaver. 2002. Coordination of circadian timing in mammals. *Nature* 418: 935–941.

Sanchez, I. and B.D. Dynlacht. 2005. New insights into cyclins, CDKs, and cell cycle control. *Seminars in Cell and Developmental Biology* 16: 311–321.

Sato, T.K., S. Panda, L.J. Miraglia, T.M. Reyes, R.D. Rudic, P. McNamara, K.A. Naik, G.A. FitzGerald, S.A. Kay and J.B. Hogenesch. 2004. A functional genomics strategy reveals Rora as a component of the mammalian circadian clock. *Neuron* 43: 527–537.

Schernhammer, E.S., F. Laden, F.E. Speizer, W.C. Willett, D.J. Hunter, I. Kawachi, C.S. Fuchs and G.A. Colditz. 2003. Night-shift work and risk of colorectal cancer in the nurses' health study. *Journal of National Cancer Institute* 95: 825–828.

Scheving L.E., T.H. Tsai, R.J. Feuers and E.L Kanabrock. 1992. Normal and abnormal cell proliferation in mice as it related to cancer. In: *Biologic Rhythms in Clinical and Laboratory Medicine*, Y. Touitou and E. Haus (eds.). Springer-Verlag, Berlin.

Sephton, S.E., R.M. Sapolsky, H.C. Kraemer and D. Spiegel. 2000. Diurnal cortisol rhythm as a predictor of breast cancer survival. *Journal of National Cancer Institute* 92: 994–1000.

Shiloh, Y. 2003. ATM and related protein kinases: safeguarding genome integrity. *Nature Reviews Cancer* 3: 155–168.

Tamai, T.K., V. Vardhanabhuti, N.S. Foulkes and D. Whitmore. 2004. Early embryonic light detection improves survival. *Current Biology* 14: R104–105.

Taylor, J.S., Y. Van de Peer and A. Meyer. 2001. Genome duplication, divergent resolution and speciation. *Trends in Genetics* 17: 299–301.

Ueda, H.R., W. Chen, A. Adachi, H. Wakamatsu, S. Hayashi, T. Takasugi, M. Nagano, K. Nakahama, Y. Suzuki, S. Sugano, M. Iino, Y. Shigeyoshi and S. Hashimoto.

2002. A transcription factor response element for gene expression during circadian night. *Nature* 418: 534–539.

Unsal-Kacmaz, K., T.E. Mullen, W.K. Kaufmann and A. Sancar. 2005. Coupling of human circadian and cell cycles by the timeless protein. *Molecular Cell Biology* 25: 3109–3116.

Vallone, D., S.B. Gondi, D. Whitmore and N.S. Foulkes. 2004. E-box function in a period gene repressed by light. *Proceedings of the National Academy of Sciences of the United States of America* 101: 4106–4111.

Vitaterna, M.H., J.S. Takahashi and F.W. Turek. 2001. Overview of circadian rhythms. *Alcohol Research Health* 25: 85–93.

Whitmore, D., N.S. Foulkes and P. Sassone-Corsi. 2000. Light acts directly on organs and cells in culture to set the vertebrate circadian clock. *Nature* 404: 87–91.

Whitmore, D., N.S. Foulkes, U. Strahle and P. Sassone-Corsi. 1998. Zebrafish Clock rhythmic expression reveals independent peripheral circadian oscillators. *Nature Neuroscience* 1: 701–707.

Yeh, K.T., M.Y. Yang, T.C. Liu, J.C. Chen, W.L. Chan, S.F. Lin and J.G. Chang. 2005. Abnormal expression of period 1 (PER1) in endometrial carcinoma. *American Journal of Pathology* 206: 111–120.

Zhao, X. and D. Mu. 1998. (6-4) photolyase: light-dependent repair of DNA damage. *Histology and Histopathology* 13: 1179–1182.

Ziv, L. and Y. Gothilf. 2006. Circadian time-keeping during early stages of development. *Proceedings of the National Academy of Sciences of the United States of America* 103: 4146–4151.

6

THE PINEAL GLAND AS A SOURCE OF MELATONIN IN FISH; INFLUENCE OF LIGHT AND TEMPERATURE

Włodzimierz Popek[CA] and *Elżbieta Ćwioro*

INTRODUCTION

The pineal gland has been known for about 2300 years. It was discovered by Herophilus of Alexandria (325-280 BC), a scientist who gained renown for the first public dissection of human cadaver and pulse diagnosis. Together with Erasistratus, he studied the human brain and showed that it is a source of all nerves. Initially, Herophilus and his contemporary philosophers considered the pineal gland as a centre of spiritual and physical life. It was only after 500 years later that Galen (AD 129–200), a physician, biologist, philosopher and philologist, first described the pineal gland in detail. After thorough study, he found the pineal gland to be similar to other glands in the human body.

The pineal gland was completely forgotten in the Middle Ages, when the development of medical sciences was hindered. It was not until the 17th century that the French philosopher and mathematician, René Descartes (1596–1650) discovered it. He expressed a view that the pineal gland is the seat of the rational soul. According to this theory, *perception* of the world arises when a sensory pattern is transmitted to the brain *via* the pineal

Department of Ichthyobiology and Fisheries, University of Agriculture in Krakow, Krakow, Poland.
[CA]Corresponding author: rzpopek@cyf-kr.edu.pl

gland. The view that the pineal gland is the seat of the soul prevailed for several hundred years.

Studies from the second half of the 19th century brought new insights into this organ, especially in ontogenetic, phylogenetic, morphological and functional aspects. The early 20th century saw the first scientific description of a tumour arising from the pineal gland of two young boys, which accelerated puberty (Askanazy, 1906). Further clinical cases suggested a hypothesis that the pineal gland is an endocrine gland that probably inhibits the genital system (Marburg, 1913). Experiments with pinealectomized mammals: rats (Sarteschi, 1913), cats (D'Amour and D'Amour, 1937) and dogs (Foa, 1928) confirmed all previous observations. After World War II, the literature described 178 cases of tumours in the pineal region of children and young people aged 1–16 years, some of whom showed premature puberty (Kitay and Altschule, 1954).

A major breakthrough in the study of pineal gland function was made by the dermatologist Aaron Lerner. The addition of bovine pineal extracts to swimming water caused blanching of the skin of tadpoles. Further studies led to the isolation of a compound known as melatonin. It was named after the pigment cells melanophores. A year later the team of Lerner determined the compound's structure and showed it to be a hormone (Lerner *et al.*, 1959).

Real progress in the study of the role of melatonin in animal bodies was made with the advent of radioimmunoassay (RIA), which determines trace levels of biological substances. This method made it possible to measure melatonin concentrations not only in the pineal gland but also in body tissues, above all in blood.

In addition to the pineal gland, melatonin can be synthesized in the retina (Gern and Ralph, 1979), intestine (Bubenik and Dhanvantari, 1989), spleen (Quay and Ma, 1976) and the lacrimal gland (Mhatre *et al.*, 1988). However, in these body parts, the hormone is produced for local needs. Only melatonin synthesized in the retina can augment the action of pineal melatonin, as is the case with birds (up to 33% in quail) (Underwood *et al.*, 1984).

In fish, melatonin was first isolated chromatographically from the pineal gland of Chinook salmon (*Oncorhynchus tshawytscha*) (Fenwick, 1970a,b), and later immunocytochemically in several other species of teleost fish.

As noted above, melatonin synthesis may occur in different parts of the body, but the most important two are the pineal gland and retina. In higher vertebrates, both these sources can release melatonin into the bloodstream. However, experiments with fish have shown that the pineal gland is the only source of melatonin in the blood. This was supported by a study with crucian carp, in which pinealectomy caused a drastic reduction in blood melatonin concentration (Iigo *et al.*, 1995). Further research with carp showed that pinealectomy would completely clear melatonin from the blood circular system (below RIA sensitivity threshold of 0.2 pg/ml). Also the transection of optic nerves and the arrest of eye circulation showed that the inability of retina-produced melatonin to penetrate into the blood and to transfer neural information from retina to the brain and pineal gland moved the acrophase of melatonin concentration 6 hours towards the beginning of the dark phase and decreased basal melatonin concentration by about 30% (Popek *et al.*, 1997) (Fig. 6.1). This procedure did not desynchronize melatonin rhythm but decreased its amplitude. Thus, fish retina is necessary for synchronizing and maintaining the proper rhythm of melatonin release into the blood.

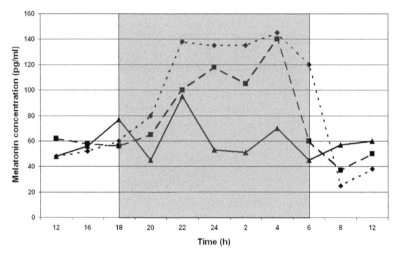

Fig. 6.1 Mean 24-h concentration of melatonin in the blood of carp (L:D=12:12). Dotted line—untreated fish, broken line—fish subjected to sham operation, continuous line—fish subjected to transection of optic nerve and arrest of eye circulation. Shadowed area—dark phase (night).

The fish pineal gland, surgically harvested and placed in biogel, is characterized by relatively long viability. This was confirmed in crucian carp, where a complete decline in the level of melatonin in the medium was obtained after 6 days of pineal gland storage under constant light (Kezuka *et al.*, 1992). Turning off the light on day 7 caused the gland to synthesize and secrete melatonin.

The above results, which show the possibility of investigating the secretory activity of the pineal gland in different environmental conditions, enabled "thermal" experiments with carp pineal glands to be carried out under "*in vitro*" conditions.

Immediately after collection, pineal glands were transferred to perifusion columns and placed in biogel. The perifusion medium, pumped through Teflon tubes, flowed through the columns with glands and was taken using a fraction collector into Eppendorf tubes. Perifusion lasted for 3 to 4 hours. The set was illuminated at light intensity of 2000 lx or isolated by aluminium foil (0 lx). The perifusion medium container was placed in thermostat to control the temperature (Fig. 6.2).

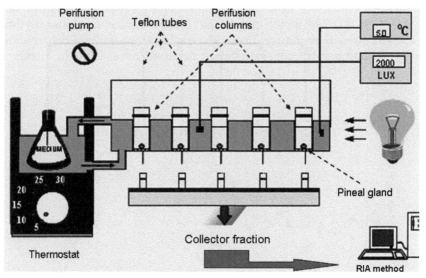

Fig. 6.2 Diagram of pineal perifusion kit. One pineal gland of carp was placed in biogel in each of five perifusion columns of 2 ml volume. The perifusion medium, pumped through Teflon tubes, flowed through the columns and was then collected using a fraction collector into Eppendorf tubes.

The results obtained showed that the carp pineal gland, similarly to crucian carp, pike, rainbow trout and sucker, responds independently to light intensity changes. No action of light on the pineal gland (total darkness) caused almost immediately strong synthesis of melatonin in the gland and release of copious quantities of melatonin to perifusion fluid (almost 400 pg/ml), while a light intensity of 2000 lx, by acting directly on the gland, rapidly inhibited the hormone secretion. Melatonin concentration in the perifusion medium remained constantly low (30 pg/ml on average) (Fig. 6.3).

These observations led us to a conclusion that under artificial conditions, the carp pineal gland shows similar secretory activity as under physiological conditions.

A further aim was to determine the rate of melatonin release from the pineal gland, i.e. to find out if after placing the gland in the medium, the hormone will be spontaneously released as in the case of the pituitary gland. The phenomenon occurs when the gland is suddenly deprived of the neural or neurohormonal control and spontaneously produces or releases the accumulated hormone. This is accompanied by several rapid sequential pulses with high and low hormone concentration in the medium. Analysis of Fig. 6.3 shows that in the medium collected during perifusion of carp pineal glands in the light phase (2000 lx), the mean concentration of melatonin was low and averaged 30 pg/ml. Because the samples were collected at very small intervals (every 5 minutes) and no rapid increase in melatonin concentration was observed in any of them, it can be safely concluded that light clearly inhibits the hormone synthesis and the gland has no accumulated stores. Turning the light off and bringing the gland to complete darkness caused a rapid release of melatonin to the medium, because already in the first fraction collected after 20 minutes of darkness, melatonin concentration increased and remained significantly ($p<0.01$) higher in the medium throughout the dark phase.

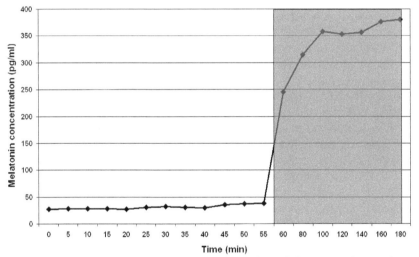

Fig. 6.3 Mean melatonin concentration in samples collected during perifusion of carp pineal glands at 20°C. Bright area (0–60 min.) 2000 lx. Dark area (60–180 min.) 0 lx.

These results clearly indicate that the pineal gland does not store melatonin. In nature, the released melatonin rapidly enters the blood and, in the case of fish, small amounts of melatonin also enter the cerebrospinal fluid. Because the half-life of melatonin in blood is very short (several to several dozen minutes), through melatonin the pineal gland keeps the body informed about the current photoperiod. It then functions as the hands of an internal clock, determining the time for the whole body. This is possible because in fish (unlike, e.g., in birds) the pineal gland is the only source of melatonin responsible for the presence of this hormone in blood (Popek *et al.*, 1997a).

The above experiments were conducted during the summer when natural day length was 16 hours (June) and temperature of the fish pond was 20°C. The same temperature was maintained in perifusion columns, so the only variable was the presence or absence of light.

As noted above, light is the main environmental factor that controls melatonin secretion. However, data from literature point to the effect of other factors such as temperature, humidity (the rainy season) and the magnetic field, which affects secretory activity of the pineal gland by interacting with photoperiod. Of these, an especially important factor in addition to light is often temperature. This is clearly evident in cold-blooded vertebrates, because temperature is an omnipresent environmental factor that changes rhythmically. Its 24-hour fluctuations (higher temperatures during the day, lower temperatures during the night) particularly affect terrestrial animals, although they can also influence the physiology of animals inhabiting bodies of water.

In warm-blooded animals, even rapidly changing ambient temperature has little effect on body temperature. In cold-blooded animals, changes in ambient temperature can be a stronger environmental factor than light that affects melatonin secretion. In mammals, the activity of enzymes taking part in melatonin biosynthesis, such as NAT and HIOMT, is inhibited by the increase of external temperature beyond 33°C (Nir and Hirschmann, 1978). Also in rats exposed to elevated (34°C) or lower (7°C) temperature, NAT activity was found to decrease (Ulrich *et al.*, 1973). Low temperature also prevents light-induced inactivation of NAT in the hamster (*Djungarian hamster*) (Stieglitz *et al.*, 1991).

In birds, experiments with chicken pineal gland demonstrated that ambient temperature affects the rhythm of melatonin release, becoming a second factor (next to light) regulating melatonin synthesis, although the biochemical mechanism of this effect is not fully understood (Zatz *et al.*, 1994).

Much more sensitive to temperature fluctuations are cold-blooded vertebrates, especially those inhabiting temperate regions, where temperature is one of the most important environmental factors. In reptiles, melatonin synthesis and release is extremely dependent on temperature. This is confirmed by a study with snakes, which showed a lower amplitude of melatonin secretion at 10°C compared to the level of melatonin secreted in animals exposed to 25°C and 35°C. Similar findings were obtained for the turtle (Viven-Roels *et al.*, 1988). Likewise, the pineal gland of the lizard, in addition to photoreceptive function, shows thermoreceptive capacity, and melatonin secretion may peak even during the light phase of the cycle if temperature is low and associated with the dark (nocturnal) phase (Underwood and Calaban, 1987). Many publications have also emphasized that temperature changes regulate the annual sexual cycles of lizards more strongly than changes in exposure duration. In some reptile species, it was shown that low environmental temperature may completely eliminate the effect of changing light duration.

Also in amphibians, reproductive activity is more strongly stimulated by environmental temperature than by light. Within this class of animals, the circadian rhythm of melatonin secretion is synchronized not only by light but also by temperature, as evidenced by a study with salamander, in which 24-hour melatonin concentration in blood was lower at 5°C compared to 15°C (Rawdings and Hutchison, 1992). Also a short photoperiod and low temperature (equivalent to the winter period) suppress the rhythmic nature of melatonin concentration in the blood, pineal gland and retina of the frog (Delgado and Viven-Roels, 1989).

Much fewer reports have been published on temperature and its role in melatonin secretion in fish, although the importance of this factor was sometimes emphasized. The effect of temperature on pineal activity was investigated in the rainbow trout (*Oncorhynchus mykiss*) and *in vitro* studies revealed that pineal response to changes in light intensity is much weaker at lower temperature, which demonstrates that it can be the second environmental factor to affect endocrine activity of the pineal gland. Perifusions of the rainbow trout pineal glands at different temperature ranges showed the importance of temperature for the regulation of melatonin synthesis in both the light and dark phase. It was also reported that elevated temperature not only increases melatonin synthesis but also makes the pineal gland more sensitive to light (Max and Menaker, 1992). The importance of external factors (temperature and light) in melatonin synthesis was also shown in the white sucker (*Catostomus commersoni*), whose pineal gland may function as a transmitter of both thermal and light environmental information (Zachmann *et al.*, 1991).

However, these data pertain to salmonids, which are typical psychrophilic species. The experiments discussed below were conducted with long-day fish (mainly Cyprinidae) that prefer warm water and spawn during the summer period.

To show conclusively the role of temperature in melatonin release from the carp pineal gland, experiments were undertaken in which variable photoperiod (2000 lx and total darkness) was paralleled by different perifusion temperatures (5°C to 30°C). In this temperature range, the entire life cycle of most Cyprinidae species takes place at medium latitudes.

The experiments were started in spring (second half of March), when the natural day length was equal to night length (L:D=12:12), and mean water temperature in ponds was 10°C. After killing the fish, their pineal glands were placed in perifusion columns (Fig. 6.2). Medium samples were collected every 30 minutes for 3 hours and, as mentioned above, different incubation temperatures were applied in each group. It was shown that for each temperature used, mean melatonin concentration in the media collected during the dark phase (0 lx) was significantly (p<0.01) higher than the concentration of this hormone in samples taken during the light phase (2000 lx). Another characteristic trait for pineal response was an increase in melatonin secretion (during the dark phase) that was almost proportional to the increase in ambient temperature. Comparison of changes in melatonin concentration in the medium at different temperatures showed that from 90 to 180 minutes of perifusion (0 lx), the concentration of this hormone at 20°C, 25°C and 30°C differed significantly (p<0.01) in relation to the concentrations observed at 5°C and 15°C. There were no significant differences during the light phase (2000 lx) (Fig. 6.4).

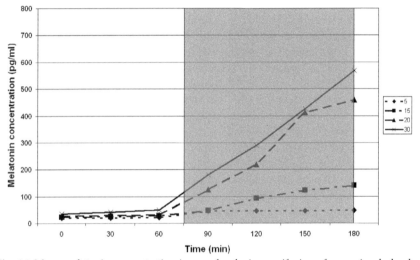

Fig. 6.4 Mean melatonin concentration in samples during perifusion of carp pineal glands at 5°C, 15°C, 20°C and 30°C. Experiment conducted in spring (L:D=12:12; pond water temperature 10°C). Bright area (0–60 min.) 2000 lx. Dark area (60–180 min.) 0 lx.

Further experiments were conducted to show possible differences in the seasonal sensitivity of the carp pineal glands to temperature. The only difference was that pineal glands were collected from carp harvested on the longest day of the year (June, L:D=16:8, pond water temperature of 20°C) (Fig. 6.5), during the autumnal equinox (September, L:D=12:12, pond water temperature of 15°C) (Fig. 6.6), and on the shortest day of the year (December, L:D=8:16, pond water temperature of 5°C) (Fig. 6.7).

In summing up the "temperature" experiments *in vitro*, it is concluded that similarly to the spring experiment, in the other seasons of the year the pineal gland responded to complete darkness with increased melatonin secretion to the perifusion fluid, which was followed by a rapid decline in secretion after the gland was illuminated. Also in these experiments, temperature played a significant role by stimulating melatonin secretion during the dark phase. Like in spring, in the other seasons of the year there was a temperature threshold in melatonin concentrations obtained at 20°C and beyond. In contrast, perifusions of pineal glands at less than 15°C resulted in clearly weaker melatonin secretion and the temperature of 5°C caused melatonin concentration in the collected media to be as low as that observed during the dark phase (Popek and Ćwioro, unpublished data).

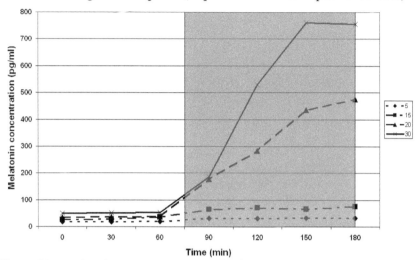

Fig. 6.5 Mean melatonin concentration in samples during perifusion of carp pineal glands at 5°C, 15°C, 20°C and 30°C. Experiment conducted in summer (L:D=16:8; pond water temperature 20°C). Bright area (0–60 min.) 2000 lx. Dark area (60–180 min.) 0 lx.

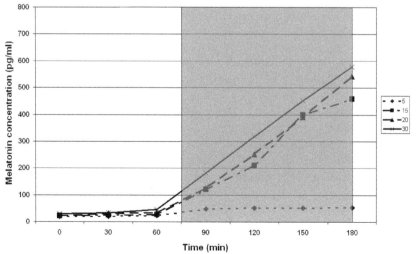

Fig. 6.6 Mean melatonin concentration in samples during perifusion of carp pineal glands at 5°C, 15°C, 20°C and 30°C. Experiment conducted in autumn (L:D=12:12; pond water temperature 15°C). Bright area (0–60 min.) 2000 lx. Dark area (60–180 min.) 0 lx.

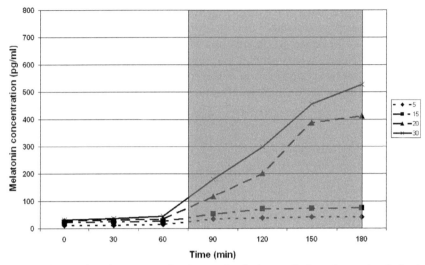

Fig. 6.7 Mean melatonin concentration in samples during perifusion of carp pineal glands at 5°C, 15°C, 20°C and 30°C. Experiment conducted in winter (L:D=8:16; pond water temperature 5°C). Bright area (0–60 min.) 2000 lx. Dark area (60–180 min.) 0 lx.

These results are in agreement with the observations of other authors, who reported that low temperatures in reptiles and amphibians disturbed or even completely suppressed 24-hour melatonin rhythms both in the pineal gland and in the blood, and when photoperiod and temperature were applied together, an interaction was observed between light and temperature cycles and melatonin secretion rhythm. It was also hypothesized that temperature has a stronger effect on the amplitude of melatonin rhythm than on the basal level of melatonin.

Similar results were found for *in vitro* research in sea lamprey (*Petromyzon marinus*), where the melatonin peak was lower at 100°C than at 20°C. What is more, at 10°C lamprey pineal glands showed a steady decline in the amount of melatonin released during the night. At 20°C, melatonin concentration was observed to increase (Bolliet *et al.*, 1993).

The adaptability of fish pineal glands, which reflects the effect of variable photoperiod and water temperature, was confirmed by *in vivo* studies with sexually mature female carp. The experiment was carried out during the summer. After harvesting from the ponds, fish were divided into groups and adapted over 10 days to different artificial conditions. It was shown that in fish exposed to long photoperiod (L:D=16:8) but different water temperatures, the circadian rhythm of blood melatonin concentration was similar. However, the amplitude of this rhythm was much lower in fish exposed to 12°C (Figs. 6.8 and 6.9) (Popek *et al.*, 2002).

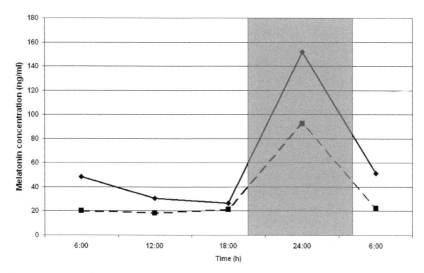

Fig. 6.8 Mean melatonin concentration in the blood of carp during long photoperiod (L:D=16:8). Continuous line—fish adapted to 20°C, fish adapted to 12°C. Shadowed area—dark phase (night).

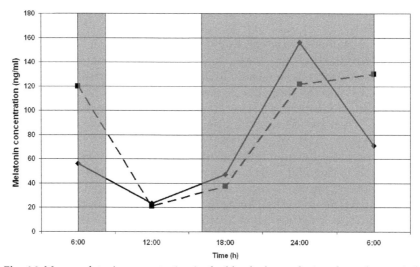

Fig. 6.9 Mean melatonin concentration in the blood of carp during short photoperiod (L:D=16:8). Continuous line—fish adapted to 20°C, fish adapted to 12°C. Shadowed area—dark phase (night).

The above data clearly indicate that the second most important environmental factor (temperature) can affect melatonin secretion from carp pineal glands by strengthening or weakening the effect of photoperiod. However, this process is not straightforward. Probably, an internal oscillator (biological clock) that maintains melatonin rhythm in the pineal gland is suppressed by low temperature. This effect occurs probably as a result of suppressed formation of cAMP and is a consequence of reduced NAT activity in the pineal gland. These suggestions are derived from a study by Falcon, who on the one hand showed thermal sensitivity of NAT and on the other demonstrated that changes in NAT activity in the retina and pineal gland of the pike are proportional according to temperature. Experiments also showed that there is a temperature threshold (about 37°C) beyond which NAT activity is observed to decrease, probably as a result of modified enzyme structure or enzyme degradation. It is worth noting that as temperature increased (from 4°C to 27°C), the NAT profile curve shown in the homogenates of pike pineal gland overlapped the profile obtained from the measurement of cyclic AMP accumulation in this gland (Falcon *et al.*, 1994). A study with rainbow trout and other fish species showed that NAT activity in the pineal gland of these fish can peak at lower temperatures (between 12°C and 17°C), when blood melatonin concentration is the highest. This can be due to different adaptation mechanisms at enzyme level (Thibault *et al.*, 1993).

When studying the effect of temperature on melatonin secretion in the rainbow trout, Max and Menaker (1992) proposed that the maximum physiological increase of melatonin concentration over basal concentration is 400%. Beyond this limit, the concentration increase is probably a kinetic effect. With this in mind, it can be safely stated that during the winter-spring period, when pond water temperature does not exceed 10°C, the physiological activity of carp pineal glands is very low. During this time, 15°C is probably the limit to which the pineal gland can still adapt.

In late spring and early summer, when the increase in day length is paralleled by a rapid increase in water temperature, the pineal gland of the carp becomes more flexible. During the period that can last until late autumn, this gland is able to fully adapt to an extreme temperature range of 5°C to 30°C.

The results obtained to date indicate that the pineal gland is able to distinguish between seasons of the year, even if photoperiods are identical. Both in the summer and in the autumn, the day length to night length ratio under natural conditions is 12:12. However, in the autumn water temperature and tolerance of the pineal gland to temperature are still rather high, which differentiates the profile of 24-hour melatonin synthesis in relation to 24-hour changes in the concentration of this hormone in the spring, when water temperature is lower.

In conclusion, light is definitely the strongest environmental factor affecting the pineal gland and thus the blood melatonin level. However, in lower vertebrates external temperature may also play an important role in regulating the rhythm of melatonin synthesis. The results obtained add to our knowledge of the role of the pineal gland in fish, because the effect of temperature changes on secretory activity of this gland in carp was shown for the first time. It is known that the internal biological clock of the pineal gland regulates rhythmic melatonin synthesis. This clock is synchronized directly by the photoperiod, where light has an inhibiting effect on melatonin production. However, it was also shown that the temperature factor may directly affect the regulation of rhythmic melatonin synthesis and secretion. Therefore, the pineal gland in fish can function as a transmitter of both heat and light information needed to regulate—through rhythmic melatonin secretion—24-hour and seasonal rhythms of activity, reproductive cycles, migrations and many other behaviours in environmental conditions that change periodically.

References

Askanazy, M. 1906. Teratom und chorionepitheliome der zilber. *Verhandlungen der Deutschen Gesellschaft für Pathologie* 58.

Bolliet, V., M.A. Ali, M. Anctil and A. Zachmann. 1993. Melatonin secretion *in vitro* from the pineal complex of the Lamprey, *Petromyzon marinus*. *General and Comparative Endocrinology* 89: 101–106.

Bubenik, G.A. and S. Dhanvantari. 1989. Influence of serotonin and melatonin on some parameters of gastrointestinal activity. *Journal of Pineal Research* 7: 333–344.

D'Amour, M.D. and F.E. D'Amour. 1937. Effect of pinealectomy over several generations. *Proceedings of the Society for Experimental Biology and Medicine* 37: 244–246.

Delgado, M.J. and B. Viven-Roels. 1989. Effect of environmental temperature and photoperiod on the melatonin levels in the pineal, lateral eye, and plasma of the frog, *Rana perezi*. Importance of ocular melatonin. *General and Comparative Endocrinology* 75: 46–53.

Falcón, J., V. Bolliet, J.P. Ravault, D. Chesneau, M.A. Ali and J.P. Collin. 1994. Rhythmic secretion of melatonin by the superfused pike pineal organ: thermo- and photoperiod interaction. *Neuroendocrinology* 60: 535–543.

Fenwick, J.C. 1970a. The pineal organ: Photoperiod and reproductive cycles in the goldfish, *Carassius auratus* L. *Journal of Endocrinology* 46: 101–111.

Fenwick, J.C. 1970b. Demonstration and effect of melatonin in fish. *General and Comparative Endocrinology* 14: 86–97.

Foa, C. 1928. Nouvi experimenti sulla fisiologia della ghiandolla pineale. *Archivio di Scienze Biologiche* (Bologna) 12: 306–321.

Gern, W.A. and L.L. Ralph. 1979. Melatonin synthesis by the retina. *Science* 204: 183–184.

Iigo, M., K. Furukawa, A. Hattoria, M. Haraa, R. Ohtani-Kanekoa, T. Suzukia, M. Tabatac and K. Aidab. 1995. Effects of pinealectomy and constant light exposure on day-night changes of melatonin binding sites in the goldfish brain. *Neuroscience Letters* 197: 61–64.

Kezuka, H., M. Iigo, K. Furukawa, K. Aida and I. Hanyu. 1992. Effect of photoperiod, pinealectomy and ophthalmectomy on circulating melatonin rhythms in the Goldfish, *Carassius auratus*. *Zoological Science* 9: 1047–1053.

Kitay, J. D. and M. D. Altschule. 1954. *The Pineal Gland*. Harvard Univ. Press, Cambridge, Mass.

Lerner, A..B., J.D. Case and R.U. Heinzelman. 1959. Structure of melatonin. *Journal of the American Chemical Society* 81: 6084–6085.

Marburg, O. 1913. Die Klinik der Zirbeldrüsenerkrankkungen. *Ergebnisse der inneren Medizin und Kinderheilkunde* 10: 180–195.

Max, M. and M. Menaker. 1992. Regulation of melatonin production by light, darkness, and temperature in the trout pineal. *Journal of Comparative Physiology* 170: 479–489.

Mhatre, M.C., A.S. van Jaarsveld and R.J. Reiter. 1988. Melatonin in the lacrimal gland: first demonstration and experimental manipulation. *Biochemical and Biophysical Research Communications* 153: 1186–1192.

Nir, I. and N. Hirshmann. 1978. Pineal N-acetyltransferase depression in rats exposed to heat. *Experientia* 34: 1645–1646.

Popek, W., P. Epler, K. Bieniarz and M. Sokołowska-Mikołajczyk. 1997. Contribution of factors regulating melatonin release from pineal gland of carp (*Cyprinus carpio* L.) in normal and in polluted enviroments. *Archives of Polish Fisheries* 5: 59–75.

Popek, W., A. Węgrocka, E. Drąg and P. Epler. 2002. Effect of the pineal gland in the hormonal regulation of sexual maturation in female carp. *Medycyna Weterynaryjna* 58: 371–374. (In Polish).

Quay, W.B. and Y.H. Ma. 1976. Demonstration of gastrointenstinal hydroxyindole-O-methyltransferase. *IRCS Medical Science* 4: 563–565.

Rawdings, R.S. and V.H. Hutchison. 1992. Influence of temperature and photoperiod on plasma melatonin in the mudpuppy, *Necturus maculosus*. *General and Comparative Endocrinology* 88: 364–374.

Sarteschi, U. 1913. La sindrome epifisaria macrogenitisomia precoce ottenuta sperimentalmente nei mammiferi. *Pathologica* 5: 707–710.

Steiglitz, A., S. Steinlechner, T. Ruf and G. Heldmaier. 1991. Cold prevents the light inducted inactivation of pineal N-acetyltransferase in the Djungarian hamster, *Phodopus sungorus*. *Journal of Comparative Physiology* 168: 599–603.

Thibault, C., J.P. Collin and J. Falcón. 1993. Intrapineal circadian oscillator(s), cyclic nucleotides and melatonin production in the pike pineal photoreceptor cells. In: *Melatonin and the Pineal Gland. From Basic Science to Clinical Application*, Y. Touitou, J. Arendt and P. Pévet (eds), Elsevier, Amsterdam, 1017: pp. 11–18.

Ulrich, R., A. Yuwiller, L. Wetterberg and D.C. Klein. 1973. Effect of light and temperature on the pineal gland in suckling rats. *Neuroedocrinology* 13: 255–263.

Underwood, H. and M. Calaban. 1987. Pineal melatonin rhythms in the lizard, *Anolis carolinensis*. I. Response to light and temperature cycles. *Journal of Biological Rhythms* 2: 179–193.

Underwood, H., T. Binkley, K. Siopes and K. Mosher. 1984. Melatonin rhythms in the eyes, pineal bodies, and blood of Japanese quail (*Conturnix conturnix japonica*). *General and Comparative Endocrinology* 56: 70–81.

Viven-Roels, B., P. Pévet and B. Claustrat. 1988. Pineal and circulating melatonin rhythms in the box turtle *Terrapene carolina triunguis*: Effect of photoperiod, light pulse, and environmental temperature. *General and Comparative Endocrinology* 69: 163–173.

Zachmann, A., S.C.M. Knijff, V. Bolliet and M.A. Ali. 1991. Effects of temperature cycles and photoperiod on rhythmic melatonin secretion from the pineal organ of the white sucker (*Catostomus commersoni*) *in vitro*. *Neuroendocrinology Letters* 13: 325–330.

Zatz, M.G., D.G. Lange and M.D. Rollag. 1994. What does changing the temperature do to the melatonin rhythm in cultured chick pineal cells? *The American Journal of Physiology* 226: 50–58.

7

CELLULAR CLOCKS AND THE IMPORTANCE OF LIGHT IN ZEBRAFISH

David Whitmore

INTRODUCTION

The use of forward, genetic mutant screens has proved to be one of the more powerful strategies in biology for working out what genes are involved in a given process (Mullins *et al.*, 1994). In fact, some might say that it is the only reasonable strategy when one knows nothing about a particular biological event. In the world of screening, *Drosophila* and mouse are "king", both of which have certain obvious advantages and disadvantages. The one alternative vertebrate model system to the mouse is zebrafish. These animals have a particular niche advantage over the mouse, especially when one is interested in examining processes involved in early embryo development. The reasons for this are clear, in that fish produce large numbers of transparent embryos, which develop outside of the female: a major advantage for studying the early stages of development. A large number of research groups have made use of this particular advantage and consequently, there has been considerable progress in the analysis of mechanisms involved in early embryogenesis using fish (Geisler *et al.*, 2007). Perhaps rather surprisingly, this model system has not been used anywhere nearly as extensively to study other biological problems. There are a few forays into the field of cancer biology, some drug screening, a little sleep research, and

University College London, Dept. of Cell and Developmental Biology, Centre for Cell and Molecular Dynamics, Rockefeller Building, 21 University Street, London, WC1E 6DE, UK.
e-mail: d.whitmore@ucl.ac.uk

of course the topic of this chapter, research on circadian clocks (Stern *et al.*, 2003; Tamai *et al.*, 2005; Prober *et al.*, 2006). This volume will contain several additional articles on clock biology in zebrafish, and so there will be some inevitable overlap, though the opinions of each author may interestingly differ. To reduce repetition to a minimum, I will focus on aspects of fish chronology especially relating to clocks in early embryos, the significance of light exposure in these early stages, and how light has been shown to influence the core clock mechanisms using zebrafish embryonic cell lines.

To begin with, however, it is probably important to start with a little history, and importantly acknowledge the effort and insight of Greg Cahill, who recently passed away far too prematurely. It was Greg who started the first examination of clocks in zebrafish back in 1996, and it was certainly his endeavours that inspired myself and several others to begin work in this system (Cahill, 1996). Many of his findings will be discussed below.

THE EMBRYONIC CLOCK

As mentioned above, the key advantage of zebrafish over mouse as a model system lies in the relative ease with which one can study changes in early development. An initial study on larval clocks was performed by Greg Cahill using an activity/video analysis system (Cahill *et al.*, 1998). Greg's aim was to establish an assay of rhythmicity that would be suitable for use in a circadian mutagenic screen. Typically rhythmic locomotor activity has been the assay of choice in both *Drosophila* and mouse, and so it made sense to begin here with an analysis of zebrafish clocks. The larvae used in these early studies were relatively old by zebrafish development standards being on the order of 10–15 days post-fertilization. But the results did clearly reveal the presence of a functional circadian pacemaker, with a population average period of 25.6 hours, and variance of between 0.5 to 1 hour. Though a little noisy as a screening assay, the Cahill laboratory was able to refine this approach, and establish it as a viable tool to successfully look for zebrafish circadian mutants (DeBruyne *et al.*, 2004).

Zebrafish larvae clearly have a functional clock. However, this study and those prior to this date were performed with the generally accepted belief that clocks in lower vertebrates were restricted to the retina and pineal gland; two structures where the clock regulates the rhythmic release of the hormone melatonin. Our understanding of circadian clock organization changed rather dramatically in 1998 and 2000 with the work of Professor Foulkes and myself (Whitmore *et al.*, 1998; Whitmore *et al.*, 2000). A series of experiments showed that cultured fish organs possessed endogenous circadian pacemakers (see Fig. 7.1). In fact, just about every tissue examined, with the exception of the testis, seemed to contain a clock. This study represented the first demonstration of peripheral tissue clocks in a vertebrate, and followed on from studies showing the same in *Drosophila*, and of course the presence of clocks in mammalian cell lines by the Schibler group (Plautz *et al.*, 1997; Balsalobre *et al.*, 1998). Further studies in zebrafish showed that these tissue and cell culture clocks could be directly entrained by exposure to a rhythmic light-dark cycle (Whitmore *et al.*, 2000). This direct cellular light sensitivity was rather unexpected for a vertebrate, even an animal as relatively translucent as a fish, and these results drew obvious comparisons with similar data in *Drosophila*. Yet there is a clear difference with higher/ placental mammals, where cells and organs appear to lack this cell autonomous light response (Tamai *et al.*, 2003).

An obvious question is whether this direct light sensitivity can also be found at the earliest stages of zebrafish development. We decided to explore this by examining expression of the zebrafish *period2* (*per2*) gene, as well as looking at changes in the DNA repair gene *6-4 photolyase* (Tamai *et al.*, 2004). Embryos were collected immediately after fertilization and placed into an incubator in complete darkness. At either 5 or 6 hours post fertilization

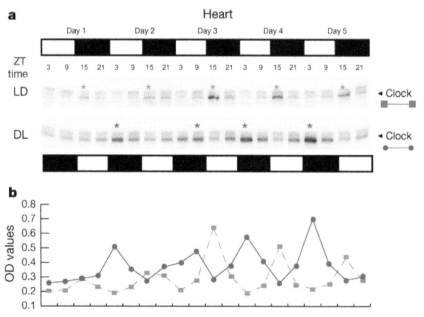

Fig. 7.1 Hearts were dissected from the same batch of zebrafish on the same light-dark cycle. Half were placed into an incubator on a forward light-dark cycle, and the other half into an adjoining incubator on a reverse light-dark cycle. Samples were collected at six-hour time points, RNA extracted, and an RNase protection assay performed to examine levels of the *clock* gene transcript (Fig. 7.1a). The dashed trace in Fig. 7.1b shows the rhythm obtained on the forward cycle, and the solid line that on a reverse light-dark regime. It is clear that the circadian oscillation rapidly re-entrains under the reverse cycle. These data show not only the presence of an endogenous clock in the zebrafish heart, but that tissues are also directly light sensitive. (From Whitmore *et al.*, 2000).

(hpf) half of the sibling embryos were given a light pulse of 3–5 hours, at an intensity of 500µW/cm² and "white" light of 400–700nm. Immediately after the light pulse, the embryos were either harvested for RNA extraction or fixed in 4% paraformaldehyde for *in situ* hybridization studies (see Fig. 7.2). The results from these studies were clear and quite dramatic. Early stage zebrafish embryos, just at the beginning of gastrulation, showed a strong

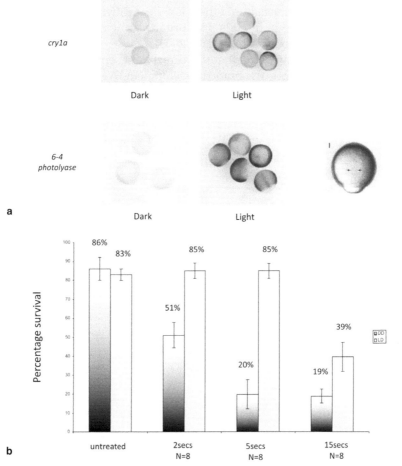

Fig. 7.2 (a) showing the clear induction of the two light responsive genes, *cry1a* and *6-4 photolyase*, by *in situ* hybridization on early gastrula embryos. Embryos were placed in the dark immediately after fertilization, and then given a five-hour light pulse after 6 hours of development at 28°C. The majority, if not all of the early embryonic cells, surrounding the yolk mass, are directly light responsive at this early stage. (b) shows the consequences of exposing early, day1 embryos to a short pulse of DNA damaging UV-light. Control sibling embryos were kept in the dark, while a matching group were raised on a light-dark cycle. Embryos raised in the light showed dramatically less death than their siblings growing in the dark, following UV exposure (From Tamai *et al.*, 2004).

transcriptional response to light exposure, with both *per2*, *cry1a* and *6-4 photolyase* being strongly induced throughout the cells of the developing animal. There were no distinct or clear light responsive regions within the embryo, but a global, diffuse light sensitivity. So embryos can respond to light well before the differentiation of any classical light responsive structures. In the earliest stages of development, from 3–6 hours post fertilization, the expression of DNA repair enzymes is high in both light and dark samples, which is undoubtedly the consequence of maternal deposition of this transcript in the zebrafish egg. The ability to detect light at around 5 hpf is interesting as, depending upon exact temperature, this corresponds approximately to the time that the embryos start to transcribe genes from their own genome, i.e., zygotic transcription begins at the mid-blastula transition. As soon as the zygotic genome becomes active, it appears that the embryo expresses the necessary cellular components, photopigments and signalling pathways, for it to be a light responsive organism. Of course, it is possible that the embryo is light responsive even prior to this developmental stage, using some post-transcriptional mechanism, but as all of our assays to date are transcriptionally based, this is hard for us to determine.

What is the purpose of this embryonic light sensitivity? Well an obvious significance could be relating to clock entrainment, and this will be discussed below. However, there are additional and very important advantages for an embryo that can "see" light. It was always a puzzle to me that zebrafish tend to mate within the immediate few hours after the lights come on at dawn. This means that the embryo, developing outside of the female, is suddenly exposed to the harmful impact of UV light exposure, and at a time in their development when one would imagine they are most sensitive to such a damaging stimulus. Surely, it would make far more sense to mate in the evening, and allow your offspring to go through rounds of critical cell division in the dark? Our study had shown that early light exposure was required to turn on expression of the DNA repair enzyme, *6-4 photolyase*. Is this a critical event in the survival of these early embryos in a "real world" situation? To test this we took two groups embryos, controls placed into constant darkness and their siblings raised on a LD cycle immediately after fertilization. Both groups were then given a short UV light pulse for a matter of seconds (Tamai *et al.*, 2004). Those embryos raised in constant darkness showed a high level of mortality, approximately 80% following a 5 second UV pulse, while their siblings, on a light-dark cycle showed no greater mortality than UV untreated control animals (see Fig. 7.2b). Furthermore, of the 20% of dark embryos that survived, many showed high levels of clear anatomical deformities, suggesting considerable levels of DNA damage. Growing up on a light-dark cycle, and "seeing" light early in development has a clear beneficial effect on the ability of these animals to survive severe environmental stress, and repair damaged DNA that might result from this.

What about the circadian clock in the developing zebrafish embryo? With the cloning of clock gene homologs in this species, it became possible to directly examine core clock, cellular oscillations in these embryos, rather than having to rely on output/downstream events such as locomotor behaviour or melatonin assays. This first circadian gene cloned in zebrafish was *zfclock* (Whitmore *et al.*, 1998). Several years later, a zebrafish homolog of *period* was cloned, *per3*, by the Thisse group as part of a large scale *in situ* hybridization screen (Delaunay *et al.*, 2000). Isolating zebrafish *period 3* (*per3*) in conjunction with the use of *in situ* hybridization then provided the Thisse group with a good opportunity to explore clock function at the earliest stages of development, prior to significant cellular differentiation, and the establishment of other known, measurable clock outputs (Delaunay *et al.*, 2000). The results of this study generated many tantalizing results, but also a great deal of confusion that has taken many years for the zebrafish clock community to "clear up". The core observation of their studies was that the circadian clock, as measured by changes in *per3* expression, started remarkably early in the development of the embryos. When eggs were collected and embryos raised on a light-dark cycle, a clear oscillation in *per3* expression was detected. In fact, it was even proposed that *per3* transcript levels oscillate prior to fertilization in the zebrafish oocyte, which of course makes interesting "demands" on transcriptional regulation in this haploid and relatively transcriptionally "quiescent" state. In addition, it raises interesting questions about how the clock mechanism could be working prior to the establishment of a diploid condition at fertilization. The result that is somewhat more controversial, was the observation that *per3* oscillations appeared to be just as robust, and significantly showed no dampening, in embryos raised under constant dark conditions. As there should be no entraining/synchronizing cues under these constant conditions, it forces one into quite a dramatic conclusion. How does a given embryo "know" the time of day? The only answer, and the title of the manuscript in question, must be that the embryo inherits not only a functional clock from the mother, presumably through maternal deposition of key transcripts into the egg, but it also must inherit circadian phase, the exact timing of the circadian rhythm. The deposition of certain important clock transcripts from the mother is certainly true (see below), but sadly clock gene oscillations under true constant conditions have not proved reproducible, at least in our hands (Dekens *et al.*, 2008). The constant dark oscillations in *per3* are most likely the result of entrainment to subtle, rhythmic changes in temperature, a hypothesis that would also explain why this rhythm failed to damp over the course of the study; a phenomenon that one never normally sees for clock gene expression under constant conditions in zebrafish.

Several groups, including my own, have attempted to re-examine this important issue, and confirm the existence of early embryonic clocks. An explicit rebuttal of the Thisse observation came from Kaneko and Cahill in 2005 (Kaneko *et al.*, 2005). The Cahill lab developed the first luminescent reporter, transgenic fish, with the *per3* promoter driving expression of luciferase. Using these transgenics they were able to follow gene expression oscillations in living embryos under a variety of lighting conditions. From this it became clear that *per3-luiferase* did not show any oscillations under constant dark, and temperature conditions. This observation fit perfectly with earlier Cahill observations that the rhythm in melatonin release, a strong circadian output in these embryos, also requires prior exposure to an entraining light-dark cycle, if oscillations are to be detected (Kaneko *et al.*, 2005). Rather unexpectedly, however, it was necessary to expose embryos to a solid 5–6 days of light-dark cycles before any significant oscillation in *per3* was observed. This seemed rather odd, for if *per3* was really a core clock component, then this would imply that the circadian oscillator was not actually starting until around day 5. This was in contradiction to Cahill's own observations of embryonic melatonin rhythmicity, which showed clear changes on day 2 of development (Kazimi *et al.*, 1999). Either the pacemaker was beginning to work initially only in highly localized, differentiated structures, such as the very early to form pineal gland (such that changes in the global signal cannot be detected), or else this luminescent reporter construct was not accurately reporting changes of the endogenous gene at this early stage. Sadly, it is our impression that the latter case is true, and possibly for complex reasons, due perhaps to the insertion site or regulation of this transgene, it does not appear to correctly represent endogenous gene function at these earliest of stages (see below).

In 2005 and 2006, the Gothilf laboratory published two excellent studies examining the role of light and light-induced *per2* expression on entrainment of the *arylalkylamine-N-acetyltransferase* (*zfaaNAT2*) transcript rhythm in the developing embryonic pineal gland (Ziv *et al.*, 2005, 2006). As I am sure that this will be discussed in great detail in a neighbouring chapter, I will not focus on these results extensively. However, it was perfectly clear from their data that early light exposure, even on the first day of development, was essential for pineal *aaNAT* rhythmicity, and therefore for entrainment of the pineal circadian clock. Interestingly, this putative light entrainment could occur even before the pineal structure itself differentiated, meaning that the phase must have been obtained by clocks within progenitor cells, and then have been retained through the process of cell differentiation. The use of specific antisense, morpholino constructs generated against *per2*, blocked this light-dependent clock entrainment in the pineal. These results not only provide evidence for a key role of *per2* in zebrafish clock entrainment, but also show that, at least in terms of pineal *aaNAT* expression, early light exposure is key for clock entrainment, and argues that embryonic oscillations are not inherited.

Several distinct questions still remained unanswered. Pineal clock function becomes apparent on the second or third day of development, but what is happening on the first day of development with cellular circadian rhythmicity throughout the embryo? Does the light response on the first day of development act to start the embryonic clock, or is it synchronizing a series of oscillators that begin spontaneously, as a normal part of the developmental process? We decided to go back and re-examine exactly what was happening with the clock mechanism during that critical first day, and try to determine the importance of light on these events (Dekens *et al.*, 2008). Our efforts were somewhat technically limited, as it is clear that the best way to gain insight into the developing clock would be to image oscillations in gene expression within single cells of the growing embryo. Though transgenic animals exist, in the form of the *per3-luciferase* animals, the number of photons released by each cell is too small to detect with current camera technology, and in addition, with every cell producing a luminescent signal, it would be difficult to obtain cellular resolution in any case. Consequently, we were limited to rather standard molecular techniques, along with every one, including RNase protection assays, and high-resolution fluorescent *in situ* hybridizations. Nevertheless, several interesting facts have become clear. Many of the critical and central clock components are maternally inherited in the very early stage embryo, when only a relatively few cells exist in the simple blastula (Ziv *et al.*, 2006; Dekens *et al.*, 2008). These include the two core transcriptional activators *clock* and *bmal*, as well as certain genes thought to be involved in the negative feedback loop of the clock mechanism, such as *period1*. Of the two strongly light-induced genes, *per2* is maternally present, but *cryptochrome 1a* (*cry1a*) is absent from the embryo within the first few hours post fertilization, and cannot be detected until the animals are exposed to light.

The first key observation, however, was that when embryos are raised in constant darkness, with tightly controlled environmental temperature, then no oscillations are seen in any of the currently identified clock components (see Fig. 7.3). This is obviously in contrast to the observation in the initial Thisse paper, but in line with measurements of clock outputs by Cahill and Gothilf. It is clear from our results that the embryo must experience light either to entrain an embryonic clock, or possibly even to start one oscillating. If one examines *per1*, or even *per3* expression during the first day of development, then one does appear to see a single cycle of "rhythmicity" either on a light-dark cycle or even in constant darkness, but this is not reflected in the expression of other clock components. It is clear that the maternal transcript for the *period* genes is rapidly depleted between 3 and 6 hours post fertilization, only to spontaneously increase again in the early evening following the onset of zygotic transcription. In the dark, *per1* then sits at a constant intermediate level of expression, compared to the very

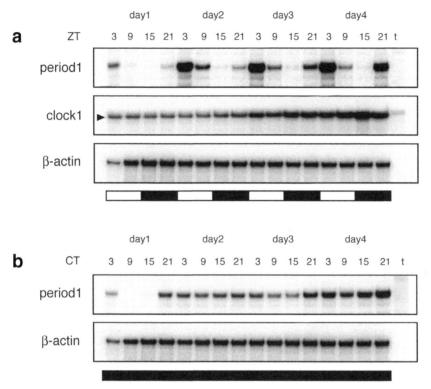

Fig. 7.3 RNase protection assays were performed on embryo samples collected over the first four days of development on either a light-dark cycle, or under constant darkness. a) Expression of *period1*, *clock* and *actin* were examined over this time course. On a light-dark cycle *per1* can be clearly seen to strongly oscillate throughout the experiment, whereas the smallest change in clock expression is only just detectable on day 4. The *actin* loading control shows no rhythmic changes. b) *period1* does not appear to oscillate in embryos raised in constant darkness. However, an interesting apparent rhythm does occur on day 1, as the maternally deposited transcript decays, followed by activation of endogenous circadian gene expression at about ZT15/21 (From Dekens *et al.*, 2008).

strong oscillation seen on the second day of development for embryos raised on a light-dark cycle.

These results suggest that the circadian oscillation undergoes a natural, spontaneous developmental activation during the middle of the first day of development, but requires early light exposure to set this embryonic rhythm. But how to examine this problem further especially, as mentioned above, as it is not yet possible to image single cell oscillations within the context of the whole embryo? One approach was to perform a high-resolution fluorescent *in situ* hybridization analysis of early embryos raised in the light or constant darkness. Using this technique, it is possible to resolve gene expression within single cells. Control embryos raised on a light-dark cycle showed a very strong rhythm in *per1* expression at the beginning of day 2 of development, with the vast majority of cells showing peak expression at ZT3, in the early morning, with almost no cells showing expression at ZT15, in the early evening (see Fig. 7.4). If the clock had not started in DD embryos, but was simply sitting at a medium level of non-oscillatory expression, as shown by RNase protection/gel analysis, then one would predict that all of these cells will express a constant mid-level of *per1*. However, if the clock has started, but is not entrained, then one might expect to see a random distribution of cells expressing *per1*, with some cells showing a high level of *per1* expression and others with negligible expression. The actual result was the second of these predictions. These data strongly support the hypothesis that embryonic clocks really have spontaneously started oscillating sometime on the first day of development, but they must experience environmental light signals for entrainment to occur and global rhythmicity to be established in the developing animal. Though perhaps far from a perfect study, this still represents the strongest evidence we have been able to obtain so far that the embryonic clock does in fact start working as part of a developmental program, and not maternal inheritance.

Additional evidence for the presence of a functional day 1 embryonic clock comes from an examination of this early light responsiveness. The first difference one sees in terms of rhythmic gene expression is that the levels of spontaneously increased *per1* expression are lower at ZT21 day 1 in light than under constant dark conditions. This means that light is able to impact core circadian oscillator function at the end of day 1, and consequently, the process of clock entrainment must have begun within the first 12 hours of development, when the embryos were exposed to light. To explore this phenomenon further we injected embryos with RNA coding for a CLOCK-dominant negative protein. When translated, this protein will compete with wild type endogenous CLOCK protein for binding with BMAL, and lead to the production of an inactive dimer. This is effectively akin to generating a transient CLOCK mutant zebrafish embryo. If one then examines *per1* expression in embryos on day 1 of development, it becomes apparent that the spontaneous increase in *period*, which is normally seen at the end of the first day, is dramatically repressed. This means that the spontaneous increase in *per1* expression at this stage is dependent upon the presence of a functional CLOCK:BMAL complex, again supporting the idea of the establishment of a functional pacemaker at this early stage of development. A rather curious observation is that the circadian oscillation in expression of *clock* and *bmal* cannot be detected until around day 3–4 of development, several days after the establishment of a strong *per1* rhythm. Now some caution needs to be taken with such data, as the natural rhythms in these two genes is inherently much lower amplitude than for the *period* genes, making accurate measurements much more challenging. Nevertheless, this result raises the interesting possibility that different aspects of the core clock

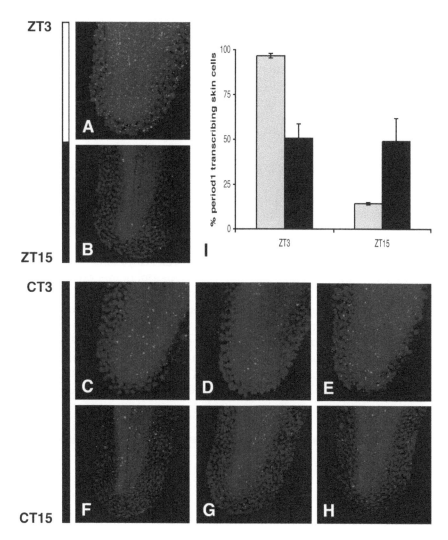

Fig. 7.4 Fluorescent *in situ* hybridization was performed in an attempt to examine gene expression changes within single cells of the developing embryo. All images of *per1* expression were made on the second day of development. It is clear, from panels A and B, that there is a dramatic rhythm in *per* expression between cells at ZT3 and ZT15 raised on a light dark cycle. Panels C to H show examples of expression for cells raised under constant dark conditions. If the clock had not started in the embryo one would predict all cells to be expressing *per1* at a mid-/average level. If, however, the clock has started, but is not synchronized between cells, then some cells will show high levels, and others low levels of expression of *per1*. Panel I shows a quantification of these data, and reveals an average number of cells expressing *per1*, supporting the idea of a population of de-synchronized clocks existing in the early embryo when raised in the dark. (From Dekens *et al.*, 2008).

Color image of this figure appears in the color plate section at the end of the book.

mechanism mature at slightly different times during early development. Perhaps a "complete" pacemaker mechanism is not truly present until day 4, a hypothesis which needs further testing. However, this is a result, which interestingly correlates with the onset of several rhythmic clock output events, such as clock gating of the cell cycle, for example (Dekens *et al.*, 2003). Maybe some of these downstream events do not become fully clock regulated until rhythmic expression of the transcriptional activators, *clock* and *bmal*, is established. Additional studies are required to "nail down" this idea.

CLOCK ENTRAINMENT WITHIN SINGLE ZEBRAFISH CELLS

It is perhaps not surprising, considering the fact that embryos appear to have a functional clock on the second day of development, that cell lines made from such animals also possess an endogenous clock (Whitmore *et al.*, 2000). The more unusual aspect of zebrafish cell lines is not so much this inherent rhythmicity, but the fact that such cells can themselves directly respond to light (Whitmore *et al.*, 2000). This phenomenon marks out zebrafish cell lines as a unique model system for the study of clock function, as they must possess not only the components of a functional clock, but also all of the elements of an entrainment pathway, including the relevant photopigments. The use of luminescent reporter gene constructs has greatly aided the examination of clock function in these cells, as can be seen from recordings of *per1*-luciferase shown below (Vallone *et al.*, 2004) (see Fig. 7.5). In addition, the direct light sensitivity opens up a unique opportunity to

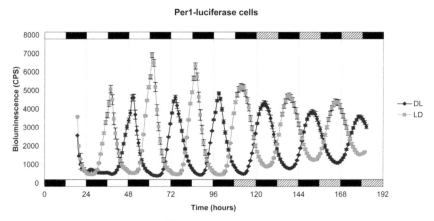

Fig. 7.5 Zebrafish cells were stably transfected with a *per1*-luminescent reporter gene construct, allowing circadian rhythms in gene expression to be followed in an automated manner. These cells survive extremely well in a 96 well plate on a luminometer. A clear oscillation in *per1* is detectable, which free-runs in constant darkness. Cells also retain the ability to directly entrain to altered light-dark cycles, confirming the presence of the required circadian photopigments and light signalling pathways.

study core clock entrainment in a dynamic manner whilst changing the immediate lighting conditions (Tamai *et al.*, 2007).

The initial studies on cellular clock function relied on a population analysis of these cellular rhythms, either by harvesting cells at set time points for RNA extraction, or following rhythms in 96 well plates where the signal is an average of about 100,000 cells or more. Consequently, it was an assumption that each cell really contained its own clock, and that light sensitivity was not spreading through the population from a subset of light responsive cells; as unlikely as that hypothesis would appear. Furthermore, if cells were maintained in constant darkness for numerous days, or even weeks, no apparent rhythm in any clock genes could be detected. Again the most likely hypothesis was that such cells simply drifted out of phase with each other, but of course it was also possible that the clocks within each cell could "wind down" after long durations in constant conditions. To explicitly examine such issues it was necessary to follow circadian oscillations within the living single cell. This means imaging gene expression within single cells over numerous days within the population. Though this has now been achieved not only in zebrafish but also mammalian cells, it still remains quite a challenging endeavour (Welsh *et al.*, 2004; Carr *et al.*, 2005).

There were two initial problems to such a study. The first was producing a cell line with sufficient photon yield to be able to image. Even the most sensitive photon counting cameras are pushed to the extreme limits by luminescent signals generated from single cells. Secondly, it would be "useful" if all cells could be oscillating within the same dynamic range. Or more simply put, that the peak and trough levels of brightness are relatively similar between cells. One reason for this is that the dynamic range of current camera technology is quite limited, and it is impossible to set a range of camera parameters that could cover cells with a wide and varied range of intensities, that are produced by cells with a variable number of reporter gene inserts. These problems were solved by generating clonal cell lines using fluorescence activated cell sorting (FACS); where many clonal cell populations were generated from individually isolated single cells. This approach has allowed us to select for clones with a range of luminescent intensities, and also to have some confidence that the population would be oscillating within a similar dynamic range.

Having solved these technical problems, then it was simply a matter of imaging cells that had been either entrained to a light-dark cycle, or maintained in constant darkness for several weeks. The one minor issue being that all of the cell culture had to be performed in complete darkness, under IR illumination. Data collected over three days for cells, following entrainment, revealed that all of the cells showed a robust circadian oscillation with a high level of synchrony (*per1* expression peaking in the early morning). In contrast, cells that were maintained in constant darkness for over a week still showed circadian oscillations in *per1* expression, but the timing of peak levels was randomly distributed throughout the day (see Fig. 7.6). Clearly, individual cells do continue to oscillate for long periods in the dark, and the role of light really is primarily as a synchronizing signal, and not a stimulus required for "starting" the clock.

If a population of cells are maintained in the dark for several weeks then, as mentioned above, the global *per1* signal averages to a flat, non-

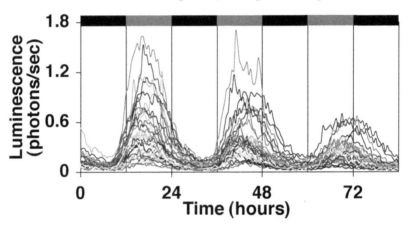

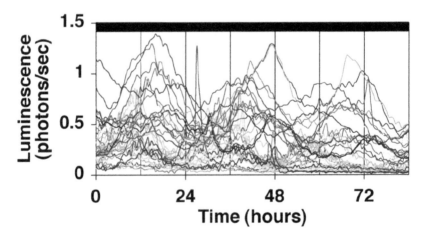

Fig. 7.6 Single cell luminescent imaging reveals the presence of a circadian clock within each individual cell. Cells that have been entrained previously to a light-dark cycle continue to oscillate in phase in the dark for several days. However, cells that have been maintained in the dark still show circadian oscillations in *per1* expression, but these have random, non-synchronized oscillations. Light is required to entrain the clock, but not maintain or start the circadian rhythm. (From Carr *et al.*, 2005).

oscillatory level. A single 15 minute light pulse applied to these cells has a dramatic synchronizing effect, in that an immediate oscillation is apparent, which has a consistent starting phase of approximately ZT3 (see Fig. 7.7). In other words, a single light pulse appears to move the cells to a "common" phase equivalent to the early morning. But what is actually happening at the singe cell level? Imaging cells prior to, and for several days following a single 15 minute light pulse revealed that all of the cells responded to the light signal by shifting their randomly distributed phases such that they became synchronized to the same circadian phase. A point from which these oscillators then began to free-run. So all of the cells within the population appear to behave identically, at least as far as we could determine, and without question each cell possesses this direct light sensitivity. One requirement for such strong resetting to light is, of course, the possession of a high amplitude, Type 0, phase response curve, as described by Pittendrigh in *Drosophila pseudobscura* many years ago. This simply means that the oscillator within these cells responds to light by producing a very large phase shift in the rhythm, data for which can be seen in the figure below. Trivially of course this means that zebrafish are superbly evolved for modern jet travel, and can shift rapidly to any new time zone.

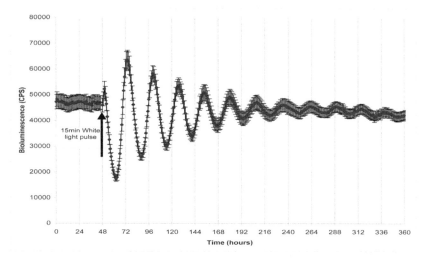

Fig. 7.7 At the population level, cells maintained in constant darkness show a flat, constant level of *per1* gene expression. If given a single, 15-minute light pulse, however, these cells rapidly synchronize and clear circadian oscillations are apparent. This dampens down over the following 12 days to a mid-value of luminescence. Interestingly, the phase of the population is always the same, with the synchronized rhythm beginning just prior to ZT3 at dawn. This movement to a common phase is the consequence, obviously, of the amplitude and shape of the underlying Type 0 phase response curve. (From Carr *et al.*, 2005).

It is clear that phase shifting to a single pulse is a rapid, near instantaneous event, but how does the oscillator entrain to complete photoperiods? The classical studies of Pittendrigh and Minis explored *Drosophila* entrainment to both full photoperiods, as well as skeleton regimes, with the aim of exploring the issue of parametric versus non-parametric entrainment (Pittendrigh *et al.*, 1964). The question was relatively simple. Was a single pulse of light at dawn, or two matching pulses at dawn and dusk, sufficient to replicate the entrainment pattern seen to a full photoperiod? Did the oscillator "jump" to a new phase each day upon light exposure, such that steady state entrainment was finally reached in response to these light pulses, or was light acting to change the angular velocity, i.e. speed of the inherent pacemaker continuously during the day? All of these experiments in *Drosophila* were performed through a series of laborious eclosion studies, where thousands of *Drosophila* pupae were examined in each study. Each eclosing pupa contributed to this large population data set, the size of which gave these studies much of their power. These classical studies clearly established that single light pulses (or two pulse skeleton photoperiods) could mimic very well the entraining effects of full photoperiods, that is complete light-dark cycles, with issues of stability only arising when the two light pulses were moved "further apart" (simulating a lengthening photoperiod). Under these conditions, a period of bi-stability can occur where individual animals might have "difficulties" in determining which pulse represents dawn and which dusk. As a population, however, eclosion activity "jumps", such that it is timed to occur in the shorter of the two inter-pulse intervals (Pittendrigh *et al.*, 1964).

Directly light responsive, clock-containing cell lines offer a unique opportunity to actually see how molecular components of a core circadian oscillator, as opposed a behavioural output, respond to such entraining stimuli. Therefore, we placed our cells onto a variety of lighting regimes, including a single 15 minute "dawn" pulse of light every 24 hours, and two similar pulses representing dawn and dusk (see Fig. 7.8). The molecular entrainment to these two conditions was then compared to that which occurred to a matching full photoperiod of light of equal intensity and wavelength composition. The results for this were quite interesting when one examined expression of both *per1* and *cry1a-luciferase* reporters. With one single entraining pulse each day, the timing of the peak expression of *per1* was very similar to that seen on a full photoperiod, as was the timing and waveform of the decline or falling phase of its expression. The oscillation established by this single pulse remarkably similar to that seen on the full photoperiod. However, the timing of the trough, or low point value, was not phase matched to that for cells receiving a full 12 hours of light, and the rising phase of the rhythm was phase advanced. In fact, a more careful analysis shows that the rising phase of *per1* expression actually phase advances slightly on each entraining cycle.

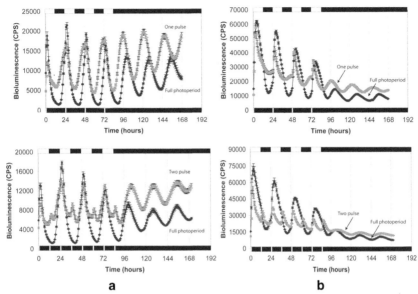

Fig. 7.8 How does entrainment to a complete photoperiod compare to that using a single, or skeleton pulse regime of 15 minutes of light? Fig. 7.8a shows the oscillation of *per1* expression. The top panel compares entrainment between a full photoperiod and single light pulse per cycle. While the lower panel compares the complete light-dark cycle to a skeleton, two pulse paradigm. Entrainment of peak *per1* phase is remarkably similar under all conditions. However, entrainment with the single pulse in not complete, as the trough value and rising phase do not achieve a stable phase relationship. There is also a dramatic after-effect on free-running period following single pulse entrainment. Using a skeleton, the stability of entrainment is much improved, though the amplitude of the oscillation is not as great as that seen on a full photoperiod. Fig. 7.8b shows results for *cry1a* expression under the same experimental paradigm. *Cry1a* is clearly always acutely induced by the light pulse. Again lack of complete entrainment to the single pulse paradigm is apparent. (From Tamai *et al.*, 2007).

Though establishment of peak phase is quite good, it would appear that one does not achieve full or complete entrainment of the molecular rhythm to a single daily pulse of light.

This differential entrainment between the accuracy of peak timing and the difference in trough/rising phase entrainment, raises the interesting possibility of there really being an "evening" and "morning" oscillator within zebrafish cells, each of which, in this case, would be sensitive to light either at dawn or dusk, respectively. If this is the case, how does the system entrain to a two pulse skeleton photoperiod, with 15-minute dawn and dusk pulses being separated by 12 hours? In this scenario, compared to the full photoperiod, the timing of peak *per1* expression and its rising phase are more accurately timed, and compare more precisely with the full light cycle situation. The actual traces are a little harder to examine as a consequence of the clear, acute induction of *per1* expression that one can detect in response to the second, or dusk light pulse. Nevertheless, the precision in phase and period of the molecular oscillation is greater in the two-pulse situation, confirming improved oscillator entrainment, and supporting, at a molecular level, the results obtained many years ago by Pittendrigh in his studies of *Drosophila* eclosion.

Several additional observations regarding this entrainment process are of interest. For one, even though the accuracy of timing using two-pulse entrainment is very close to that seen on a full photoperiod, the amplitude of the rhythm, that is the peak to trough values of the oscillation, is significantly smaller. One clear consequence, therefore, of entrainment to a complete light-dark cycle is the establishment of a robust high amplitude rhythm. The sustained or parametric effect of experiencing a complete light-dark cycle is the extended repression of the *per1* rhythm during the day, and this molecular event is key to the generation of this high amplitude oscillation. So, the one or two pulse scenario can clearly mimic phase, but parametric aspects of light exposure are necessary for a more "reliable" or robust molecular oscillation. A second interesting observation relates to the generation of after effects on the free-running period of the clock when the cells enter constant darkness, following entrainment to the various lighting conditions. The free-running period of the *per1* rhythm in DD, following single pulse entrainment is very short compared to that on a full photoperiod. This difference is in the order of 3–4 hours, which leads to a rapid phase difference between cells in the two conditions after several days in the dark. This after effect is not apparent following entrainment to the skeleton two-pulse situation, where the free-running periods between skeleton and full photoperiods are identical. This observation clearly reflects, therefore, a long lasting consequence on the period of the cellular clock depending upon prior light history. Though a common observation at the level of animal behaviour, such a phenomenon is less well described at the cellular/molecular level.

What does one observe if one examines expression of *cry1a*, using a *cry1a-luciferase* reporter construct? Initially, it is clear that when the cells are exposed to light that *cry1a* expression is increased. So in the two-pulse situation there are two clear, small peaks in *cry1a* levels, matching each light pulse. In the case of the single daily light stimulus, the lack of complete entrainment is even more apparent, as *cry1a* expression clearly phase advances each day following the single light pulse per day treatment. This acute induction of *cry1a*, of course, suggests a key role for this gene in clock entrainment, a topic that will be discussed more fully below.

Another remarkable observation regarding light entrainment in zebrafish is the action of sustained light exposure on the molecular clock. It had been reported many years ago again by Pittendrigh that sustained light exposure appeared to stop the circadian pacemaker, as measured by alterations in the *Drosophila* eclosion rhythm (Pittendrigh *et al.*, 1964). What do long light pulses actually change within the zebrafish molecular pacemaker? Again looking at the levels of *per1*-luciferase expression, it is clear that the circadian pacemaker does in fact "stop" when cells are exposed to long light pulses, typically from 24 up to 60 hours. *Per1* is tonically repressed by this extended light treatment, and remains at this flat, low level until the light stimulus is removed. To explore this phenomenon further, we performed a series of varying long pulse, or "wedge" experiments (see Fig. 7.9). In this case, different plates of cells are exposed each to light of varying durations. A "forward wedge" is when the light pulse begins at the same circadian phase, but then ends at a variety of times, depending upon its duration. While a "backward wedge" simply means starting these long pulses at different phases, but ending them all at the same time. Both approaches reveal interesting details about the nature of this light-clock stopping phenomenon. The "forward wedge" clearly shows that the circadian oscillation is held motionless for the length of the light pulse. Cells that are exposed to light for 24 hours show a delay in the rhythm, to the next *per1* peak, of about 36 hours, while cells exposed to a 60 hour light pulse show a delay of about 72 hours. The "backward wedge" clearly shows that, upon release from the sustained light condition, all of the cellular pacemakers begin their "motion" from a phase roughly equivalent to dusk, or ZT12. It is for this reason that the delay in response to a light pulse is equivalent to the duration of that light pulse plus approximately an additional 12 hours. The circadian clock is held motionless, or perhaps more correctly phrased, the amplitude of its oscillation is reduced greatly and very close to the singularity, until the light treatment is terminated. At this point, the clock begins to oscillate again from dusk. It is remarkable in some ways in which this molecular observation fits so well with the description of clock "behaviour" in eclosing *Drosophila* shown by Pittendrigh many years ago.

What are the underlying molecular mechanisms behind this cellular light entrainment, and clock "stopping" process? Well at this time there are more unknowns than there are identified signalling components. However, several things are quite clear. Light exposure causes the acute induction of at least two, and undoubtedly more, central clock genes. This includes an increase in *cryptochrome 1a* (*cry1a*), as mentioned above, and *period2* (*per2*). Both of these appear to have important roles in clock entrainment. *Per2* was discussed above, in the context of entrainment of pineal NAT rhythms in the developing zebrafish pineal gland, work performed by the Gothilf group (Ziv *et al.*, 2006). This will undoubtedly be discussed in more detail in a neighbouring chapter. As far as I am aware, at this time, there is little molecular understanding of how a light increase in this gene leads to changes in oscillator phase, and certainly my group has failed to elucidate this mechanism. Consequently, we have focussed more on the role and action of light induced *cry1a* expression. As mentioned before, *cry1a* levels always increase when cells or tissues are

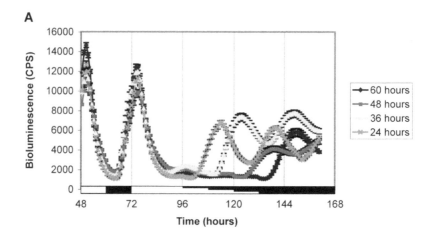

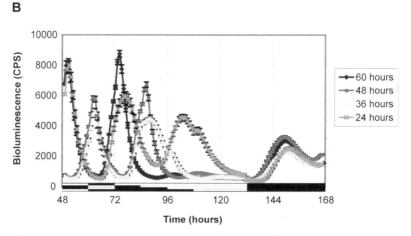

Fig. 7.9 To explore the light "stopping" action of sustained light exposure, *per1* expression was compared under a series of long-pulse experiments. The top trace shows a "forwards wedge", with light pulses beginning at the same circadian phase. The circadian oscillation in *per1* is clearly held motionless during the sustained light exposure, and begins again upon return to constant dark conditions. A "reverse wedge" (lower panel) shows that the pacemaker begins from approximately the same phase upon release into DD, about ZT12 or dusk. The clock appears to be held motionless by light pulses longer than 12 hours (From Tamai *et al.*, 2007).

Color image of this figure appears in the color plate section at the end of the book.

given an entraining light pulse. To explore this further we decided to explore the correlation between levels of *cry1a* induction, light intensity, and size of induced phase shift. *Cry1a-luciferase* cells were given a single light pulse, of varying light intensity, for 15 minutes at CT16. The light used was "white" light of exactly 400–700nm, and intensities ranged from 100 to 10,000

$\mu W / cm^2$. At the same time, and in the same experimental plate, changes in the timing or phase of *per1* were measured. As light intensity is increased across this range, so the size of the phase shift also increases. Light induced phase shifts range from approximately 6 hours at the lowest intensity up to about 12 hours for bright light. The levels of *cry1a* induction strongly correlate with both light intensity, and more importantly with the size of the resulting phase shift in the rhythm.

All circadian systems have distinct phase response curves (PRCs), which underlie their entrainment or phase shifting characteristics. In zebrafish, the PRC is composed mainly of large phase delays, with a narrow region of phase advances around dawn. Due to the considerable size of the light induced phase shifts, this response would be classified as a Type 0 PRC. To examine the role of *cry1a* in this process further, we decided to explore the correlation between the phase-dependent size of light-induced phase shift with the induction levels of *cry1a*. Light pulses of fixed intensity were given at a variety of circadian phases, and the size of resulting phase shift was measured by examining the resultant timing in the *per1*-luciferase rhythm. At the same time, the amount of increased *cry1a* expression was determined using *cry1a-luciferase* cells. The largest phase delays to light are seen in the late subjective night, around CT20, when a single light pulse can cause more than a 15-hour shift in the rhythm. Interestingly, this is also the phase of the cycle when light induces the largest amount of *cry1a* transcript. Contrast this to light pulses given at CT4, in the early subjective day, when almost no phase shift is induced by light, and one can see that this corresponds to the time on the circadian cycle when light induces the least amount of *cry1a* transcript.

Clearly, there are a series of very strong correlations between phase shifting, entrainment and the light dependent increase in *cry1a* expression. But how could this molecule be working to regulate clock timing in response to light? To explore this question further we have performed a large series of protein interaction assays. The dimerization of CLOCK and BMAL proteins is a key event in forming a transcriptionally active dimer. As has been well established in numerous circadian model systems, this CLOCK:BMAL dimer then binds to E-boxes (CACGTG) in the promoters of target genes, where it activates their transcription (King *et al.*, 2000). A yeast two-hybrid analysis of CLOCK and BMAL protein interactions in zebrafish shows that these two proteins physically interact at two key domains within their protein structure (Tamai *et al.*, 2007). Both the PAS B and basic helix-loop-helix (bHLH) domains are important areas where these proteins physically make contact in order to form an active dimer (see Fig. 7.10). A close examination of the interaction behaviour of CRY1a shows that this protein also forms a complex with both CLOCK and BMAL, and in fact, interacts with the PAS B region of CLOCK, and the PAS B, bHLH and C-terminals regions of BMAL.

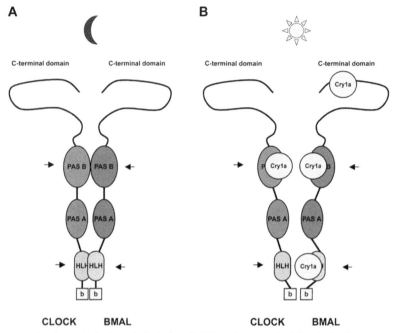

Fig. 7.10 A schematic for how light-induced CRY1a disrupts CLOCK:BMAL function in a light-dependent manner. CLOCK and BMAL have been shown to interact through their PAS B and bHLH domains. CRY1a in the light binds to these regions and prevents the formation of a transcriptionally active dimer. In this way, long light exposures can "stop" circadian pacemaker function, until the CRY1a protein is degraded (From Tamai *et al.*, 2007).

Color image of this figure appears in the color plate section at the end of the book.

Effectively, CRY1a can bind to the key regions necessary for the formation of a CLOCK:BMAL dimer, as well as the part of the protein, the C-terminus, necessary for downstream transcriptional activation. Consequently, we believe that CRY1a competes within the cell for binding to these important protein domains, and as such, can block the formation of an active CLOCK and BMAL dimer. Of course, CRY1a levels are high during light exposure, and so this protein binding/competition will only occur in the presence of light. The "more" light, then the more CRY1a is likely to be present and the greater the disruptive effect on CLOCK and BMAL dimerization is likely to be. It is through this mechanism that we believe light, through *cry1a* induction, can have a repressive effect on *per1* transcription, and lead to such strong phase shifts in the circadian oscillation.

From our yeast protein interaction studies, an important aspect of this protein competition appears to be the fact that CRY1a can only interact with CLOCK or BMAL monomers, but has little inhibitory effect when the active CLOCK:BMAL dimer has already formed. This is an important detail, and in part goes to explain how light can "stop" the circadian oscillator in a phase dependent manner. Sustained light treatments lead to an increase in *cry1a* expression, which remains at a raised level of expression as long as the light stimulus is present. During this treatment, the *per1* levels of expression are highly repressed. Over-expression of *cry1a* in these cells mimics this strong repressive action of light. But why does the clock stop at ZT12 or dusk? Our working model to explain this phenomenon is that the dawn peak of *per1* expression (peaking around ZT3) is generated by an active CLOCK:BMAL dimer acting on the E-boxes within the *per1* promoter. The natural drop in *per1* expression during the day is likely to be due to both the degradation of active CLOCK and BMAL, as well the repressive action of the endogenous negative feedback loop. Light at this time has a relatively small phase shifting effect, possibly in part because light-induced CRY1a cannot disrupt the active CLOCK:BMAL dimer. As the day progresses, then transcription of new *clock* and *bmal* begins in the late afternoon to early evening, reaching a peak around ZT15. We believe that this mRNA is rapidly turned into CLOCK and BMAL proteins. If no light is present during the early evening, as for example on a 12-hour day, then these proteins can dimerize efficiently, and then act on the *per1* promoter to start the cycle over again. If, however, light extends past ZT12 into the early night, then the levels of cellular CRY1a will be increased, and be able to bind to CLOCK and BMAL monomers prior to the formation of a transcriptionally active dimer. In this way light will act to block CLOCK:BMAL function, stopping the pacemaker (if enough CRY1a is present) until the light signal is removed and CRY1a rapidly degrades. Though far from complete, this model of light driven protein competition explains a great deal of the rhythmic cellular biology that we observe in response to light.

There are many questions and unknown aspects of this signalling pathway still to be explored. Perhaps the first and most obvious relates to the identity of the photopigment, or pigments, involved in this cell and tissue-based light response. The truth, at this time, is that no one definitively knows the identity of the molecules critical for this response, but there are and have been a few suggestions. The most common approach used to characterize a photopigment, whether circadian or one involved in visual responses, is to produce an action spectrum for that response. Simply put, one scans across the range of light wavelengths, typically from the high UV just below 400nm up to the extreme red just above 700nm, and then examines the response of some circadian response or phenotype. This response is typically the size of the phase shift to light, but can also be a "physical change" such as pupil size, or of course a change in the transcription of a specific gene. There is only one published action spectra for the circadian response in zebrafish cell lines, though a number of groups have unpublished data for the same response (Cermakian *et al.*, 2002). The work of Cermakian *et al.*, on a particular zebrafish cell line called the Z3 cells, has argued that the photopigment involved in the light induction of *per2* is primarily sensitive to blue light, with an apparent peak at 380nm in the high UV. The response at all other wavelengths is remarkably similar, with a slight drop off in response above 450nm. Surprisingly this response neither fits the predicted response curve for an opsin, nor the predicted action spectra one would expect from a cryptochrome/flavin photopigment. Of course, the fact that cryptochromes are thought to be important circadian photopigments in *Drosophila* (Stanewsky *et al.*, 1998), if not in mammals, has been used as an argument for their role in teleosts. These data would seem to fit that hypothesis. However, it must be stressed that this conclusion is really only based on a single point of the action spectra and great caution must be taken in interpreting this result. Clearly, an action spectrum of much higher quality and resolution is necessary. In the case of the light effect on melatonin suppression this data now exists, and is likely to be discussed extensively in another chapter of this book. However, from these excellent studies, it is clear that multiple pigments appear to play a role in the light response involved in zebrafish melatonin regulation (Ziv *et al.*, 2007). Interestingly, two peaks of optimal sensitivity suggest not only a role for exo-rhodopsin, which is highly expressed in the pineal, but also a pigment more tuned to the red wavelengths. If multiple photopigments are involved in the regulation of zebrafish melatonin, then perhaps it should not come as a surprise that the same is likely to be true for the light response within the cells and peripheral tissues of this species.

A search for novel opsins in zebrafish has led to the identification of several rather novel molecules in this species. Of these, probably the cloning of *teleost multiple tissue* or *tmt* opsin has provided the most interesting candidate to date. The original isolation of tmt opsin came from the screening of a gridded *Fugu* cosmid genomic library using a probe generated against Atlantic salmon VA opsin. However, a zebrafish homolog was eventually isolated and full-length sequence obtained through the use of RACE PCR. What is interesting about this particular opsin, however, is its unusual tissue expression pattern. Initial studies showed that it can be found throughout the whole of the adult brain when examined by RNase protection assays on samples prepared of fore-, mid- and hindbrain (Moutsaki *et al.*, 2003). In addition, it can be detected in a range of tissues throughout the body, including heart. But perhaps more interestingly, it can also be detected in zebrafish cell lines, as well as very early stage blastula embryos. Importantly, no other known opsins share this wide tissue distribution. Though, of course, some caution needs to be taken in considering this comment, as this obviously only represents negative data at this stage of our studies. Clearly, the expression pattern of *tmt* opsin fits that expected of a circadian photopigment in zebrafish, where all of these tissues and cells are directly light responsive. However, a large amount of functional data remains to be collected, and the definitive proof that *tmt* is the key photopigment is still lacking.

In mammals, a great deal of attention has been focussed on the role of melanopsin as the critical circadian photopigment necessary not only for circadian entrainment, but also other non-visual roles such as pupillary constriction (Panda *et al.*, 2002; Lucas *et al.*, 2003). As one might expect, zebrafish possess quite a wide number of melanopsin isoforms, but to date, there is no description that they show a wide range of tissue expression patterns (Bellingham *et al.*, 2006). Nor is there any functional data in teleosts as to exactly what physiological roles they may be playing in this species.

Information on the signal transduction cascade from photopigment through to the oscillator, or at least down to the induction of *cry1a* and *per2* induction, is remarkably confusing, considering the small number of labs working in this area. It would appear that cAMP-dependent signalling events are not involved in this response, but a role for mitogen-activated protein kinase (MAPK) has been suggested, based on a series of pharmacological studies (Cermakian *et al.*, 2002). In addition, the activation of the fos/AP1 signalling pathway may also be involved, and a role for the activation of oxidation/reduction pathways has been suggested (Hirayama *et al.*, 2005, 2007). Much of these data remain to be confirmed, and their specific cellular relevance to clock entrainment is yet to be fully tested. However, it is safe to say, and using the vernacular, "a no brainer", that pathways involving post-translational modifications, such as phosphorylation are likely to play an important role in this signalling cascade. Furthermore, it is extremely likely that the process of protein degradation will also be key to the light-signalling event, as it is in the function of the core clock mechanism itself. Clearly, much remains to be done on the exploration of the light input pathway to the zebrafish circadian pacemaker.

THOUGHTS ON FUTURE ISSUES RELATING TO THE ZEBRAFISH CIRCADIAN CLOCK

Several of the other chapters in this book will deal with additional aspects of the circadian biology of clocks in zebrafish. One of the major areas of clock research, not only in this model organism, but also across the field in general, is the study of clock regulated output processes. Why do most, if not all cells possess a circadian pacemaker, and what aspects of physiology are rhythmically regulated? Of course, the mammalian arena of clock biology has already made great progress on this topic using global transcriptome-type approaches to study rhythmic changes in tissues, such as the liver and other organs (Reddy *et al.*, 2006). Though interesting data, this has not really led to any great synthesis, other than to acknowledge that on average about 15% of the genome is rhythmically controlled. I am sure that the same will be true for zebrafish cells and tissues as well. Of course, such information, if processed correctly, does have great value in pointing the way towards the key regulatory events within the cell. The trick will be in recognizing the key molecules from the lists that are produced. Nevertheless, a major area of future circadian research will be, of course, determining how the clock controls rhythmic outputs. A major reason for this being that such information may provide some targets that one can then manipulate in order to control downstream physiology. This will be an important area in the study of clock-cell cycle interactions for example, and zebrafish has an important role to play in this area.

There are many unknowns in the area of zebrafish light sensitivity and signal transduction events, as discussed above. The nature of the photopigments, and their clearly unusual biochemistry, is of considerable interest to those of us that work on them. Of course, these molecules may be species, or perhaps teleost, specific, which makes it interesting basic biology, but one can already see the mouse people "leaving the auditorium". Again, however, a full understanding of these input events will be essential if one is to have a significant impact on aspects of fish biology, especially relating to details of fish reproduction and husbandry.

The topic of the core circadian clock mechanism in zebrafish is one that is rarely addressed, and we have partially skirted around this issue in this review. In essence many of the key clock molecules are conserved from *Drosophila* to mouse, and zebrafish are no exception. *Clock* and *bmal* play a key role as positive transcriptional regulators, and the numerous *cryptochromes* clearly act in a variety of repressor roles, both in the negative feedback loop of the clock, but also in the light input pathway. However, there are numerous differences as well between mammals and fish. It has been argued, by some colleagues, that the additional number of copies of many clock genes, as a consequence of genome duplication events, means that there is a great deal of redundancy in the zebrafish clock mechanism. I disagree with this view. The fact that genes have been retained, and not lost over time, strongly implies that they have specific, not redundant roles, and our data on *cryptochromes* (sadly much of which remains to be published) tends to support this hypothesis. What these genome duplication events have provided certain teleost species is a greater number of molecules on which evolution can act. This is not redundancy, but it is increased complexity. Fish have "made use" of these extra copies of clock genes to produce subtle levels of control and regulation that are not seen in more simple models, such as the mouse and human. Unravelling all of their specific roles will be challenging, but from an evolutionary perspective, quite fascinating. It is highly likely that the clocks of different teleost species have adapted specifically to their local environments, more rapidly and with more flexibility than seen in the mammals. This diversification can, of course, be seen in many other aspects of the teleosts, including body plan, reproductive physiology, etc., and the same is certainly true also for the clock mechanism. From an evolutionary perspective this is fun, but from a competitive research point of view with mouse, it is daunting.

One area where we do need to pull together as a fish community, however, is in the development of many of the tools necessary to exploit these teleost systems. In part, because of the size of the mouse community and investment in this preparation, many tools and "tricks" are possible in the mouse that don't exist in zebrafish. This needs to be rectified to some extent as a community by the development of new key technologies. Clearly morpholinos and "tilling" are useful approaches, but they have their limitations. Rapid and efficient "knock-out" approaches must be established, and there are some developments in this direction, but this technology needs to be more widely disseminated amongst research groups. With this in place, the future of circadian studies in zebrafish and other teleost species is very promising, and many new discoveries will be made over the next few years, taking advantage of the unique attributes of the fish circadian system.

References

Balsalobre, A., F. Damiola and U. Schibler. 1998. A serum shock induces circadian gene expression in mammalian tissue culture cells. *Cell* 93: 929–937.

Bellingham, J., S.S. Chaurasia, Z. Melyan, C. Liu, M.A. Cameron, E.E. Tarttelin, P.M. Iuvone, M.W. Hankins, G. Tosini and R.J. Lucas. 2006. Evolution of melanopsin photoreceptors: discovery and characterization of a new melanopsin in nonmammalian vertebrates. *PLoS Biology* 4: e254.

Cahill, G.M. 1996. Circadian regulation of melatonin production in cultured zebrafish pineal and retina. *Brain Research* 708: 177–181.

Cahill, G.M., M.W. Hurd and M.M. Batchelor. 1998. Circadian rhythmicity in the locomotor activity of larval zebrafish. *Neuroreport* 9: 3445–3449.

Carr, A-J. and D. Whitmore. 2005. Imaging of single light-responsive clock cells reveals fluctuating free-running periods. *Nature Cell Biology* 7: 319–321.

Cermakian, N., M.P. Pando, C.L. Thompson, A.B. Pinchak, C.P. Selby, L. Gutierrez, D.E. Wells, G.M. Cahill, A. Sancar and P. Sassone-Corsi. 2002. Light induction of a vertebrate clock gene involves signaling through blue-light receptors and MAP kinases. *Current Biology* 12: 844–848.

DeBruyne, J., M.W. Hurd, L. Gutiérrez, M. Kaneko, Y. Tan, D.E. Wells and G.M. Cahill. 2004. Isolation and phenogenetics of a novel circadian rhythm mutant in zebrafish. *Journal of Neurogenetics* 18: 403–428.

Dekens, M.P. and D. Whitmore. 2008. Autonomous onset of the circadian clock in the zebrafish embryo. *The EMBO Journal* 27: 2757–2765.

Dekens, M.P., C. Santoriello, D. Vallone, G. Grassi, D. Whitmore and N.S. Foulkes. 2003. Light regulates the cell cycle in zebrafish. *Current Biology* 13: 2051–2057.

Delaunay, F., C. Thisse, O. Marchand, V. Laudet and B. Thisse. 2000. An inherited functional circadian clock in zebrafish embryos. *Science* 289: 297–300.

Geisler, R., G.J. Rauch, S. Geiger-Rudolph, A. Albrecht, F. van Bebber, A. Berger, E. Busch-Nentwich, R. Dahm, M.P. Dekens, C. Dooley, A.F. Elli, I. Gehring, H. Geiger, M. Geisler, S. Glaser, S. Holley, M. Huber, A. Kerr, A. Kirn, M. Knirsch, M. Konantz, A.M. Küchler, F. Maderspacher, S.C. Neuhauss, T. Nicolson, E.A. Ober, E. Praeg, R.

Ray, B. Rentzsch, J.M. Rick, E. Rief, H.E. Schauerte, C.P. Schepp, U. Schönberger, H.B. Schonthaler, C. Seiler, S. Sidi, C. Söllner, A. Wehner, C. Weiler and C. Nüsslein-Volhard. 2007. Large-scale mapping of mutations affecting zebrafish development. *BMC Genomics* 8: 11.

Hirayama, J., L. Cardone, M. Doi and P. Sassone-Corsi. 2005. Common pathways in circadian and cell cycle clocks: light-dependent activation of Fos/AP-1 in zebrafish controls CRY-1a and WEE-1. *Proceedings of the National Academy of Sciences of the United States of America* 102: 10194–10199.

Hirayama J., S. Cho and P. Sassone-Corsi. 2007. Circadian control by the reduction/oxidation pathway: catalase represses light-dependent clock gene expression in the zebrafish. *Proceedings of the National Academy of Sciences of the United States of America* 104: 15747–15752.

Kaneko, M. and G.M. Cahill. 2005. Light-dependent development of circadian gene expression in transgenic zebrafish. *PLoS Biology* 3: e34.

Kazimi, N. and G.M. Cahill. 1999. Development of a circadian melatonin rhythm in embryonic zebrafish. *Brain Research Develpoment Brain Research.* 117: 47–52.

King, D.P. and J.S. Takahashi. 2000. Molecular genetics of circadian rhythms in mammals. *Annual Review of Neuroscience* 23: 713–742.

Lucas, R.J., S. Hattar, M. Takao, D.M. Berson, R.G. Foster and K.W. Yau. 2003. Diminished pupillary light reflex at high irradiances in melanopsin-knockout mice. *Science* 299: 245–247.

Moutsaki, P., D. Whitmore, J. Bellingham, K. Sakamoto, Z.K. David-Gray and R.G. Foster. 2003. Teleost multiple tissue (tmt) opsin: a candidate photopigment regulating the peripheral clocks of zebrafish? *Brain Research. Molecular Brain Research* 112: 135–145.

Mullins, M.C., M. Hammerschmidt, P. Haffter and C. Nüsslein-Volhard. 1994. Large-scale mutagenesis in the zebrafish: in search of genes controlling development in a vertebrate. *Current Biology* 4: 189–202.

Panda, S., T.K. Sato, A.M. Castrucci, M.D. Rollag, W.J. DeGrip, J.B. Hogenesch, I. Provencio and S.A. Kay. 2002. Melanopsin (Opn4) requirement for normal light-induced circadian phase shifting. *Science* 298: 2213–2216.

Pittendrigh, C.S. and D.H. Minis. 1964. The entrainment of circadian oscillations by light and their role as photoperiodic clocks. *The American Naturalist* 98: 261–294.

Plautz, J.D., M. Kaneko, J.C. Hall and S.A. Kay. 1997. Independent photoreceptive circadian clocks throughout *Drosophila*. *Science* 278: 1632–1635.

Prober, D.A., J. Rihel, A.A. Onah, R.J. Sung and A.F. Schier. 2006. Hypocretin/orexin overexpression induces an insomnia-like phenotype in zebrafish. *Journal of Neuroscience* 26: 13400–13410.

Reddy, A.B., N.A. Karp, E.S. Maywood, E.A. Sage, M. Deery, J.S. O'Neill, G.K. Wong, J. Chesham, M. Odell, K.S. Lilley, C.P. Kyriacou and M.H. Hastings. 2006. Circadian orchestration of the hepatic proteome. *Current Biology* 16: 1107–1115.

Stanewsky, R., M. Kaneko, P. Emery, B. Beretta, K. Wager-Smith, S.A. Kay, M. Rosbash and J.C. Hall. 1998. The cryb mutation identifies cryptochrome as a circadian photoreceptor in *Drosophila*. *Cell* 95: 681–692.

Stern, H.M and L.I. Zon. 2003. Cancer genetics and drug discovery in the zebrafish. *Nature Reviews Cancer* 3: 533–539.

Tamai, T.K., V. Vardhanabhuti, S. Arthur, N.S. Foulkes and D. Whitmore. 2003. Flies and fish: birds of a feather. *Journal of Neuroendocrinology* 15: 344–349.

Tamai, T.K., V. Vardhanabhuti, N.S. Foulkes and D. Whitmore. 2004. Early embryonic light detection improves survival. *Current Biology* 14: 104–105.

Tamai, T.K., A.J. Carr and D. Whitmore. 2005. Zebrafish circadian clocks: cells that see light. *Biochemical Society Transactions* 33: 962–966.

Tamai, T.K., L.C. Young and D. Whitmore. 2007. Light signaling to the zebrafish circadian clock by Cryptochrome 1a. *Proceedings of the National Academy of Sciences of the United States of America* 104: 14712–14717.

Vallone, D., S.B. Gondi, D. Whitmore and N.S. Foulkes. 2004. E-box function in a *period* gene repressed by light. *Proceedings of the National Academy of Sciences of the United States of America* 101: 4106–4111.

Welsh, D.K., S.H. Yoo, A.C. Liu, J.S. Takahashi and S.A. Kay. 2004. Bioluminescence imaging of individual fibroblasts reveals persistent, independently phased circadian rhythms of clock gene expression. *Current Biology* 14: 2289–2295.

Whitmore, D., N.S. Foulkes and P. Sassone-Corsi. 2000. Light acts directly on organs and cells in culture to set the vertebrate circadian clock. *Nature (London)* 404: 87–91.

Whitmore, D., N.S. Foulkes, U. Strahle and P. Sassone-Corsi. 1998. Zebrafish *Clock* rhythmic expression reveals independent peripheral circadian oscillators. *Nature Neuroscience* 1: 701–707.

Ziv, L. and Y. Gothilf. 2006. Circadian time-keeping during early stages of development. *Proceedings of the National Academy of Sciences of the United States of America* 103: 4146–4151.

Ziv, L., S. Levkovitz, R. Toyama, J. Falcon and Y. Gothilf. 2005. Functional development of the zebrafish pineal gland: Light-induced expression of *period 2* is required for onset of the circadian clock. *Journal of Neuroendocrinology* 17: 314–320.

Ziv, L., A. Tovin, D. Strasser and Y. Gothilf. 2007. Spectral sensitivity of melatonin suppression in the zebrafish pineal gland. *Experimental Eye Research* 84: 92–99.

8

FEEDING RHYTHMS IN FISH: FROM BEHAVIORAL TO MOLECULAR APPROACH

Jose Fernando López-Olmeda and *Francisco Javier Sánchez-Vázquez*

INTRODUCTION

Why feeding rhythms?

First, we shall wonder why we should consider feeding rhythms after all. On one hand, we should note that food is hardly constantly available in the natural environment, being usually restricted to a particular time when food availability is highest. On the other hand, fish must trade off between food availability and the occurrence of predators, which is also highest at a certain time. Under such a cyclic environment, fish have evolved time keeping mechanisms to predict feeding time and so their physiological processes can be activated in advance, allowing the animal to avoid risk or exploit a given food source more efficiently (Madrid *et al.*, 2001). Indeed, fish are not active throughout the 24 hours, but they usually display most of their activity along the light or the dark phase. These behavioral patterns in many species have been fixed genetically due to the pressure generated by stable selective forces such as avoidance of predators, the availability of preys or the optimization of food (Daan, 1981).

Some of the forces acting on living organisms are unpredictable, but others occur at times that may be quite predictably associated with certain phases of daylight, tides, lunar cycle or seasons. Most of feeding rhythms in

Department of Physiology, Faculty of Biology, University of Murcia, 30100 Murcia, Spain.
e-mail: jflopez@um.es

fish have been described as circadian rhythms, but the existence of tidal, lunar and seasonal rhythms has also been observed.

How to study feeding rhythms?

This is certainly not a simple question; because to study when fish feed, we need to allow fish feed whenever (and as much) as they want, and use some sort of recording device to register continuously their feeding behavior. In the last two decades, feeding rhythms in fish have been successfully studied using self-feeding devices coupled to computers. Self-feeders are based in the development of an operant conditioning by fish, which learn to activate a sensor to obtain a food reward (Sánchez-Vázquez *et al.*, 1994). A typical self-feeding system (Fig. 8.1) is made of three elements: I) a demand sensor, which should be easily activated voluntarily by the fish, and should avoid accidental activations, i.e., as a consequence of the movement of the water; II) a food dispenser, which would provide a certain amount of food in response to the activation of the sensor; III) a recording device, which would register the number and time of food demands, and eventually control

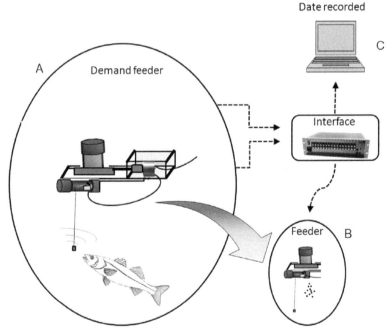

Fig. 8.1 Schematic representation of the main components of a demand feeder. Schematic representation of the switch and the string sensor (1A), the fish bites and pulls the sensor, which triggers the switch and thus provides the food (1B). In addition, the demand feeder is connected to an interface and a computer, which stores the number and time of food demands for its analysis (1C). (Figure provided by R. Fortes-Silva).

whether fish receive the food reward. In most of the cases, the recording device is a computer, which is coupled to the sensor and the feeder through an interface.

From the three components of the self-feeder described above, the demand sensor is the most important part because it must be adapted to the characteristic feeding behavior of the fish species under investigation. Early models of food-demand sensors consisted of a hanging rod, which had to be pushed by fish to trigger the feeder. However, the possible occurrence of involuntary activations generated by the agitation of fish during food delivery or accidental movements due to water turbulence generated by waves or wind, led to the design and development of a stronger model, a 'string sensor' (Rubio *et al.*, 2004). The string sensor comprises a switch, a string and a lure. The fish bites the lure and thus pulls the string, which triggers the switch. The utilization of string sensors to study the demand-feeding rhythms in fish has been successfully performed in many laboratory studies in a number of fish species, such as the European sea bass (*Dicentrarchus labrax*), gilthead sea bream (*Sparus aurata*), rainbow trout (*Oncorhynchus mykiss*), sharpsnout sea bream (*Diplodus puntazzo*), Senegal sole (*Solea senegalensis*), goldfish (*Carassius auratus*) and tench (*Tinca tinca*) (Sánchez-Vázquez *et al.*, 1994, 1995a, 1996; Sánchez-Vázquez and Tabata, 1998; Velázquez *et al.*, 2004; Herrero *et al.*, 2005; Vera *et al.*, 2006; Boluda Navarro *et al.*, 2009). Since this 'pull' sensor requires certain strength to activate the feeder, a new type of food-demand sensor had to be developed for small fish. Recently, a highly sensitive sensor has been successfully tested in zebrafish (*Danio rerio*) and sea bass post-larvae, which otherwise would have been unable to trigger a string sensor (unpublished data). This new sensor consists of an infrared photocell located in a corner of the aquarium, so that each time fish approach that particular area, they interrupt the infrared light-beam and food is dispensed.

FEEDING RHYTHMS IN FISH

Periodicity

Daily rhythms

Most animals are active either during the day or at night, but rarely throughout the 24 hours. The animals have acquired these activity patterns as the result of long periods of evolution under stable selective forces such as the avoidance of predators, the optimization of feeding and the reproductive success. In most of the species, the diurnal or nocturnal feeding behavior has been fixed genetically and it is often conditioned by the existence of special sensory requirements such as the dependence on the vision for the capture of preys (Madrid *et al.*, 2001).

First research on daily feeding rhythms in fish was reported by Hoar (1942), who described that the Atlantic salmon (*Salmo salar*) and the brook trout (*Salvelinus fontinalis*) displayed preferences for feeding along the light phase of the cycle during summer. Some years later, the first self-feeder device was designed, offering an important tool for the study of feeding rhythms in fishes (Rozin and Mayer, 1961). Since these first studies, the daily patterns of feeding behavior have been characterized in a broad variety of fish species (Madrid *et al.*, 2001; Reebs, 2002).

Generally, activity patterns in mammals and birds are easy to classify because of its stability (Helfman, 1993). However, this task can be difficult in teleost fish since individuals from the same species can display a high flexibility in their daily feeding and locomotor activity patterns. Some fish species can display diurnal and nocturnal behaviors, shifting from one phase to the other, along their life cycle. This ability is known as dualism, and has been reported to be a common feature among fish (Eriksson, 1978). The mechanisms which drive dualism are unknown, and it has been suggested to be related to a highly flexible circadian system, which allows the animals to adapt quickly to environmental changes (Sánchez-Vázquez *et al.*, 1996; López-Olmeda and Sánchez-Vázquez, 2009).

In the light of the reports presented above, it seems clear that rhythmic feeding activity is widespread in fishes, with examples of arrhythmicity being scarce or controversial. For instance, Rozin and Mayer (1961) did not find any rhythmicity in the feeding activity of goldfish, whereas other authors did find self-sustained rhythmic patterns in this species (Sánchez-Vázquez *et al.*, 1996). Some authors have suggested that arrhythmic feeding may occur in larvae (Pérez and Buisson, 1986), though personal observations in sea bass post larvae pointed to the existence of well-defined daily feeding rhythms, and in fish that consume low-energy food sources such as the grass carp (*Ctenopharyngodon idella*) (Cui *et al.*, 1993).

Tidal and lunar rhythms

In nature, these rhythms are tightly related to vertical and horizontal movements of small preys that use the tides to migrate and synchronize their reproductive rhythms. First studies by Brown (1946) reported a periodicity of 3–4 weeks in the food intake of brown trout (*Salmo trutta*). Later, rhythms of food intake and growth with a periodicity related to the lunar cycles were described in coho salmon (*Oncorhynchus kisutch*), rainbow trout and Arctic charr (*Salvelinus alpinus*) (Farbridge and Leatherland, 1987a, 1987b; Dabrowski *et al.*, 1992). In studies by Farbridge and Leatherland (1987a, 1987b) in coho salmon and rainbow trout, fish were fed *ad libitum* four times a day under a LD 12:12 photoperiod and constant water temperature. However, despite the constant conditions, rhythms in food intake and growth were detected with a periodicity of 14–15 days, with peaks located between the new moon and full moon in the case of salmon, and 4–5 days preceding the new moon or full moon in the case of rainbow trout. Nevertheless, the signals related to the lunar cycles that act as synchronizers are still unknown.

Seasonal rhythms

As poikilothermic animals, food intake of fish is directly influenced by the seasonal variations in water temperature. In addition, seasonal changes in photoperiod also influence feeding in fish. In general, under a similar water temperature, an increase in food intake can be observed in spring, coinciding with the lengthening of the photoperiod, and a decrease in autumn, coinciding with the shortening of the photoperiod (Komourdjian *et al.*, 1976; Higgins and Talbot, 1985).

The seasonal variations in photoperiod and temperature can decisively influence the phase of the daily rhythms of feeding behavior. Early evidences on such seasonal changes were reported in Arctic fishes. These fish species are subjected in their natural environment to extreme conditions of photoperiod and light intensity during the Arctic summer and winter solstices. For instance, Müller (1978) reported annual variations of the daily activity patterns of the burbot (*Lota lota*) and the sculpin (*Cottus carolinae*), and Jørgensen and Jobling (1989, 1990) reported seasonal phase inversions in the daily feeding patterns of Arctic charr.

According to Eriksson (1978), fish that display seasonal phase inversions of their daily behavioral patterns can be classified into two categories:

- Crepuscular fishes, such as Atlantic salmon and rainbow trout, in which the diurnalism or nocturnalism is the result of the fusion of the activities related with dusk and dawn.
- Biphasic fishes, such as burbot and sculpin, in which the seasonal phase inversion is a true phenomenon of inversion of their circadian system.

The first observations from Eriksson (1978) might conclude that the seasonal phase inversions are a characteristic phenomenon of species from high latitudes. Nevertheless, seasonal phase inversions were reported later in fish species from temperate latitudes such as the European sea bass or the gilthead sea bream (Fig. 8.2) (Sánchez-Vázquez *et al.*, 1998a; Velázquez *et al.*, 2004). For instance, the European sea bass displays diurnal feeding behavior from spring to autumn and shifts to nocturnal behavior in winter (Sánchez-Vázquez *et al.*, 1998a). Despite the number of recent studies deepen into the knowledge of seasonal changes, the biological processes that induce such behavior are unknown to date. In the case of Arctic fish species,

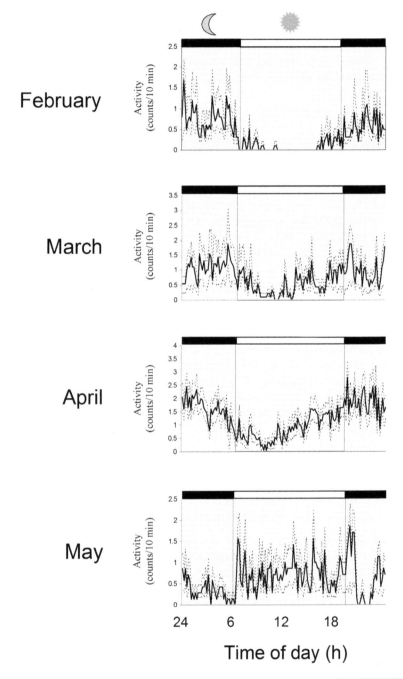

Fig. 8.2 contd....

submitted to extreme environmental conditions, the seasonal variations in their behavior can be explained by the wide variations in photoperiod. For fish species from temperate latitudes such as the sea bass, however, biotic factors such as food availability, both in quantity and quality, together with other changes in abiotic factors, could be the driving forces of this behavior. In addition, seasonal phase inversions could be related to the reproduction. In the case of sea bass and sea bream, the inversion from diurnalism to nocturnalism coincides with the reproductive season, the winter, and could be tightly related to daily rhythms of reproduction and spawning (Meseguer *et al.*, 2008). In the case of sticklebacks (*Gasterosteus aculeatus*), these fish display diurnal feeding and locomotor behavior, but during the reproductive season they constrain some behaviors mainly to the dark phase such as egg fanning in males (Reebs, 2002). Therefore, seasonal inversions in these species could be an adaptive response that would increase either parental or spawn survivorship, or both.

Dualism

Fish may show dual behavior, which is the ability of animals to display diurnal and nocturnal behavioral patterns, shifting from one phase to the other, along their life (Figs. 8.2 and 8.3). Moreover, their locomotor and feeding behavior can present independent phasing, as the same individual can display diurnal locomotor activity and nocturnal feeding activity or *vice versa* (Sánchez-Vázquez *et al.*, 1995a; Vera *et al.*, 2006). Dualism has been reported in a wide number of fish species (Table 8.1), thus being a common feature among fish (Eriksson, 1978). The mechanisms which drive dualism are unknown, and it has been suggested to be related to a highly flexible circadian system, which allows the animals to adapt quickly to environmental changes (Sánchez-Vázquez *et al.*, 1996; López-Olmeda and Sánchez-Vázquez, 2009). The biological significance of dualism is unknown to date though it may be related with one or several of the following factors: a) avoid predators; b) take advantage of the changes in food availability and c) reproduction.

Fig. 8.2 contd....

Fig. 8.2 Mean waveforms of feeding behavior of European sea bass along a four-month period. Sea bass (130 g of mean body weight) were kept in net cages of 125 m^3 placed in the open sea and maintained under natural photoperiod and water temperature, from the month of February to May. Fish were allowed to self-feed by means of demand feeders (Fig. 8.1). The seasonal phase inversion that appears in this fish species in wild conditions can be observed. The daily profile of food demands was nocturnal in February, occurring a progressive phase inversion from nocturnal to diurnal feeding behavior between the months of March and May. Feeding activity has been represented as the mean of food demands during the experimental days of each month, for periods of 10 min along the 24 h cycle. White and black bars above the waveforms represent the light and dark phase, respectively, of the LD cycle. (From Sánchez-Vázquez *et al.*, unpublished data).

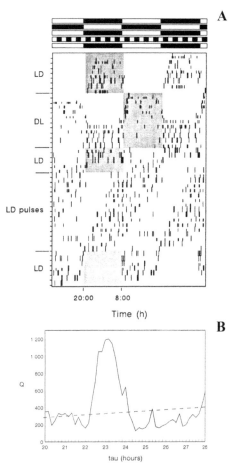

Fig. 8.3 Actogram of feeding activity records of European sea bass (8.3A) under laboratory conditions and allowed to self-feed by means of demand-feeders (Fig. 8.1). In the first part of the experiment, sea bass were maintained under a LD 12:12 h cycle, displaying nocturnal feeding behavior. The photoperiod was reversed and sea bass reserved the phase of their feeding rhythm, maintaining the nocturnal behavior for the first days, but then they shifted to diurnal feeding behavior. In the next part of the experiment, the photoperiod was reversed again, and the feeding behavior of sea bass remained phase-locked, shifting from diurnal to nocturnal behavior. Finally, to test the endogenous origin of feeding rhythms in sea bass, the photoperiod was set to light and dark pulses of 45 minutes (LD 45:45 min), appearing free-running rhythms of a period different than 24 h. Chi-square periodogram analysis (confidence level of 95%) of the free-running feeding rhythms under the LD pulses cycle has been represented below the actogram (8.3B). Data in the actogram have been double-plotted on a time scale of 48 h, the height of each point representing the number of infrared light-beam interruptions per 10 min. Each horizontal line corresponds to one experimental day, represented along the ordinate axis. White and black bars above the actograms represent the light and dark phases, respectively, of the LD cycle and LD pulses. (From Sánchez-Vázquez *et al.*, 1995a).

Table 8.1 Fish species capable of displaying dual feeding behavior.

Fish Species	Keeping Conditions	Type of inversions	References
Salmo salar	Single fish	Dual phasing	Fraser *et al.*, 1995
Salmo salar	Groups	Seasonal inversions	Smith *et al.*, 1993
Salmo trutta	Single fish	Seasonal inversions	Heggenes *et al.*, 1993
Salvelinus alpinus	Groups	Seasonal inversions	Jørgensen & Jobling, 1989, 1990
Salvelinus alpinus	Single fish	Dual phasing	Alanärä & Brännäs, 1997
Oncorhynchus mykiss	Single fish	Dual phasing	Landless, 1976;Alänarä & Brännäs, 1997
Dicentrarchus labrax	Single fish	Dual phasing	sánchez -Vázquez *et al.*, 1995a,b
Dicentrarchus labrax	Groups	Dual phasing	sánchez -Vázquez *et al.*, 1995a,b
Dicentrarchus labrax	Groups	Seasonal inversions	sánchez -Vázquez *et al.*, 1998a
Carassius auratus	Single fish	Dual phasing	sánchez -Vázquez *et al.*, 1996, 1998b
Sparus aurata	Groups	Seasonal inversions	Velázquez *et al.*, 2004
Sparus aurata	Groups	Dual phasing	López-Olmeda *et al.*, 2009b
Danio rerio	Groups	Dual phasing	Del Pozo & Sánchez-Vázquez, unpublished results
Oreochromis niloticus	Single fish	Dual phasing	Fortes-Silva *et al.*, unpublished results

The selection of one phase of the LD cycle to synchronize the feeding behavior or other type of behavior seems to depend essentially on two factors, a) an endogenous circadian pacemaker that generates oscillations in the behavioral variables and b) the masking effect that exerts the light on the behavior of fish. In the diurnal species, the peacemaker is synchronized with the light phase (photophase) of the light/dark cycle and they also show a positive masking by the light, what means that light stimulates or allows their feeding behavior. On the other hand, in the nocturnal fish species the light exerts a negative masking effect, while their circadian pacemaker is synchronized to the dark phase (scotophase). In the case of dual species, their circadian pacemaker would be synchronized with one of the phases of the light/dark cycle, while the light could exert a stimulating or an inhibitory effect depending whether fish display diurnal or nocturnal behavior, respectively.

Curiously, it is difficult to find a dual species among the fish species described traditionally as nocturnal such as the tench, Senegal sole or the European catfish (*Silurus glanis*) (Boujard, 1995; Herrero *et al.*, 2003; Bayarri *et al.*, 2004; Herrero *et al.*, 2005; Vera *et al.*, 2005; Boluda Navarro *et al.*, 2009). In contrast, many fish species considered traditionally as diurnal display dual behavior (Table 8.1), i.e., the Atlantic salmon, rainbow trout, Arctic charr, goldfish, European sea bass, gilthead sea bream, sharpsnout sea bream and zebrafish (Landless, 1976; Fraser *et al.*, 1995; Sánchez-Vázquez *et al.*, 1995a, 1996 ; Alanärä and Brännäs, 1997; Fraser and Metcalfe, 1997; Sánchez-Vázquez and Tabata, 1998; Sánchez-Muros *et al.*, 2003; Velázquez *et al.*, 2004; Vera *et al.*, 2006; López-Olmeda and Sánchez-Vázquez, 2009; López-Olmeda *et al.*, 2009b; Del Pozo and Sánchez-Vázquez, unpublished data) (Fig. 8.2, Fig. 8.3).

Under laboratory conditions, European sea bass from the same spawn, maintained in a recirculating water system and demand-fed, can display nocturnal feeding behavior in some aquaria whereas, simultaneously, in other aquaria sharing the same water, they can display diurnal feeding behavior (Sánchez-Vázquez *et al.*, 1995a,b). Under these controlled environmental conditions, it was possible to reverse the phase of feeding behavior in some fish by restricting the food reward to the opposite phase to the feeding phase selected spontaneously by the fish, although some fish maintained their demands in the phase of no-reward, which indicated a certain resistance of fish to feed out of their spontaneously selected feeding phase (Sánchez-Vázquez *et al.*, 1995b). In addition, the fish shifted from one behavior to the other, i.e. from nocturnal to diurnal, when the photoperiod was reversed (Fig. 8.3) (Sánchez-Vázquez *et al.*, 1995a).

The phase inversions of sea bass also appeared when they were allow to self-feed restricted to only three hours a day, distributed in one-hour periods, two diurnal periods and one nocturnal period (Azzaydi *et al.*, 1998). Surprisingly, the phase inversions also appeared in sea bass fed automatically, distributing the food (provided in an amount higher than their physiological needs) in three meals, two diurnal and one nocturnal. In the latter case, the waste of food varied along the seasons, with the nocturnal meal being rejected in summer.

Since dualism in wild conditions is related to seasons, such phase inversions have been tried to induce under laboratory conditions by manipulating both photoperiod and water temperature. The photoperiod was modified in a wide range, from LD 2:22 to LD 22:2, but this experimental design failed to induce inversions of behavioral patterns in the European sea bass (Aranda *et al.*, 1999a). In contrast, in the case of the brown bullhead (*Ictalurus nebulosus*), changes in light intensity effectively induced inversions of the feeding behavior (Eriksson, 1978). Fish exposed to LD 12:12 cycles of different light intensities showed nocturnal patterns under high light intensities and diurnal patterns under low intensities.

Under natural environmental conditions, temperature fluctuates in parallel to changes in photoperiod. In the Atlantic salmon, temperatures below 10°C induce nocturnal patterns of behavior (Fraser *et al.*, 1995). In contrast, in demand-fed sea bass, gradual increases of temperature from 22 to 28°C or decreases from 22 to 16°C were not able to induce phase inversions (Aranda *et al.*, 1999b). Moreover, the combination of long and short photoperiods (16L:8D and 8L:16D, respectively) with warm and cold water temperatures (28 and 16°C, respectively), simulating the environmental conditions of the Mediterranean sea in summer and winter, failed to modify the temporal patterns of food intake (Aranda *et al.*, 1999b).

Although the mechanisms that regulate the phase inversions of feeding patterns are still unknown, it is possible to hypothesize that, in addition to external cues, an endogenous annual clock may be involved, inducing the appearance of such behaviors in some seasons of the year. The existence of this endogenous annual clock may explain the failure of some synchronizers to exert phase inversions at refractory times of the clock. Supporting this hypothesis, Eriksson (1978) demonstrated that after transferring the brown trout from the natural environment to a laboratory with a LD 12:12 photoperiod, a higher number of fish showed diurnal behavior when they were transferred in summer when compared to winter.

Endogenous origin

If an external cue associated with feeding was being used by animals to develop their feeding rhythms, they would quickly lose the rhythms when such external cue is removed. Nevertheless, this is not the case and a hypothesis based on external cueing can be discarded as animals maintained under a controlled environment, without variations in the environmental factors, still display feeding rhythms (Aranda *et al.*, 2001). However, some environmental signals such as light or temperature may influence or modulate the expression of feeding rhythms.

In the light of the current evidences, a model based on an internal timing mechanism can be proposed. In this case, two different models have been suggested: an hourglass mechanism and a self-sustained clock or endogenous pacemaker entrainable by feeding (Mistlberger, 1994). The hourglass mechanism has been usually suggested by classical physiologist and nutritionists, who hypothesize that the rhythms related to feeding are driven by the cycles in energy depletion and repletion. Nevertheless, this hourglass model fails to explain why many rhythms such as feeding behavior or food anticipatory activity persist in starved animals.

The second hypothesis, the endogenous pacemaker mechanism, is demonstrated by the persistence of a rhythm under constant environmental conditions. Under such conditions the period of the rhythm deviates slightly from the environmental cycle to which it is synchronized, displaying a free-running period that matches the period of the endogenous oscillator. When the free-running rhythm persists for many periods without attenuation, it can be assumed that the rhythm belongs to the type of systems that are capable of self-sustaining oscillations (Aschoff, 1981). In order to consider that a feeding rhythm has endogenous origin, or whether by contrast it is exclusively regulated in an exogenous manner by the environmental factors, two conditions must be fulfilled: i) the rhythm must persist under LD cycles out of the circadian range (e.g. ultradian cycles of LD 45:45 min) (Fig. 8.3), and ii) the rhythm must persist in the absence of external synchronizers in the circadian range (Madrid *et al.*, 2001). The endogenous origin of feeding rhythms in fish, following these two conditions, have been demonstrated in several species such as the European sea bass (Fig. 8.3), goldfish, rainbow trout and zebrafish (Sánchez-Vázquez *et al.*, 1995a, 1996; Sánchez-Vázquez and Tabata, 1998; Del Pozo and Sánchez-Vázquez, unpublished data).

Self-sustained oscillations can be then explained by the existence of an endogenous pacemaker. In mammals, the existence of two different oscillators has been hypothesized: a light-entrainbale oscillator (LEO), located in the suprachiasmatic nuclei (SCN) of the hypothalamus (Meijer and Rietveld, 1989); and a food-entrainable oscillator (FEO), independent from the LEO and whose anatomical location is still unknown (Stephan *et al.*, 1979; Stephan, 2002; Davidson, 2006). In fish, data on properties of feeding entrainment support the hypothesis of the existence of a FEO, although it is still uncertain whether fish FEO and LEO are independent. Feeding entrainment could be explained in mechanistic terms based on the properties of a complex single oscillator. This is not the case for higher vertebrates since the existence of a separate FEO has been demonstrated. In fish, however, it cannot be excluded that they may use other mechanisms based on the information supplied by a master LEO, influencing their feeding rhythms.

Therefore, in fish, two possible mechanisms can be suggested: 1) the existence of separate by tightly coupled light- and food-entrainable oscillators, or 2) a single oscillator entrainable by both light and food, one synchronizer being stronger than the other (Sánchez-Vázquez *et al.*, 1997; Aranda *et al.*, 2001).

FEEDING ENTRAINMENT

Food Anticipatory Activity

Therefore, food can act as a synchronizer and entrain a number of overt rhythms in animals when food is available in a periodic manner. Most animals presented with food on a periodic basis display an increase in locomotor activity in anticipation of the forthcoming meal within a few days after the feeding cycle is established, a phenomenon known as food anticipatory activity (FAA) (Mistlberger, 1994). Anticipation of timed meals has been reported for a wide variety of animals, from bees to higher vertebrates such as monkeys (Stephan, 2002). In mammals, SCN lesions did not abolish feeding rhythms such as FAA, which remained entrained to a fixed daily meal. These data support the idea of the existence of a FEO independent of the LEO (Stephan, 2002). FAA can be defined as the increase of activity, two-fold or higher above a baseline of activity, maintained for at least 30 minutes and displayed just before feeding time (Mistlberger, 1994).

The ability of animals to anticipate feeding time was demonstrated for the first time in rats (Richter, 1922). The first evidence in fish was observed in the bluegill (*Lepomis macrochirus*) and largemouth bass (*Micropterus salmoides*), which showed prefeeding activity consisting of increasing swimming activity 1 to 3 hours before the fixed mealtime (Davis, 1963). Later, Davis and Bardach (1965) reported the appearance of FAA in killifish (*Fundulus heteroclitus*) maintained under constant light (LL). In recent years, the occurrence of FAA in fish has been described in a number of species such as the goldfish, tench, golden shiner (*Notemigonus crysoleucas*), greenback flounder (*Rhombosolea tapirina*), European sea bass, loach (*Misgurnus anguillicaudatus*), rainbow trout, European catfish, inanga (*Galaxias maculatus*), Arctic charr and zebrafish (Naruse and Oishi, 1994; Sánchez-Vázquez *et al.*, 1997; Reebs, 1999; Lague and Reebs, 2000; Reebs and Lague, 2000; Aranda *et al.*, 2001; Bolliet *et al.*, 2001; Chen and Purser, 2001; Purser and Chen, 2001; Brännäs *et al.*, 2005; Herrero *et al.*, 2005; Azzaydi *et al.*, 2007; López-Olmeda, 2009; López-Olmeda *et al.*, 2009a). In the case of goldfish and golden shiner, FAA appeared in animals fed either during the light or during the dark phase of the LD cycle, with a relative influence of the LD cycle on the FAA patterns being observed in animals fed near dusk or dawn (Reebs and Lague, 2000; López-Olmeda *et al.*, 2009a) (Fig. 8.4). Besides animals maintained under a LD cycle, FAA also occurs in animals with feeding time as the only synchronizing signal, i.e. fish maintained under constant lighting conditions (Aranda *et al.*, 2001; Vera *et al.*, 2007).

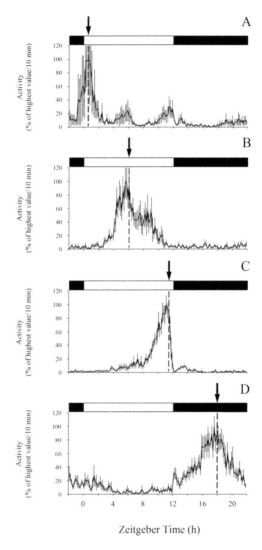

Fig. 8.4 Mean waveforms of locomotor activity of goldfish fed periodically once a day with a complete meal containing dextrin as the source of carbohydrate at four different times: at the beginning of the light phase (8.4A), at mid light (8.4B), at the end of the light phase (8.4C), and at mid dark (8.4D), indicated by the black arrow at the top of the actograms and the dashed line. Goldfish displayed food-anticipatory activity at all of the feeding times tested. Locomotor activity has been represented as the mean of interruptions of the lightbeam during the experimental days, for each period of 10 min along the 24 h cycle. To standarize the values, data have been calculated in each group as the percentage of the maximum value of activity. White and black bars above the waveforms represent the light and dark phase, respectively, of the LD cycle. (From López-Olmeda *et al.*, 2009a).

The synchronization to feeding time involves advantages for the fish similar to that related to light synchronization. When food availability is predictable, the animal can use this information to anticipate it and maximize food utilization. For instance, maintaining a continuous active state, as happens when food is delivered randomly along the 24 h (Vera *et al.*, 2007), involves a high waste of energy. In contrast, the anticipation of a meal prepares the animal and improves its food intake and nutrient utilization (Sánchez-Vázquez and Madrid, 2001).

One of the main characteristics of FAA is its gradual development, with several feeding cycles being required for its significance appearance. Moreover, after a shift of the feeding time, the time that fish require to resynchronize their FAA is directly related to the quantity of hours that the mealtime has been shifted. These facts, together with the observation of FAA under constant photoperiodic conditions (constant light or constant darkness) and its persistence for several days in fasted animals (after suppressing the scheduled feeding), support the hypothesis of FAA as a process directed by an endogenous pacemaker which is directed by feeding time in usual conditions (Sánchez-Vázquez and Madrid, 2001).

The development of FAA is also influenced by the energy contents and size of food provided periodically. For instance, fish can anticipate up to three different meals a day, although the anticipatory activity is lower than the obtained with a single daily meal. Furthermore, food of small size induces a higher anticipatory activity than meals of high size (Sánchez-Vázquez *et al.*, 2001). Finally, fish are capable of anticipating not only feeding time but also the places in which food is delivered, achieving a temporal and spatial integration to optimize food intake.

Gut physiology

Feeding entrainment has been observed in many behavioral and metabolic variables in mammals such as locomotor activity, feeding behavior, water intake, body temperature and plasma corticosterone. Feeding entrainment in fish, however, is less known than in mammals.

Among the endocrine variables, feeding entrainment of cortisol rhythms is the most studied. Cortisol is a hormone that displays many functions in the organism, among them this hormone plays a key role in metabolism, increasing the plasma glucose levels by enhancing the hepatic gluconeogenesis, and increasing the glycolytic potential in peripheral tissues such as the brain (Mommsen *et al.*, 1999). Cortisol displays circadian rhythms in vertebrates, with the phase of cortisol tightly locked to the activity phase of the animal: the peak of cortisol occurs in the early morning in diurnal animals and in the early night in nocturnal animals (Dickmeis, 2009). In fish, daily variations of cortisol have been described for several species such as European sea bass, carp (*Cyprinus carpio*), brown trout, rainbow trout, goldfish, Atlantic salmon and gilthead sea bream (Fig. 8.5) (Singley and Chavin, 1975; Rance *et al.*, 1982; Pickering and Pottinger, 1983; Kühn *et al.*, 1986; Thorpe *et al.*, 1987; Cerdá-Reverter *et al.*, 1998; López-Olmeda *et al.*, 2009b), with the acrophase of the rhythm varying depending on the fish species. In goldfish maintained under LL, a single daily meal provided at a fixed time effectively entrained the daily rhythm of cortisol, which peaked several hours before mealtime. In addition, the shifts of feeding time shifted the phase of cortisol in the goldfish under LL, and were able to entrain other hormones such as neuropeptide Y in the brain (Fig. 8.6) (Spieler and Noeske, 1981, 1984; Vera *et al.*, 2007). The rhythm of cortisol in plasma seems to be dependent of feeding as it was suppressed in fasted rainbow trout when compared with fed animals (Polakof *et al.*, 2007a). Moreover, in some cases, food can elicit a postprandial increase in cortisol, although this increase is lower than the daily peak observed during the acrophase (Small, 2005). Recent studies performed in our laboratory showed a rhythm of cortisol in the gilthead sea bream both under LD and LL. This rhythm was influenced by feeding and feeding behavior, shifting the acrophase of the rhythm depending on feeding conditions (Fig. 8.5) (López-Olmeda *et al.*, 2009b).

As regards to enzymes, circadian rhythms in enzymes involved in metabolism have been reported for several fish species. In rainbow trout, daily rhythms have been reported for metabolic enzymes in the brain, liver, white muscle and gills (Polakof *et al.*, 2007a,b). Some of the daily variations of these enzymes were dependent on feeding since they disappeared in fasted animals, while the daily rhythms of other enzymes were maintained even in fasted fish. For instance, some of the enzymes dependent of feeding were hexokinase in the brain, liver, gills and muscle; glucose 6-phosphatase in liver; lactate dehydrogenase in liver and gills; pyruvate kinase in brain and muscle; and α-glycerophosphate dehydrogenase in liver and muscle (Polakof *et al.*, 2007a,b). In gilthead sea bream, glucose 6-phosphate dehydrogenase displays both daily and seasonal rhythms in the liver, which could be influenced by both feeding and water temperature (Gómez-Milán and Sánchez-Muros, 2007). In goldfish maintained under LL, a fixed mealtime synchronized the rhythm of amylase in the gut and, what is more, the increase in amylase activity was observed 2 hours before feeding, which indicates the anticipation to mealtime in the production of this enzyme and highlights the adaptive response of feeding anticipation to maximize the digestion and food utilization (Fig. 8.6) (Vera *et al.*, 2007).

As many hormones involved in metabolism display circadian rhythmicity, it is logic that the metabolism in turn may show circadian rhythms. In fish, one of the most studied metabolic rhythms is the daily variation of blood glucose, which has been reported for several species such

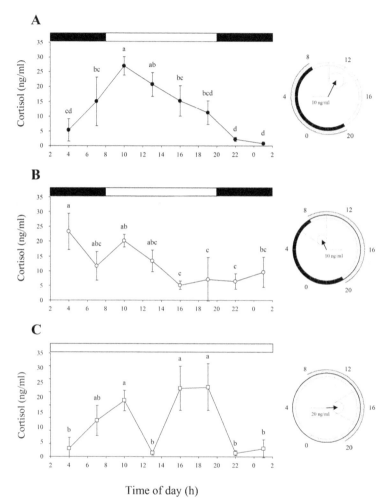

Time of day (h)

Fig. 8.5 Daily profiles of plasma cortisol in gilthead sea bream from animals with nocturnal self-feeding behavior under a LD cycle (8.5A), sea bream with diurnal self-feeding behavior under a LD cycle (8.5B), and sea bream maintained under constant light and self-feeding but only rewarded with food along a 12 hour period (8.5C). White and black bars above each graph represent light and darkness, respectively. Different letters indicate statistically significant differences between time points (one-way ANOVA, p<0.05). At the right side, the Cosinor analysis has been represented for each daily rhythm of cortisol. Time scale of Cosinor is in hours, with a 24-h cycle in each graph. Bars around the perimeter of the circle represent the light (white bars) and the dark phase (black bars) of the light cycle, and the phase of restricted feeding (dotted bar). The arrow inside the graphic represents the amplitude and points to the acrophase of the rhythm, the circle around the arrow head represents the confidence ellipse of the amplitude, and the two bars from the ellipse to the edges of the graph represent the fiducial limits of the acrophase. The half radius inside each graph represents the scale of the arrow of the amplitude in ng/ml of cortisol. (From López-Olmeda *et al.*, 2009b).

A

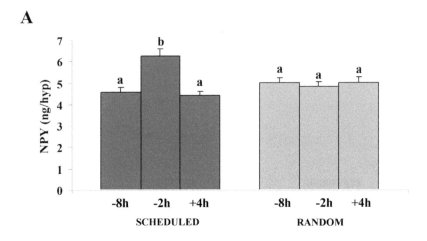

B

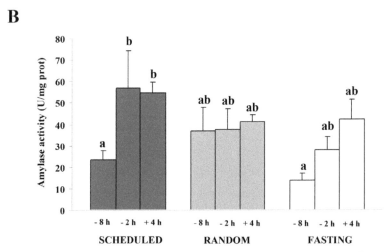

Fig. 8.6 Hypothalamic Neuropeptide Y (NPY) content (8.6A) and gut amylase activity (8.6B) from goldfish submitted to scheduled feeding (dark grey bars), random feeding (light grey bars) and two days fasting (white bars). Samples were collected 8 hours before (–8 h), 2 hours before (–2 h) and 4 hours after (+4 h) mealtime of the scheduled feeding group. Different letters indicate statistically significant differences between sampling points (ANOVA, p<0.05). On the sampling day, mealtime in the randomly fed group was programmed to occur at the same time as the scheduled fed group. NPY was not measured in the fasted animals as fasting inhibits the production of this hormone. As can be observed, both NPY and amylase activity displayed anticipation to feeding time in the scheduled fed animals, but not in the other groups. (From Vera *et al.*, 2007).

as carp, European sea bass, gilthead sea bream, common dentex (*Dentex dentex*), red porgy (*Pagrus pagrus*), tench and rainbow trout (Kühn *et al.*, 1986; Pavlidis *et al.*, 1997; Cerdá-Reverter *et al.*, 1998; Pavlidis *et al.*, 1999a,b; De Pedro *et al.*, 2005; Polakof *et al.*, 2007a). In addition, other metabolic compounds found in the blood such as proteins, amino acids, lactate, triglycerides, cholesterol and electrolytes also display daily rhythms in fish (Kühn *et al.*, 1986; Pavlidis *et al.*, 1997, 1999a,b; De Pedro *et al.*, 2005; Polakof *et al.*, 2007a). In the rainbow trout, the daily rhythms of some of these metabolic compounds such as plasma glucose and triglycerides, as well as brain glycogen, glucose and lactate, were dependent on feeding as their diel variations disappeared in fasted animals (Polakof *et al.*, 2007a). Recently, a daily rhythm in carbohydrate tolerance in goldfish has been reported (Fig. 8.7) (López-Olmeda *et al.*, 2009a). In goldfish fed a diet with dextrin, significant differences in the increases in postprandial glucose were observed depending on the mealtime, with the highest increases being found around mid-light and the lowest values around mid-dark (López-Olmeda *et al.*, 2009a). The mechanisms that drive these daily rhythms in glucose tolerance are still unknown in fish, while in mammals they seem to be related with a rhythm in peripheral resistance to insulin (Van Cauter *et al.*, 1997).

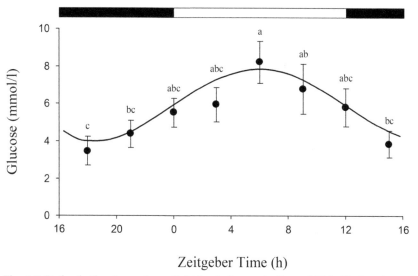

Zeitgeber Time (h)

Fig. 8.7 Daily rhythm in postprandial blood glucose in the goldfish. Each point was measured 2 hours after feeding with a complete meal with dextrin as the source of carbohydrate. In the experiment, blood glucose concentrations 2 hours after mealtime were measured in eight groups of fish fed once a day at eight different times along the 24 hour light cycle (every three hours: ZT0, 3, 6, 9, 12, 15, 18, and 21). Different letters indicate statistically significant differences between feeding times (one-way ANOVA, $p<0.05$). (From López-Olmeda *et al.*, 2009a).

Clock genes entrainment

All oscillators have three basic requirements: positive input to drive a change, feedback, and a delay in the execution of the feedback (Dunlap, 2004). The nature of an oscillation is that it describes a system that tends, in a regular manner, to move away from equilibrium before returning. To achieve this, all that is needed is a process whose product feeds back to slow down the rate of the process itself (negative element), and a delay in the execution of the feedback. A further necessity for an oscillator is a positive element that keeps the oscillator from winding down (Dunlap, 1999). All known circadian oscillators at the molecular level use loops that close within cells, and that rely on positive and negative elements in which transcription of clock genes yields clock proteins that block the action of positive elements whose function is to active clock genes. Therefore, the positive element in the loop is the transcriptional activation of one or more clock genes, mediated by transcriptional activators that are paired via interactions of PAS domains within the proteins. These positive elements activate then the transcription of clock genes, which in turn generate clock proteins that provide the negative element in the loop. These negative elements feed back to block activation of clock genes, with the result that the amount of clock gene mRNA and proteins declines (Dunlap, 2004). The molecular clock and its elements have been widely studied in recent years in several model organisms. In fish, zebrafish has been the most studied species and its molecular clock has been described by several reports (Cahill, 2002).

However, the influence of feeding on the molecular clock in fish is unknown to date. In mammals, a number of peripheral oscillators have been described in peripheral organs such as the heart, kidney, liver or the intestine (Yamazaki *et al.*, 2000). When peripheral organs such as liver and lung of transgenic rats with the *per1* promoter linked to the luciferase reporter were cultured, endogenous oscillations in *per1* were observed for several days under culture, although the rhythms of *per1* expression in these peripheral tissues dampened after several cycles (Yamazaki *et al.*, 2000). Later on, the study of *per2* linked to the luciferase reporter in transgenic mice revealed that the rhythms of clock gene expression in peripheral tissues in culture could be maintained for several weeks, much longer than it was thought first (Yoo *et al.*, 2004). Phase shifts due to changes in the LD cycle elicit rapid phase shifts in the SCN, whereas peripheral tissues and overt rhythms such as locomotor activity require more time, so it was suggested that the SCN in mammals functions as a self-sustained pacemaker capable of exerting phase control on damped peripheral oscillators, but a number of transient cycles are required to establish phase control (Stephan, 2002). However, two recent reports (Damiola *et al.*, 2000; Stokkan *et al.*, 2001) revealed that the phase of peripheral oscillators can be uncoupled from the phase of the SCN by means of restricted feeding and independently of lighting conditions. Indeed, restricted feeding can entrain the rhythms of clock gene expression in the liver of SCN-ablated mice (Hara *et al.*, 2001). Despite the relative independence of peripheral oscillators and the ability of food to reset their phase, the SCN exerts a control on these oscillators as several cycles are required to shift the phase. Time for resetting the phase in the liver was reduced considerably in knock-out mice for the glucocorticoid receptor, pointing to a role of glucocorticoids in the phase-locking of peripheral oscillators to the LD cycle (le Minh *et al.*, 2000). In addition, administration of a glucocorticoid analog, the dexametasone, induced phase response curves (PRC) similar of those obtained with light pulses (Balsalobre *et al.*, 2000). A role for metabolic rhythms in the control of circadian rhythms of gene expression in peripheral tissues has also been suggested (Schibler and Naef, 2005), as factors such as the ratio $NAD(P)^+/NAD(P)H$ (redox state) or the intracellular concentration of carbon monoxide (CO) influence clock gene expression and CLOCK/NPAS2 binding (Rutter *et al.*, 2001; Dioum *et al.*, 2002). In addition, several molecular factors related with cellular metabolism such as the transcriptional coactivator PGC-1α or the NAD^+-dependent deacetylase SIRT1 have been reported to influence and regulate the action of clock proteins (Liu *et al.*, 2007; Asher *et al.*, 2008; Nakahata *et al.*, 2008).

In zebrafish, rhythms in clock genes expression have also been found in the nervous system and several peripheral oscillators. The first observations were reported by Whitmore *et al.* (1998), who cloned the *clock* gene of zebrafish and observed its daily variations under DD in the eye, pineal, heart and kidney. In addition to *clock*, other clock genes such as *bmal1, bmal2, per1* and *per3* have also been reported to oscillate in both central and peripheral tissues in culture under DD (Cermakian *et al.*, 2000; Pando *et al.*, 2001; Kaneko *et al.*, 2006). Interestingly, several reports performed on the circadian rhythms of clock genes in zebrafish revealed that clock gene expression in cells and tissues could be directly entrained by the light (Whitmore *et al.*, 2000; Pando *et al.*, 2001), which suggested that the circadian system of zebrafish may be between the hierarchical system of mammals and the circadian system of *Drosophila*. However, it must be considered that previous reports on direct light responsiveness in zebrafish have been performed using cells and tissues in culture. What is more, a recent paper by Dickmeis *et al.* (2007) reported that the rhythms of cell division in peripheral tissues are lost in zebrafish mutants that present an absence of production of cortisol. Therefore, and in parallel to what is found in mammals, glucocorticoids seem to be an important signal for the entraining of peripheral rhythms in zebrafish, which could also constitute a link between the central and peripheral oscillators in fish. Recent research performed in our laboratory showed that feeding can reset the phase of clock gene expression in peripheral tissues of zebrafish (Fig. 8.8) (López-

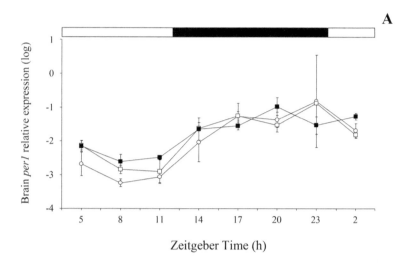

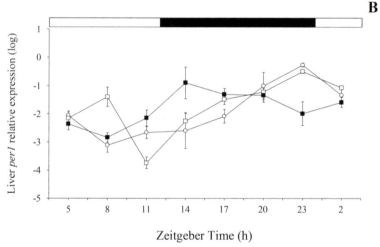

Fig. 8.8 Daily variations of *per1* expression in the zebrafish brain (8.8A) and liver (8.8B) of three experimental groups of zebrafish fed once a day in three different ways: randomly along the light phase (white circles); and at a fixed time: at mid light (white squares), and at mid dark (black squares). Gene expression was measured by means of quantitative PCR (TaqMan probes), and data were calculated as relative expression by using *β-actin* as the housekeeping gene. A significant daily rhythm was observed in all groups in the brain, with the acrophases located at ZT 20.7–21.6 (Cosinor, p<0.05). In liver, for each group, a significant circadian rhythm was observed (Cosinor, p<0.05), but the acrophases differed among treatments: both random and scheduled fed at ML peaked at the end of the dark phase (ZT22.8 and ZT22.6, respectively). However, the phase of *per1* expression in the liver of animals fed at MD was displaced to ZT15.9, almost 7 hours prior to the acrophases of the groups fed during daytime, indicating the resetting ability of feeding time in this peripheral oscillator. (From López-Olmeda, 2009).

Olmeda, 2009). In that experiments, zebrafish were divided into three groups fed once a day at different times: randomly, at mid-light, and at mid-dark. When the expression levels of *per1* in central (brain) and peripheral (liver) oscillators were analyzed, it could be observed that the phase of *per1* in the brain remained synchronized to the LD cycle. In contrast, the phase of *per1* expression in the liver was shifted in the group fed at mid-dark (López-Olmeda, 2009). Finally, another recent paper described the influence of metabolism on circadian rhythms of clock gene expression. As reported by Hirayama *et al.* (2007), the cellular redox state regulates the expression of several clock genes in zebrafish, an effect mediated by the enzyme catalase.

CONCLUSIONS

In conclusion, we may summarize that feeding activity of fish, as most physiological functions, displays rhythms which can be daily rhythms, seasonal or even tidal or lunar rhythms in some species. Moreover, the daily feeding patterns of fish may be diurnal or nocturnal depending on the species, or even more flexible in those species showing dual feeding behavior, a characteristic that seems to be quite common in fish and which provides them with a highly flexible circadian system. In addition, feeding time entrains a number of variables of gut physiology, such as endocrine, enzymatic and metabolic activities. The adaptive role of food anticipatory activity would be to prepare the animal physiology for the forthcoming meal, increasing the efficiency of the digestive processes. At the molecular level, scheduled-feeding entrains clock gene expression in the liver but not in the brain. These results, together with the observations in feeding behavior, point to the existence of two different oscillators in fish, one entrainable by the light (LEO) and the other entrainable by food (FEO).

Acknowledgements

This work was supported by the Spanish Ministry of Science and Innovation by projects AGL2007-66507-C02-02 and AQUAGENOMICS 28502 (Consolider-Ingenio Program), and by the Seneca Foundation by project 05690/PI/07.

References

Alanärä, A. and E. Brännäs. 1997. Diurnal and nocturnal feeding activity in Arctic charr (*Salvelinus alpinus*) and rainbow trout (*Oncorhynchus mykiss*). *Canadian Journal of Fisheries and Aquatic Sciences* 54: 2894–2900.
Aranda, A., J.A. Madrid and F.J. Sánchez-Vázquez. 2001. Influence of light on feeding anticipatory activity in goldfish. *Journal of Biological Rhythms* 16: 50–57.

Aranda, A., J.A. Madrid, S. Zamora and F.J. Sánchez-Vázquez. 1999a. Synchronizing effect of photoperiod on the dual phasing of demand-feeding rhythms in sea bass. *Biological Rhythm Research* 30: 392–406.

Aranda, A., F.J. Sánchez-Vázquez and J.A. Madrid. 1999b. Influence of water temperature on demand feeding-rhythms in sea bass. *Journal of Fish Biology* 55: 1029–1039.

Aschoff, J. 1981. *Handbook of Behavioral Neurobiology 4: Biological Rhythms*. Plenum Press, New York.

Asher, G., D. Gatfield, M. Stratmann, H. Reinke, C. Dibner, F. Kreppel, R. Mostoslavsky, F.W. Alt and U. Schibler. 2008. SIRT1 regulates circadian clock gene expression through PER2 deacetylation. *Cell* 134: 317–328.

Azzaydi, M., J.A. Madrid, S. Zamora, F.J. Sánchez-Vázquez and F.J. Martínez. 1998. Effect of three feeding strategies (automatic, ad libitum demand-feeding and time-restricted demand-feeding) on feeding rhythms and growth in European sea bass (*Dicentrarchus labrax* L.). *Aquaculture* 163: 285–296.

Azzaydi, M., V.C. Rubio, F.J.M. López, F.J. Sánchez-Vázquez, S. Zamora and J.A. Madrid. 2007. Effect of restricted feeding schedule on seasonal shifting of daily demand-feeding pattern and food anticipatory activity in European sea bass (*Dicentrarchus labrax* L.). *Chronobiology International* 24: 859–874.

Balsalobre, A., S.A. Brown, L. Marcacci, F. Tronche, C. Kellendonk, H.M. Reichardt, G. Schütz and U. Schibler. 2000. Resetting of circadian time in peripheral tissues by glucocorticoid signaling. *Science* 289: 2344–2347.

Bayarri, M.J., J.A. Muñoz-Cueto, J.F. López-Olmeda, L.M. Vera, M.A. Rol de Lama, J.A. Madrid and F.J. Sánchez-Vázquez. 2004. Daily locomotor activity and melatonin rhythms in Senegal sole (*Solea senegalensis*). *Physiology & Behavior* 81: 577–583.

Bolliet, V., A. Aranda and T. Boujard. 2001. Demand-feeding rhythm in rainbow trout and European catfish—synchronisation by photoperiod and food availability. *Physiology and Behavior* 73: 625–633.

Boluda Navarro, D., V.C. Rubio, R.K. Luz, J.A. Madrid and F.J. Sánchez-Vázquez. 2009. Daily feeding rhythms of Senegalese sole under laboratory and farming conditions using self-feeding systems. *Aquaculture* (In press).

Boujard, T. 1995. Diel rhythms of feeding activity in the European catfish, *Silurus glanis*. *Physiology and Behavior* 58: 641–645.

Brännäs, E., U. Berglund and L.O. Eriksson. 2005. Time learning and anticipatory activity in groups of Arctic charr. *Ethology* 111: 681–692.

Brown, M.E. 1946. The growth of brown trout (*Salmo trutta*, L.). II. The growth of two-year-old trout at a constant temperature of 11.5°C. *Journal of Experimental Biology* 22: 130–144.

Cahill, G.M. 2002. Clock mechanisms in zebrafish. *Cell and Tissue Research* 309: 27–34.

Cerdá-Reverter, J.M., S. Zanuy, M. Carrillo and J.A. Madrid. 1998. Time-course studies on plasma glucose, insulin, and cortisol in sea bass (*Dicentrarchus labrax*) held under different photoperiodic regimes. *Physiology and Behavior* 64: 245–250.

Cermakian, N., D. Whitmore, N.S. Foulkes and P. Sassone-Corsi. 2000. Asynchronous oscillations of two zebrafish Clock partners reveal differential clock control and function. *Proceedings of the National Academy of Sciences of the United States of America* 97: 4339–4344.

Chen, W.M. and G.J. Purser. 2001. The effect of feeding regime on growth, locomotor activity pattern and the development of food anticipatory activity in greenback flounder. *Journal of Fish Biology* 58: 177–187.

Cui, Y., S. Chen, S. Wang and X. Liu. 1993. Laboratory observations on the circadian feeding pattern in the grass carp (*Ctenopharyngodon idella* Val.) fed three different diets. *Aquaculture* 113: 57–64.

Daan, S. 1981. Adaptive daily strategies in behavior. In: *Handbook of Behavioral Neurobiology 4: Biological Rhythms,* J. Aschoff (ed.). Plenum Press, New York, pp. 275–298.

Dabrowski, K., G. Krumschnabel, M. Paukku and J. Labanowski. 1992. Cyclic growth and activity of pancreatic enzymes in alevins of Arctic charr (*Salvelinus alpinus* L.). *Journal of Fish Biology* 40: 511–521.

Damiola, F., N. Le Minh, N. Preitner, B. Kornmann, F. Fleury-Olela and U. Schibler. 2000. Restricted feeding uncouples circadian oscillators in peripheral tissues from the central pacemaker in the suprachiasmatic nucleus. *Genes and Development* 14: 2950–2961.

Davidson, A.J. 2006. Search for the feeding-entrainable circadian oscillator: a complex proposition. *American Journal of Physiology Regulatory Integrative and Comparative Physiology* 290: R1524–R1526.

Davis, R.E. 1963. Daily 'predawn' peak of locomotion in fish. *Animal Behaviour* 12: 272–283.

Davis, R.E. and E. Bardach. 1965. Time-co-ordinated prefeeding activity in fish. *Animal Behaviour* 13: 154–162.

De Pedro, N., A.I. Guijarro, M.A. López-Patiño, R. Martínez-Álvarez and M.J. Delgado. 2005. Daily and seasonal variations in haematological and blood biochemical parameters in the tench, *Tinca tinca* Linnaeus, 1758. *Aquaculture Research* 36: 1185–1196.

Dickmeis, T. 2009. Glucocorticoids and the circadian clock. *Journal of Endocrinology* 200: 3–22.

Dickmeis, T., K. Lahiri, G. Nica, D. Vallone, C. Santoriello, C.J. Neumann, M. Hammerschmidt and N.S. Foulkes. 2007. Glucocorticoids play a key role in circadian cell cycle rhythms. *PLoS Biology* 5: 854–864.

Dioum, E.M., J. Rutter, J.R. Tuckerman, G. González, M.A. Gilles-González and S.L. McKnight. 2002. NPAS2: a gas-responsive transcription factor. *Science* 298: 2385–2387.

Dunlap, J.C. 1999. Molecular bases for circadian clocks. *Cell* 96: 271–290.

Dunlap, J.C. 2004. Molecular biology of circadian pacemaker systems. In: *Chronobiology. Biological Timekeeping,* J.C. Dunlap, J.J. Loros and P.J. DeCoursey (eds.). Sinauer Associates, Sunderland, pp. 213–253.

Eriksson, L.O. 1978. Nocturnalism versus diurnalism—dualism within individuals. In: *Rhythmic Activity of Fishes,* J.E. Thorpe (ed.). Academic Press, London, pp. 69–89.

Farbridge, K.F. and J.F. Leatherland. 1987a. Lunar cycles of coho salmon, *Oncorhynchus kisutch.* I. Growth and feeding. *Journal of Experimental Biology* 128: 165–178.

Farbridge, K.F. and J.F. Leatherland. 1987b. Lunar periodicity of growth cycles in rainbow trout, *Salmo gairdneri* Richardson. *Journal of Interdisciplinary Cycle Research* 18: 169–177.

Fraser, N.H.C. and N.B. Metcalfe. 1997. The cost of becoming nocturnal: feeding efficiency in relation to light intensity in juvenile Atlantic salmon. *Functional Ecology* 11: 760–767.

Fraser, N.H.C., J. Heggenes, N.B. Metcalfe and J.E. Thorpe. 1995. Low summer temperatures cause juvenile Atlantic salmon to become nocturnal. *Canadian Journal of Zoology* 73: 446–451.

Gómez-Milán, E. and M.J. Sánchez-Muros Lozano. 2007. Daily and annual variations of the hepatic glucose 6-phosphate dehydrogenase activity and seasonal changes in the body fats of the gilthead seabream *Sparus aurata. Journal of Experimental Zoology* A307: 516–526.

Hara, R., K. Wan, H. Wakamatsu, R. Aida, T. Moriya, M. Akiyama and S. Shibata. 2001. Restricted feeding entrains liver clock without participation of the suprachiasmatic nucleus. *Genes to Cells* 6: 269–278.

Heggenes, J., O.M.W. Krog, O.R. Lindås, J.G. Dokk and T. Bremnes. 1993. Homeostatic behavioural responses in a changing environment: Brown trout (*Salmo trutta*) become nocturnal during winter. *Journal of Animal Ecology* 62: 295–308.

Helfman, G.S. 1993. Fish behaviour by day, night and twilight. In: *Behaviour of Teleost Fishes*, T.J. Pitcher (ed). Chapman and Hall, London, pp. 479–512, 2nd Edition.

Herrero, M.J., J.A. Madrid and F.J. Sánchez-Vázquez. 2003. Entrainment to light of circadian activity rhythms in tench (*Tinca tinca*). *Chronobiology International* 20: 1001–1017.

Herrero, M.J., M. Pascual, J.A. Madrid and F.J. Sánchez-Vázquez. 2005. Demand-feeding rhythms and feeding-entrainment of locomotor activity rhythms in tench (*Tinca tinca*). *Physiology & Behavior* 84: 595–605.

Higgins, P.J. and C. Talbot. 1985. Growth and feeding in juvenile Atlantic salmon (*Salmo salar* L.). In: *Nutrition and Feeding in Fish*, C.B. Cowey, A.M. Mackie and J.G. Bell (eds.). Academic Press, London, pp. 243–263.

Hirayama, J., S. Cho and P. Sassone-Corsi. 2007. Circadian control by the reduction/oxidation pathway: Catalase represses light-dependent clock gene expression in the zebrafish. *Proceedings of the National Academy of Sciences of the United States of America* 104: 15747–15752.

Hoar, W.S. 1942. Diurnal variations in feeding activity of young salmon and trout. *Journal of the Fisheries Research Board of Canada* 34: 1655–1669.

Jørgensen, E.H. and M. Jobling. 1989. Patterns of food intake in Arctic charr, *Salvelinus alpinus*, monitored by radiography. *Aquaculture* 81: 155–160.

Jørgensen, E.H. and M. Jobling. 1990. Feeding modes in Arctic charr, *Salvelinus alpinus* L: the importance of bottom feeding for the maintenance and growth. *Aquaculture* 86: 379–385.

Kaneko, M., N. Hernández-Borsetti and G.M. Cahill. 2006. Diversity of zebrafish peripheral oscillators revealed by luciferase reporting. *Proceedings of the National Academy of Sciences of the United States of America* 103: 14614–14619.

Komourdjian M.P., R.L. Saunders and J.C. Fenwick. 1976. Evidence for the role of growth hormone as part of the 'light pituitary axis' in growth and smoltification of Atlantic salmon (*Salmo salar*). *Canadian Journal of Zoology* 54: 544–551.

Kühn, E.R., S. Corneille and F. Ollevier. 1986. Circadian variations in plasma osmolality, electrolytes, glucose, and cortisol in carp (*Cyprinus carpio*). *General and Comparative Endocrinology* 61: 459–468.

Lague, M. and S.G. Reebs. 2000. Phase-shifting the light-dark cycle influences food-anticipatory activity in golden shiners. *Physiology and Behavior* 70: 55–59.

Landless, P.J. 1976. Demand-feeding behaviour of rainbow trout. *Aquaculture* 7: 11–25.

Le Minh, N., F. Damiola, F. Tronche, G. Schütz and U. Schibler. 2001. Glucocorticoid hormones inhibit food-induced phase-shifting of peripheral circadian oscillators. *EMBO Journal* 20: 7128–7136.

Liu, C., S. Li, T. Liu, J. Borjigin and J.D. Lin. 2007. Transcriptional coactivator PGC-1a integrates the mammalian clock and energy metabolism. *Nature (London)* 447: 477–482.

López-Olmeda, J.F. 2009. Circadian modulation of the biological clock and behavioral, endocrine and metabolic rhythms of teleost fishes. Doctoral Thesis, University of Murcia, Murcia.

López-Olmeda, J.F. and F.J. Sánchez-Vázquez. 2009. Zebrafish temperature selection and synchronization of locomotor activity circadian rhythm to ahemeral cycles of light and temperature. *Chronobiology International* 26: 200–218.

López-Olmeda, J.F., M. Egea-Álvarez and F.J. Sánchez-Vázquez. 2009a. Glucose tolerance in fish: Is the daily feeding time important? *Physiology and Behavior* 96: 631–636.

López-Olmeda, J.F., A. Montoya, C. Oliveira and F.J. Sánchez-Vázquez. 2009b. Synchronization to light and restricted-feeding schedules of behavioral and humoral daily rhythms in gilthead sea bream (*Sparus aurata*). *Chronobiology International* 26: 1389–1408.

Madrid, J.A., T. Boujard and F.J. Sánchez-Vázquez. 2001. Feeding rhythms. In: *Food Intake in Fish*, D.F. Houlihan, T. Boujard and M. Jobling (eds.). Blackwell Science, Oxford, pp. 189–215.

Meijer, J.H. and W.J. Rietveld. 1989. Neurophysiology of the suprachismatic circadian pacemaker in rodents. *Physiological Reviews* 69: 671–707.

Meseguer, C., J. Ramos, M.J. Bayarri, C. Oliveira and F.J. Sánchez-Vázquez. 2008. Light synchronization of the daily spawning rhythms of gilthead bream (*Sparus aurata* L.) kept under different photoperiod and after shifting the LD cycle. *Chronobiology International* 25: 666–679.

Mistlberger, R. 1994. Circadian food-anticipatory activity: Formal models and physiological mechanisms. *Neuroscience Biobehavioral Research* 18: 171–195.

Mommsen, T.P., M.M. Vijayan and T.W. Moon. 1999. Cortisol in teleosts: dynamics, mechanisms of action, and metabolic regulation. *Reviews in Fish Biology and Fisheries* 9: 211–268.

Müller, K. 1978. The flexibility of the circadian system of fish at different latitudes. In: *Rhythmic Activity of Fishes*, J.E. Thorpe (ed). Academic Press, London, pp. 91–104.

Nakahata, Y., M. Kaluzova, B. Grimaldi, S. Sahar, J. Hirayama, D. Chen, L.P. Guarente and P. Sassone-Corsi. 2008. The NAD+-dependent deacetylase SIRT1 modulates CLOCK-mediated chromatin remodeling and circadian control. *Cell* 134: 329–340.

Naruse, M. and T. Oishi. 1994. Effects of light and food as zeitgebers on locomotor activity rhythms in the loach *Misgurnus anguillicaudatus*. *Zoological Science* 11: 113–119.

Pando, M.P., A.B. Pinchak, N. Cermakian and P. Sassone-Corsi. 2001. A cell-based system that recapitulates the dynamic light-dependent regulation of the vertebrate clock. *Proceedings of the National Academy of Sciences of the United States of America* 98: 10178–10183.

Pavlidis, M., M. Berry, P. Divanach and M. Kentouri. 1997. Diel pattern of haematocrit, serum metabolites, osmotic pressure, electrolytes and thyroid hormones in sea bass and sea bream. *Aquaculture International* 5: 237–247.

Pavlidis, M., L. Greenwood, M. Paalavuo, H. Mölsa and J.T. Laitinen. 1999a. The effect of photoperiod on diel rhythms in serum melatonin, cortisol, glucose, and electrolytes in the common dentex, *Dentex dentex*. *General and Comparative Endocrinology* 113: 240–250.

Pavlidis, M., M. Paspatis, M. Koistinen, T. Paavola, P. Divanach and M. Kentouri. 1999b. Diel rhythms of serum metabolites and thyroid hormones in red porgy held in different photoperiod regimes. *Aquaculture International* 7: 29–44.

Pérez, E. and B. Buisson. 1986. Research on the origin of the circadian activities in the course of the ontogenesis of the trout, *Salmo trutta* L.: the activities of the eggs and vesiculed alevins in constant conditions. *Biology Zentralblatt* 105: 609–613.

Pickering, A.D. and T.G. Pottinger. 1983. Seasonal and diel changes in plasma cortisol levels of the brown trout, *Salmo trutta* L. *General and Comparative Endocrinology* 49: 232–239.

Polakof, S., R.M. Ceinos, B. Fernández-Durán, J.M. Míguez and J.L. Soengas. 2007a. Daily changes in parameters of energy metabolism in brain of rainbow trout: Dependence on feeding. *Comparative Biochemistry and Physiology* A 146: 265–273.

Polakof, S., J.M. Míguez and J.L. Soengas. 2007b. Daily changes in parameters of energy metabolism in liver, white muscle, and gills of rainbow trout: Dependence on feeding. *Comparative Biochemistry and Physiology* A 147: 363–374.

Purser, G.J. and W.M. Chen. 2001. The effect of meal size and meal duration on food anticipatory activity in greenback flounder. *Journal of Fish Biology* 58: 188–200.

Rance, T.A., B.I. Baker and G. Webley. 1982. Variations in plasma cortisol concentrations over a 24-hour period in the rainbow trout *Salmo gairdneri*. *General and Comparative Endocrinology* 48: 269–274.

Reebs, S.G. 1999. Time-place learning based on food but not on predation risk in a fish, the inanga (*Galaxias maculatus*). *Ethology* 105: 361–371.

Reebs, S.G. 2002. Plasticity of diel and circadian activity rhythms in fishes. *Reviews in Fish Biology and Fisheries* 12: 349–371.

Reebs, S.G. and M. Lague. 2000. Daily food-anticipatory activity in golden shiners: A test of endogenous timing mechanisms. *Physiology and Behavior* 70: 35–43.

Richter, C.P.A. 1922. A behavioristic study of the activity of the rat. *Comparative Psychology Monographs* 1: 1–55.

Rozin, P. and J. Mayer. 1961. Regulation of food intake in the goldfish. *American Journal of Physiology* 201: 968–974.

Rubio, V.C., M. Vivas, A. Sánchez-Mut, F.J. Sánchez-Vázquez, D. Covès, G. Dutto and J.A. Madrid. 2004. Self-feeding of European sea bass (*Dicentrarchus labrax*, L.) under laboratory and farming conditions using a string sensor. *Aquaculture* 233: 393–403.

Rutter, J., M. Reick, L.J. Wu and S.L. McKnight. 2001. Regulation of Clock and NPAS2 DNA binding by the redox state of NAD cofactors. *Science* 293: 510–514.

Sánchez-Muros, M.J., V. Corchete, M.D. Suárez, G. Cardenete, E. Gómez-Milán and M. de la Higuera. 2003. Effect of feeding method and protein source on *Sparus aurata* feeding patterns. *Aquaculture* 224: 89–103.

Sánchez-Vázquez, F.J. and M. Tabata. 1998. Circadian rhythms of demand-feeding and locomotor activity in rainbow trout. *Journal of Fish Biology* 52: 255–267.

Sánchez-Vázquez, F.J. and J.A. Madrid. 2001. Feeding anticipatory activity in fish. In: *Food Intake in Fish*, D.F. Houlihan, T. Boujard and M. Jobling (eds.). Blackwell Science, Oxford, pp. 216–232.

Sánchez-Vázquez, F.J., A. Aranda and J.A. Madrid. 2001. Differential effects of meal size and food energy density on feeding entrainment in goldfish. *Journal of Biological Rhythms* 16: 58–65.

Sánchez-Vázquez, F.J., J.A. Madrid and S. Zamora. 1995a. Circadian rhythms of feeding activity in sea bass, *Dicentrarchus labrax* L.: dual phasing capacity of diel demand-feeding pattern. *Journal of Biological Rhythms* 10: 256–266.

Sánchez-Vázquez, F.J., S. Zamora and J.A. Madrid. 1995b. Light-dark and food restriction cycles in sea bass: effect of conflicting zeitgebers on demand-feeding rhythms. *Physiology and Behavior* 58: 705–714.

Sánchez-Vázquez, F.J., F.J. Martínez, S. Zamora and J.A. Madrid. 1994. Design and performance of an accurate demand feeder for the study of feeding behaviour in sea bass, *Dicentrarchus labrax* L. *Physiology and Behavior* 56: 789–794.

Sánchez-Vázquez, F.J., J.A. Madrid, S. Zamora and M. Tabata. 1997. Feeding entrainment of locomotor activity rhythms in the goldfish is mediated by a feeding-entrainable circadian oscillator. *Journal of Comparative Physiology* A181: 121–132.

Sánchez-Vázquez, F.J., J.A. Madrid, S. Zamora, M. Iigo and M. Tabata. 1996. Demand feeding and locomotor circadian rhythms in the goldfish, *Carassius auratus*: dual and independent phasing. *Physiology and Behavior* 60: 665–674.

Sánchez-Vázquez, F.J., M. Azzaydi, F.J. Martínez, S. Zamora and J.A. Madrid. 1998a. Annual rhythms of demand-feeding activity in sea bass: evidence of a seasonal phase inversion of the diel feeding pattern. *Chronobiology International* 15: 607–622.

Sánchez-Vázquez, F.J., T. Yamamoto, T. Akiyama, J.A. Madrid and M. Tabata. 1998b. Selection of macronutrients by goldfish operating self-feeders. *Physiology and Behavior* 65: 211–218.

Schibler, U. and F. Naef. 2005. Cellular oscillators: rhythmic gene expression and metabolism. *Current Opinion in Cell Biology* 17: 223–229.

Singley, J.A. and W. Chavin. 1975. Serum cortisol in normal goldfish (*Carassius auratus* L.). *Comparative Biochemistry and Physiology* A50: 77–82.

Small, B.C. 2005. Effect of fasting on nychthemeral concentrations of plasma growth hormone (GH), insulin-like growth factor I (IGF-I), and cortisol in channel catfish (*Ictalurus punctatus*). *Comparative Biochemistry and Physiology* B142: 217–223.

Smith, I.P., N.B. Metcalfe, F.A. Huntingford and S. Kadri. 1993. Daily and seasonal patterns in the feeding behaviour of Atlantic salmon (*Salmo salar* L.) in a sea cage. *Aquaculture* 117: 165–178.

Spieler, R.E. and T.A. Noeske. 1981. Timing of a single daily meal and diel variations of serum thyroxine, triiodothyronine and cortisol in goldfish. *Life Sciences* 28: 2939–2944.

Spieler, R.E. and T.A. Noeske. 1984. Effects of photoperiod and feeding schedule on diel variations of locomotor activity, cortisol, and thyroxine in goldfish. *Transactions of the American Fisheries Society* 113: 528–539.

Stephan, F.K. 2002. The "other" circadian system: food as a zeitgeber. *Journal of Biological Rhythms* 17: 284–292.

Stephan, F.K., J.M. Swann and C.L. Sisk. 1979. Entrainment of circadian rhythms by feeding schedules in rats with suprachiasmatic nucleus lesions. *Behavioral and Neural Biology* 25: 545–554.

Stokkan, K.A., S. Yamazaki, H. Tei, Y. Sakaki and M. Menaker. 2001. Entrainment of the circadian clock in the liver by feeding. *Science* 291: 490–493.

Thorpe, J.E., M.G. McConway, M.S. Miles and J.S. Muir. 1987. Diel and seasonal changes in resting plasma cortisol levels in juvenile Atlantic salmon, *Salmo salar* L. *General and Comparative Endocrinology* 65: 19–22.

Van Cauter, E., K.S. Polonsky and A.J. Scheen. 1997. Roles of circadian rhythmicity and sleep in human glucose regulation. *Endocrine Reviews* 18: 716–738.

Velázquez, M., S. Zamora and F.J. Martínez. 2004. Influence of environmental conditions on demand-feeding behaviour of gilthead seabream (*Sparus aurata*). *Journal of Applied Ichthyology* 20: 536–541.

Vera, L.M., J.A. Madrid and F.J. Sánchez-Vázquez. 2006. Locomotor, feeding and melatonin daily rhythms in sharpsnout seabream (*Diplodus puntazzo*). *Physiology and Behavior* 88: 167–172.

Vera, L.M., J.F. López-Olmeda, M.J. Bayarri, J.A. Madrid and F.J. Sánchez-Vázquez. 2005. Influence of light intensity on plasma melatonin and locomotor activity rhythms in tench. *Chronobiology International* 22: 67–78.

Vera, L.M., N. De Pedro, E. Gómez-Milán, M.J. Delgado, M.J. Sánchez-Muros, J.A. Madrid and F.J. Sánchez-Vázquez. 2007. Feeding entrainment of locomotor activity rhythms, digestive enzymes and neuroendocrine factors in goldfish. *Physiology and Behavior* 90: 518–524

Whitmore, D., N.S. Foulkes, U. Strähle and P. Sassone-Corsi. 1998. Zebrafish *Clock* rhythmic expression reveals independent peripheral circadian oscillators. *Nature Neuroscience* 1: 701–707.

Whitmore, D., N.S. Foulkes and P. Sassone-Corsi. 2000. Light acts directly on organs and cells in culture to set the vertebrate circadian clock. *Nature* 404: 87–91.

Yamazaki, S., R. Numano, M. Abe, A. Hida, R. Takahashi, M. Ueda, G.D. Block, Y. Sakaki, M. Menaker and H. Tei. 2000. Resetting central and peripheral circadian oscillators in transgenic rats. *Science* 288: 682–685.

Yoo, S.H., S. Yamazaki, P.L. Lowrey, K. Shimomura, C.H. Ko, E.D. Buhr, S.M. Siepka, H.K. Hong, W.J. Oh, O.J. Yoo, M. Menaker and J.S. Takahashi. 2004. PERIOD2::LUCIFERASE real-time reporting of circadian dynamics reveals persistent circadian oscillations in mouse peripheral tissues. *Proceedings of the National Academy of Sciences of the United States of America* 101: 5339–5346.

9

REPRODUCTION RHYTHMS IN FISH

Catarina Oliveira[1] *and Francisco Javier Sánchez-Vázquez*[2]

INTRODUCTION

Earth is not a static environment, since there are cyclic changes occurring all the time. Some of these environmental changes can be predictable, when they are associated with geophysical cycles, or unpredictable, when related with a meteorological phenomenon (such as rain). Environmental cycles occur within different periodicities: annual periodicity in association with the rotation of the Earth around the Sun, appearing seasons in temperate regions; lunar periodicity associated with the rotation of the moon around the Earth; and finally, daily periodicity associated with the rotation of the Earth on its axis, originating the daily alternation of light and darkness. All these cycles acting on living organisms through evolution, promoted the appearance of molecular oscillators or biological clocks, which can be used by animals to keep time and anticipate recurrent events, in order to optimize their survival. Such a multioscillatory pacemaker, which endogenously controls biological rhythms, is located in the hypothalamus, pineal organ and eyes of vertebrates.

Periodic time cues, which are called "*zeitgebers*", are used by the animals to reset their clock and synchronize their biological rhythms, modulating the period, amplitude and phase (Aschoff, 1981). These synchronizers can be classified in two types: abiotic or biotic factors. Among the abiotic factors, light and temperature cycles are the main synchronisers controlling daily

---

[1,2] Department of Physiology, Faculty of Biology, University of Murcia, 30100 Murcia, Spain.
e-mails: [1] oliveira@um.es; [2] javisan@um.es

behavioural and molecular rhythms (Aschoff, 1981; Carr *et al.*, 2006; Ziv and Gothilf, 2006). Actually, temperature changes of 1–2°C are observed within a 24h cycle, and they synchronise biological rhythms (Rensing and Ruoff, 2002). Both light and temperature cycles have also a crucial role in timing seasonal rhythms, such as reproduction in fish (Bromage *et al.*, 2001; Anguis and Cañavate, 2005; Clark *et al.*, 2005). Light may act through the seasonal pattern of changing daylength, while temperature may act by the naturally fluctuating thermo-cycle. As to biotic factors, periodic feeding is a strong synchronizer, since fish fed at the same time during several days develop food anticipatory activity, presenting higher locomotor and digestive activities prior to feeding time (Sánchez-Vázquez and Madrid, 2001; Vera *et al.*, 2007).

Daily rhythms have been profusely investigated in fish during the last two decades, with special emphasis on behavioural rhythms (locomotor and feeding activities) or rhythms of secretion of neuroendocrine factors. In what concerns to behavioural rhythms, fish species have been usually classified as diurnal, nocturnal or crepuscular, according to the time of day they are more active and perform their activities (locomotor, feeding, reproduction, etc). This preference for being active in certain time of the day is usually species specific and could be determined genetically or associated with adaptations to the habitats (food availability, predation, etc) or sensorial requirements (dependence on vision for food capture). However, there can be some plasticity in the activity patterns (Ali, 1992), as happens for example when the activity patterns change over the year according to the seasons, like in the case of the pike, *Esox lucius,* which passes from crepuscular in the summer for being diurnal during winter (Cook and Bergersen, 1988); or the juveniles of cod, *Gadus morhua,* which present nocturnal activity during summer and change to diurnal in winter (Clark and Green, 1990). However, there are species which maintain firmly their behavioural pattern. For example, the tench is strictly nocturnal even under extremely short photoperiods, with only two hours of darkness per day (Herrero *et al.*, 2003).

In addition to behavioural rhythms, fish may show daily and seasonal rhythms in their neuroendocrine system. In the brown trout (*Salmo trutta*) the existence of a marked rhythm of plasma concentration of cortisol, with higher values during the night, was observed for the most part of the year. However, in summer and beginning of autumn, there was a delay of the rhythm, with maximum values being registered 4h later (Pickering and Pottinger, 1983). In the European sea bass, daily rhythms of plasma concentration of cortisol, glucose and insulin were observed, with insulin rhythm presenting evidences of being controlled by the feeding time, and glucose rhythm showing more alike to be controlled by the photoperiod (Cerdá-Reverter *et al.*, 1998).

As to annual (seasonal) rhythms, the most deeply studied in fish are the reproduction rhythms, which synchronisation pathways will be discussed onwards.

MELATONIN RHYTHMS

Phototransduction

The highly photosensitive pineal organ of fish transduces the environmental information (photoperiod, temperature, salinity, etc), into a hormonal signal: melatonin. This hormone is produced in high amounts during the night and in almost undetectable levels during the day, and is immediately secreted to the bloodstream. Light, through the daily alternation of light and darkness, is known to be the most important environmental factor controlling melatonin rhythms. Since melatonin is produced by the pineal almost exclusively at night, this clock hormone provides the animal with information about the time of day. Besides, the duration of the nocturnal elevation is longer under a short photoperiod than under a long photoperiod, mediating photoperiod signals and providing the animal with seasonal information, thus also working as a calendar hormone (Reiter, 1993). Furthermore, melatonin rhythms may also transduce the lunar phase because the teleost pineal organ is so sensitive to light that very dim light during the night can inhibit melatonin production. In some fish species a light pulse of very low intensity applied during the night can inhibit the nocturnal melatonin production down to daytime values, as observed for example in the European sea bass (Bayarri *et al.*, 2002), in the tench (Vera *et al.*, 2005) and in the Senegal sole (Oliveira *et al.*, 2007). This high sensitivity to light could also be useful for fish to detect the presence of dim light at night in the natural environment, such as moon light, and thus transduce the lunar phase. Actually, in the genus *Siganus* sp. (rabbitfishes) there are several studies on the influence of the moon light cycle on their melatonin rhythms (Rahman *et al.*, 2004a,b; Takemura *et al.*, 2004b, 2006). The influence of the lunar phase, however, has not been observed exclusively in tropical fishes, since in the Mediterranean region Senegal sole exposed to full moon nights presented inhibited melatonin production (Oliveira *et al.*, 2009b). Taken together, these evidences suggest that melatonin rhythms supply accurate information about the time of day, season of the year and moon cycle.

In vitro studies have suggested that most teleost species possess endogenous intrapineal oscillators driving the rhythm of melatonin production (Bolliet *et al.*, 1994, 1996; Cahill, 1996; Iigo *et al.*, 1991, 2003, 2004; Bayarri *et al.*, 2004a; Oliveira *et al.*, 2009d). However, such an endogenous control is lacking in salmonids, suggesting that ancestral salmonids lost the circadian regulation of melatonin production after the divergence from osmerid teleots (Iigo *et al.*, 2007). In a recent study, Migaud *et al.* (2007) suggested a new theory to explain the change suffered by the circadian control of melatonin production, during teleost evolution. According to this author, the regulation of pineal activity would have evolved from an independent light sensitive pineal organ, without pacemaker activity (like in salmonids) (Fig. 9.1A) to an intermediate state where the pineal organ remains light sensitive and could possess a circadian pacemaker, but is also regulated by photic information perceived by the retina (as seen in and in cod) (Fig. 9.1B) to finally reach a third type, where pineal light sensitivity would be dramatically reduced, approaching to what happens in higher vertebrates (Fig. 9.1C). In this case, melatonin synthesis activity would be primarily regulated by a circadian pacemaker (with an unknown location) entrained by photic information perceived by the retina (Migaud *et al.*, 2007).

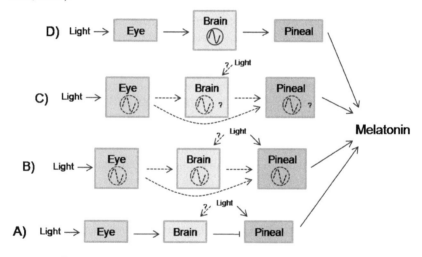

Fig. 9.1 Different systems of circadian control of melatonin production suggested by Migaud *et al.* (2007): (A) Atlantic Salmon, (B) Sea Bass, (C) Nile Tilapia, (D) Mammals.

Melatonin (N-acetyl-5-methoxytryptamine) (Fig. 9.2) is an indoleamine, whose two functional groups are not only decisive for specificity of receptor binding, but also for allowing the molecule to enter any cell, compartment or body fluid and, surprisingly for its oxidation chemistry (Poeggeler *et al.*, 2002).

The melatonin precursor is the essential amino acid L-tryptophan (L-TRP), which is captured from the blood torrent by the pinealocytes through a mechanism of active transport under adrenergic control, as seen in lampreys, teleosts, lizards, birds and mammals (Collin *et al.*, 1986, 1989; Falcón *et al.*, 1992). Inside the cell, the L-TRP gives place to the 5-

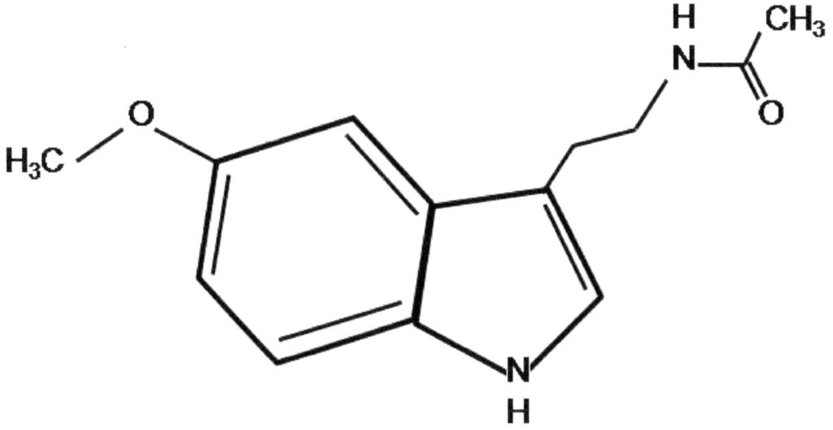

Fig. 9.2 Structure of the molecule of melatonin—adapted from Hardeland *et al.* (2006).

hydroxytryptophan through the action of the enzyme tryptophan hydroxylase (TPOH). The enzyme L-aromatic amino acid decarboxylase (AAAD) acts on the 5-hydroxytryptophan to form 5-hydroxytryptamine or serotonin. The serotonin concentration in the pineal is very high, superior to any other organ. From this moment on, the most important step in pineal metabolism of serotonin is its transformation in N-acetyl-serotonin through the action of the arylamine N-acetyltransferase (AA-NAT), enzyme which constitutes the limiting factor in the melatonin synthesis. Finally, the N-acetyl-serotonin is transformed by the hydroxyl-O-methyltransferase in melatonin which is liberated in the circulating blood torrent by the pinealocytes (Falcón, 1999).

The pineal organ does not only transduce seasonal information on photoperiod, since the melatonin production is also influenced by the temperature. If photoperiod determines the length of the melatonin elevation, the temperature determines the amplitude of the rhythm. Thus, the shape of plasma melatonin profile changes according to seasons, presenting short duration and high amplitude in the summer, and long duration and low amplitude during winter, with intermediate situations occurring in the spring and autumn (Falcón *et al.*, 2007). In sea bass this pattern has been described by García-Allegue and co-workers (2001): under natural conditions the pineal organ showed capacity to integrate seasonal information (progressive changes in the environment) generating for each season a specific melatonin profile, which may supply precise calendar information. In Senegal sole, the annual natural fluctuating temperature modulated the nocturnal production of melatonin, with higher melatonin values observed with increasing temperature. Furthermore, the annual rhythm of nocturnal melatonin disappeared under constant temperature

(Vera *et al.*, 2007). In the goldfish the daily melatonin rhythms were also markedly influenced by temperature, with clearly higher night-time levels in high than in low temperature (Iigo and Aida, 1995), as happened in sole. *In vitro* studies have demonstrated similar relation between temperature and nocturnal melatonin production. In the case of the lamprey, *Lampetra japonica* (Samejima *et al.*, 2000), or the pike, *Esox lucius* (Falcón *et al.*, 1994), under LD conditions only at 20°C a nocturnal increase in melatonin was registered, while under decreasing temperature (10°C/7°C) the rhythm disappeared. Furthermore, in the case of pike, under constant darkness (DD) temperature cycles synchronized the melatonin rhythm, which peaked with the highest temperature. In the rainbow trout pineal organs, melatonin release at 20°C was higher than at 15 and 25°C (Iigo *et al.*, 2007). The influence of temperature on melatonin production should not be surprising since fish are ectotherm animals and temperature influences biochemical reactions such as the conversion of serotonin into N-acetylserotonin by the arylalkylamine N-acetyltransferase (AANAT) (Sugden *et al.*, 1983).

Moreover, the circulating melatonin in fish can also be influenced by salinity, since the role of osmoregulatory processes in this indoleamine synthesis has been suggested (Kulczykowska, 2002). In species like the European sea bass (López-Olmeda *et al.*, 2009) or the gilthead sea bream (Kleszczynska *et al.*, 2006) this influence has been observed, with the lowest values of melatonin being recorded at the highest salinities. In contrast, melatonin levels in coho salmon augmented with increasing salinity (Gern *et al.*, 1984). This influence of salinity on plasma melatonin could be related with the migrations between freshwater and marine environments that these euryhaline species experience along their life cycle. Thus, such relation would explain the opposite results between sea bass or gilthead sea bream, diadromous species, and the Coho salmon, an anadromous species (López-Olmeda *et al.*, 2009).

MELATONIN AND REPRODUCTION

In teleost fish, breeding is timed to occur during increasing or decreasing photoperiod in temperate regions, in order to achieve maximal survival of the offspring. The photoperiod synchronizes seasonal reproduction and the pineal organ appears to be the mediator (Bromage *et al.*, 2001). There are many evidences that the pineal organ, through the production of the neurohormone melatonin, plays a key role in the regulation of daily and seasonal rhythmic patterns, including reproduction (Amano *et al.*, 2000; Bromage *et al.*, 2001; Bayarri *et al.*, 2004c; Falcón *et al.*, 2007). It is though that melatonin acts on the hypothalamic-pituitary-gonad axis, which in turn, controls reproduction timing through neuroendocrine signals (gonadotropins), the production of sex steroids and growth factors in the gonads. However, the mechanism of melatonin regulation is yet unknown, since there seems to be inconsistencies in the results of studies that have been carried out to date (Falcón *et al.*, 2007). For instance, pinealectomy has in many cases clear effects on reproduction, which can be both inhibited and stimulated. Generally, in long-day breeders stimulates reproduction in winter, whereas it inhibits reproduction in summer while the opposite happens in short-day breeders. As to melatonin injections, this procedure may have either no effects on fish maturation or suppress reproduction in fishes where long photoperiods stimulate breeding, while the studies using melatonin implants do not fully support the concept of melatonin being of major importance for photoperiod effects in teleost (Mayer *et al.*, 1997).

According to Amano *et al.* (2000) melatonin treatment had a stimulatory effect on the GSI and pituitary gonadotropin I (GtH-I) contents and testosterone (T) levels in underyearling precocious male masu salmon (*Oncorhynchus masou*). These results suggested that mimicking a short photoperiod by melatonin administration stimulated testicular development but did not completely activate the brain-pituitary-gonadal axis, in these fishes. Thus, melatonin was suggested to be one of the factors that mediate the transduction of photoperiod information to the brain-pituitary-gonadal axis. In the case of Senegal sole, both photoperiod and temperature affected melatonin production, transducing seasonal information and controlling annual reproductive rhythms: plasma melatonin rhythms showed seasonal differences under natural conditions. Seasonal melatonin rhythms, however, disappeared when sole were kept at constant temperature throughout the year, spawning and sex steroids rhythms being also disrupted (Vera *et al.*, 2007; Oliveira *et al.*, 2009c). Furthermore, for fish species like the Atlantic salmon (Randall *et al.*, 1995) or the European sea bass (García-Allegue *et al.*, 2001) the duration of the melatonin rise faithfully represented the length of the night, and hence the seasonally changing pattern of daylength. This way, all body tissues including those involved in the control and timing of reproduction would be provided with information on seasonal changes of daylength and hence, calendar time. In a study on caged European sea bass (*Dicentrarchus labrax*), artificial light regimes were shown to influence the circulating melatonin levels and to affect the daily rhythm of luteinizing hormone (LH), maintaining a fixed relationship between the first nocturnal rise of melatonin and the nocturnal peak of plasma LH in both groups, suggesting a strong relation between these two hormones (Bayarri *et al.*, 2004c). For the same species, a recent study (Servili *et al.*, 2008) suggested a role for gonadotropin releasing hormone (GnRH-II and GnRH-III) in the modulation of pineal and retinal functions, respectively. Furthermore, melatonin effects on the expression of GnRH-I, GnRH-III, and several GnRH-R subtypes were demonstrated, suggesting that these interactions between melatoninergic and GnRH systems could represent a substrate of photoperiod effects on reproductive and other rhythmic physiological events in the European sea bass. For the Indian carp, *Catla catla,* the importance of photoperiod and melatonin (linking the environment with the endocrine system) in the regulation of temporal pattern of reproductive events has also been discussed in a recent review (Maitra and Chattoraj, 2007).

Briefly, we may conclude that the major role of the pineal organ in the control of reproduction is to transduce environmental time cues into hormonal (melatonin) neural outputs that ultimately trigger reproduction. However, more studies are needed to determine the exact neuroendocrine pathways of melatonin on fish reproduction.

REPRODUCTION RHYTHMS

Reproduction in fish is widely known as a seasonal phenomenon. However, the existence lunar or daily rhythms of reproduction should not be dismissed. Each of the multiple environmental cycles (the light/dark, tidal, semi-lunar, lunar and seasonal cycles) may have independent, specific effects on reproductive timing of fish, and they may alone or in combination produce a variety of temporal reproductive patterns.

SEASONALITY

Most fish species reproduce once a year, within a species specific season, in order to ensure that the offspring is produced when environmental conditions are most suitable for its survival. Most species reproduce during spring, to allow fish larvae grow through the summer, when temperature is higher and there is more food availability. The fact that males and females synchronize and are sexually mature at a given time of the year favors the reproductive contacts between them and optimizes the spread of their genetic material. Since gonadal development and subsequent gamete and embryo formation take a long time to complete, it is clear that gonadal recrudescence must have initiated many months earlier. The seasonality of reproduction is entrained by the cyclical changes occurring in the environment during the year, as it is photoperiod, temperature, climate conditions and food availability. However, it is the seasonally-changing pattern of daylength which is probably responsible for the cuing and timing of reproduction in the majority of fish from temperate latitudes (Fig. 9.3). In tropical and sub-tropical fish the photoperiod plays a minor role, since the annual changing pattern of daylength is not so pronounced as in other latitudes. These species use other cues to synchronize reproduction season, as it is temperature or the increased productivity which follows seasonal rainfall or movements in oceanic currents. However, maturation and spawning almost certainly rely on photoperiod as a measure of daily and possibly calendar time (Bromage *et al.*, 2001).

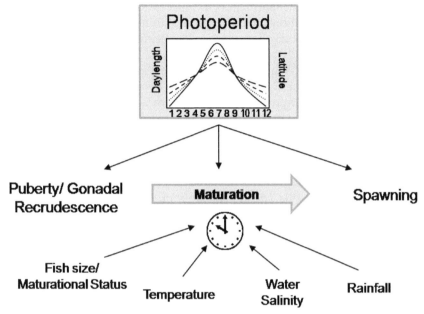

Fig. 9.3 Diagrammatic representation of the entrainment by photoperiod of the endogenous processes or clocks which time reproduction—adapted from Bromage *et al.* (2001). The other environmental factors probably act in a permissive manner to 'gate' these endogenous processes.

According to their reproductive behavior teleosts from temperate latitudes can be classified as long or short day-breeders, according to the time of year they prefer to reproduce. The Senegal sole, for example, presents market seasonality in its reproduction rhythms and is considered a long-day breeder. The natural spawning season in the Mediterranean occurs in spring (from March until June), with some occasional spawning during autumn and winter (Cabral and Costa, 1999; Cabral, 2000). This pattern of spawning is also observed in captivity, for both wild and F1 Senegal sole under natural temperature regimes, with animals laying eggs from February to May, with a secondary season in autumn. However, spawnings obtained from F1 individuals do not result in hatched larvae as they are, most of the times, unfertilized (Anguis and Cañavate, 2005; Agulleiro *et al.*, 2006; Guzmán *et al.*, 2008). Sex steroids rhythms presented higher concentration values at pre-spawning, and a minor rise around autumn, and were observed to be in phase with each other, as well as with the vitellogenin (VTG) rhythm in the case of females (García-López *et al.*, 2006a; Guzmán *et al.*, 2008). Senegal sole exposed to attenuated temperature cycle during the year lost their sex steroids rhythms, as well as the melatonin rhythms, and failed to spawn (García-López *et al.*, 2006b; Vera *et al.*, 2007; Oliveira *et al.*, 2009c), while fish maintained under conditions of constant short photoperiod from winter solstice on, did successfully spawn during spring (unpublished observations), what leads us to think that this species mostly relies on temperature then on photoperiod to synchronize their reproduction seasonality.

Other species with high commercial interest are the gilthead seabream, *Sparus aurata*, and the European sea bass, *Dicentrarchus labrax*, which represents the most important targets of nowadays aquaculture industry and research. Reproduction season of both these species occur during winter, unlike what happened in the case of Senegal sole. The gilthead sea bream is a true hermaphrodite and functions as a male during the first 2 years of life, while in the third year, there is a testicular degeneration and ovary development, resulting in transformation of the fish to functional females. In wild conditions this species matures in September/October, with spawning starting in November (Zohar and Gordin, 1979; Zohar *et al.*, 1984). The levels of sex steroids E_2 (estradiol) and T in females have been observed to be elevated during the spawning period, and to be at their lowest levels at the end of this period. The constant E_2 levels observed during spawning suggested ongoing vitellogenesis during the entire breeding season (Jerez *et al.*, 2006). Indeed, changes in plasma VTG were seen to be correlated with those of E_2 in both pre-spawning and spawning females. A slight decrease in VTG observed between pre- to spawning period fitted well the fact that plasma VTG reflects a balance between that produced by the liver and that sequestered by growing oocytes (Mosconi *et al.*, 1998). Furthermore, a recent study (Mosconi *et al.*, 2002) demonstrated that basal VTG production has seasonal changes and that the responsiveness to E_2 and pituitary hormones of sea bream liver varies during the reproductive cycle. During pre-spawning, when E_2 receptor levels are highest, the liver produces the highest levels of VTG in response to E_2. These findings demonstrate a seasonal variation in hormone-induced VTG secretion response correlated with the E_2 receptor level in sea bream liver.

The European sea bass also reproduces during the winter, but spawnings have been observed to occur mostly in late winter. Maturation started in September/October as in sea bream, and post-vitellogenic oocytes were first observed in December. Ovulation lasted from the January to mid-March (Asturiano *et al.*, 2000) which can be considered the main spawning period for the European sea bass. Plasma concentrations of E_2 and T presented a peak between December and February followed by a decrease in March and April. In the case of T, however, a second peak in plasma profile was observed in May–June, but its physiological function remains unknown (Prat *et al.*, 1990; Asturiano *et al.*, 2000). The increase in VTG concentration in plasma paralleled the elevation in E_2, with highest levels being registered in December. This peak in post-vitellogenesis period may be due to the continued synthesis of VTG accompanied by a reduction in absorption of the protein, resulting in accumulation in plasma. On the other hand, this peak may reflect the fact that when the most advanced oocytes reach the post-vitellogenic stage, oocytes in subsequent clutches are in the vitellogenic stage (Mañanós *et al.*, 1994; Mañanós *et al.*, 1997; Asturiano *et al.*, 2000).

The Atlantic cod, as all the species we have been previously discussing, has a high commercial interest. Was introduced as a candidate for diversification of the Norwegian aquaculture industry in the early 1980s, when its reproductive biology started to be investigated. In its natural environment, in Norwegian waters, the Atlantic cod normally spawns between February and May with a peak around early April (Pedersen, 1984). Similar spawning times have been found in studies performed in individuals in captivity conditions, under natural photoperiod regimens (Kjesbu, 1994; Hansen *et al.*, 2001). In captivity, the first evidences of oocyte maturation were observed around November/December, and GSI increased during the course of gonadal development and reached a maximum during spawning in both sexes. However, in the case of females, a marked elevation was only detected immediately before spawning. These changes were associated with changes in sex steroids: in females, E_2 increased during gonadal development and reached maximum plasma levels during spawning, when the T levels were also maximum, while in males, 11keto-testosterone (11KT) increased rapidly during spermiation and T increased at earlier testicular stages (Dahle *et al.*, 2003). Furthermore, in another study on this species, E_2 was observed to be responsible for the induction of the synthesis of VTG in females (Silversand *et al.*, 1993).

Salmonids are a group of fish inhabiting cold water habitats, which present a particular reproduction behavior rhythm; they spend most of their life at sea and migrate to the home rivers only to reproduce, thus being anadromous species. In the specific case of the Atlantic salmon, *Salmo salar*, research on the timing of runs of adult individuals into rivers remount to the early years of the last century (Nordqvist, 1924). Wild Atlantic salmon smolts leave fresh water in spring and move quickly into the ocean. After living in the ocean for 1–4 years, the majority of adults return to spawn in the same rivers they left as smolts. Most adult wild salmon enter Norwegian coastal waters between May and September, and they travel o a stream or pond high in oxygen, where they can lay their eggs during autumn (Hansen *et al.*, 1993; Thorpe and Stradmeyer, 1995; Holm *et al.*, 2000). Most salmons die shortly after spawning. Concerning sex steroids, T in males and females, 11KT in males, and E_2 in females increased from basal levels in summer to peak values just before spawning in late autumn and descending after this period (Lacroix *et al.*, 1997).

Tropical fishes also posse annual rhythms of reproduction, although they live in latitudes with minimal annual environmental cycles. Rabbitfish species exhibit a reproductive season in nature, which may differ among the tropical areas, but it is mostly located between late spring/early summer (from May to July) for species such as the forktail rabbitfish, the seagrass rabbitfish, the spiny rabbitfish and the golden rabbitfish. The changes observed in the reproductive season of those species according to tropical region, suggests that reproductive season is affected by regional variation of environmental factors such as water temperature and photoperiod, and food abundance (Takemura *et al.*, 2004a). Actually, this group of fishes is known to synchronize reproduction with other environmental factors, most potent on these latitudes, as the moon cycle, as we will discuss on the next section.

A remarkable fact concerning fish reproduction seasonality is the fact that some species presented annual rhythms of reproduction in the absence of any obvious changes in the environmental cues (constant conditions of light, temperature, food, etc), suggesting an endogenous control of these rhythms, thus being called as circannual rhythms (Bromage *et al.*, 2001). The rainbow trout, for example, when maintained under constant conditions of 6 hr light and 18 hr darkness (LD 6:18), constant temperature (8.5–9.0°C), and constant feeding rate, exhibited free-running circannual rhythms of gonadal maturation and ovulation, which were self-sustaining for up to three cycles. The periodicity of the rhythm showed variation between fish and in successive cycles for the same fish, ranging from approximately 11 to 15 months (Duston and Bromage, 1991). Self-sustaining circannual rhythms of spawning have also been observed in species like the Indian catfish (*Heteropneustes fossilis*) (Sundararaj *et al.*, 1982) or the European sea bass. This last sustained rhythmicity under both LD 15:9 and LD 9:15 photoperiods (Prat *et al.*, 1999). Under natural conditions, the circannual reproduction rhythms driven by the endogenous clock would be fully in phase with the ambient light cycle as a result of ongoing re-entrainment by the seasonally-changing daylength. Thus, accepting that endogenous mechanisms can trigger the timing of reproduction in fish makes easier to understand how the altered seasonal and constant photoperiod regimes exert their effects (Bromage *et al.*, 2001).

Within each species reproduction season, fish may present lunar and daily rhythms of spawning, sex steroids, and/or gonad development, as we will see straight away.

LUNAR PHASE

Fish species inhabiting tropical zones of the planet are commonly known to poses lunar reproduction rhythms and this phenomenon has been widely investigated recently. These latitudes have a relative stable environment, with little seasonal photoperiod and temperature changes, what obligates them to use lunar cues to synchronize many biological and physiological processes (Takemura *et al.*, 2004a). The perception of the moon cues by fishes may be either by the lightning cycles (changes in the intensity of moonlight, time of moonrise, and movement of the moon across the night sky) or the changes in earth–moon–sun gravitational forces, which are repeated at an interval of approximately 2 weeks, and are responsible for tides (Leatherland *et al.*, 1992). Thus, the lunar synchronized reproduction rhythms can be of different periodicities: when the reproduction is repeated once a month around a species specific lunar phase, the rhythm is synchronized with the lunar cycle and is called lunar rhythm. Otherwise, if the reproduction rhythm repeats twice a month, it is mostly related with the spring/neap tidal cycle, and is called semi-lunar rhythm. Furthermore, the reproduction rhythms can even be synchronized with the daily cycle of tides (Takemura *et al.*, 2009) but such rhythms will be discussed in the next section, since it is more suitable to be a "daily rhythm".

In what concerns to lunar spawning rhythms, a group of fish which has been deeply investigated recently are the rabbitfishes (genus *Siganus*). Fishes from this genus exhibit a relationship between reproduction and environmental changes induced by the moon. Spawning occurs once a month, around a species-specific lunar phase, as well as the peaks of E_2 and T concentrations (Takemura *et al.*, 2004a). The golden rabbitfish, *Siganus guttatus* spawned around the first quarter moon at the same time the gonadosomatic index (GSI) showed a peak in females and yolky oocytes were observed. Plasma steroid hormones and VTG levels changed in parallel with changes in GSI, with the peak of plasma VTG occurring slightly prior to spawning. Histological observations revealed that the vitellogenic oocytes appeared again 1 week after spawning and developed synchronously, what suggested that this species is a multiple spawner and the oocyte development is in a group-synchronous manner (Rahman *et al.*, 2000b). In males, plasma steroid hormones reached their peaks 1 week before spawning coinciding with the peak of testicular development (Rahman *et al.*, 2000a). To this species, the presence of the natural night conditions seemed to be crucial to the synchronous gonadal development and spawning, since fish under constant darkness and lightness of night failed to sawn (Takemura *et al.*, 2004b). As to the production of sex steroids by ovarian follicles and testis incubated *in vitro*, the response to the addition of human chorionic gonadotropin (hCG) differed according to the lunar phase. Steroidogenic activity was only increased toward the specific lunar phase: the ovarian follicles produced E_2 around the new moon and DHP around the first lunar quarter (Rahman *et al.*, 2002) and the testis produced 11KT towards the first lunar quarter (Rahman *et al.*, 2001). Like the *S. guttatus*, the pencil-streaked rabbitfish, *S. doliatus*, also showed peaks of GSI and spawned around the first quarter moon. Similar changes in the ovarian features were repeated three times in phase with the lunar cycle, thus exhibiting a synchronous pattern in gonadal development and spawning with the lunar cycle (Park *et al.*, 2006b). Otherwise, Spiny rabbitfish *S. spinus* spawns during the new moon, when higher GSI has been observed for both females and males, and higher sperm motility and quality in males (Harahap *et al.*, 2001, 2002; Park *et al.*, 2006a). Another species from this group, the seagrass rabbitfish *S. canaliculatus, also showed* major spawning occurring around the new moon, with GSI peak coinciding with spawn, and serum VTG presenting higher values at pre-spawning (last quarter moon) (Hoque *et al.*, 1999). Finally, the forktail rabbitfish *S. argenteus spawned around the last quarter moon, contrary to the species seen above.* Histological observations showed that gametes developed towards this moon phase, correlated with an increase in sperm motility and quality and weekly changes in plasma steroid hormones, which peaked at pre-spawning (full moon). Also when gametes were stimulated *in vitro* with hCG, the production of steroid hormones was observed at pre-spawning. In this species the GtH and DHP seemed to be related to the final stage of testicular maturation, and GtH to act through the production of DHP in the testis. These results suggest that the concomitant physiological events in the gonads strictly changes with the lunar cycle in this species (Rahman *et al.*, 2003a,b,c; Park *et al.*, 2006a). All the evidences stated in this last paragraph clearly show that lunar periodicity is the major factor in stimulating reproductive activity of *Siganus sp.*

In this group of fishes the melatonin rhythm is thought to play a key role in the synchronization of reproduction with lunar cues, since in the species discussed above, besides the lunar reproduction rhythms there are also several evidences of lunar rhythms of melatonin. For example, in *S. guttatus* plasma melatonin concentration at the new moon was higher than that at the full moon. These results show that the fish possibly perceive moonlight intensity and plasma melatonin fluctuates according to 'lightness' at a point of night (Takemura *et al.*, 2004b). Furthermore, cultured pineal organs respond to environmental light cycles as in the whole fish (Takemura *et al.*, 2006), and even rf*Per2* (a light-inducible clock gene) expression was affected by moonlight, presenting monthly variation related with the change in amplitude between the full and new moon periods (Sugama *et al.*, 2008). These evidences suggest that the pineal organ could be transducing the lunar cues into melatonin rhythms, and thus providing information on the lunar cycles to the brain-pituitary-gonad axis.

In another groups of fishes, as Serranidae and Cichlidae families, evidences of lunar rhythms on reproduction activities have also been observed. Some studies have reported in reef species of groupers a lunar-related spawning behaviour. For example, both *Epinephelus guttatus* and *E. striatus* aggregated and spawned around the full moon. Males patrolled territories containing several females and spawning was observed low in the water column one day after full moon. This moon phase may be a superior cue to aggregation and spawning for these species, particularly in areas of weak tidal currents (Colin *et al.*, 1987; Aguilar-Perera and Aguilar-Dávila, 1996). The honeycomb grouper *E. merra* also spawned around the same lunar phase, with GSI increasing with the approach of the full moon (Lee *et al.*, 2002), while the leopard coral grouper, *pleuctropomus leopardus*, aggregated and spawned towards the new moon (Samoilys and Squire, 1994). But not only in seawater fishes could reproduction be synchronized with the moon. In the case of cichlids, which are freshwater fishes, lunar spawning rhythms have also been observed: *Neolamprologus morii* laid their eggs during the full moon (Rossiter, 1991) while species such as *Cyprichromis leptosoma* and *Lepidiolamprologus profundicola* spawned during the first quarter moon. The synchronization of *C. leptosoma* reproduction to lunar cues could be related to the fact that the juveniles of this species use the breeding territory of *L. Profundicola* as a safety zone from their potential predators since the female *L. profundicola* guards the nesting site driving away approaching piscivorous fish (Watanabe, 2000a,b).

In spite of all the evidences seen above about the lunar synchronization of reproduction in fishes from zones with a tropical climate, due to the stable characteristics of their habitat, lunar spawning rhythms have also been registered in species from temperate zones. For instance, in gilthead seabream, *S. aurata,* the effect of moon phase on spawning behaviour led to more eggs being laid during the full moon (Saavedra and Pousão-Ferreira, 2006), while Senegal sole preferred to spawn towards the darkest nights, during the last quarter and the full moon (Oliveira *et al.*, 2009a) and presented significant higher levels of sex steroids during full moon when compared with new moon (Oliveira *et al.*, 2009b). There are only few studies on this subjected, and more investigation is needed, but the influence of the lunar cycles on the reproduction rhythms should not be put aside in these regions.

Besides lunar synchronization, the reproduction rhythms in some species can also synchronize with the spring neap tidal cycle, being called semi-lunar rhythms. Species which present such rhythms are likely to perceive and utilize changes in tidal cycle occurring regularly at approximately two-week intervals. The apparent adaptive advantage is related with the fact that the fertilized eggs deposited in the supralittoral during high spring tides, remain in this zone until hatching around the next high spring tide. During development the eggs remain protected from aquatic predators and hatching larvae are rapidly dispersed out of the predator rich area at the time of hatching (Gibson, 1992). Examples of this strategy are the puffer *Takifugu niphobles* and the California grunion, *Leuresthes tenuis* (Yamahira, 2004). Other species which have a well-established semi-lunar rhythm and have been observed to lay fertilized eggs during the high spring tides are the mummichog *Fundulus hetelocritus* and the gulf killifish *F. grandis* (Hsiao and Meier, 1989). The Asian sea bass, *Lates calcarifer* (Bloch), however, did present semi-lunar spawning rhythms, but the spawn occurred at low tides during the neat tides period, coinciding with the quarter moons. During this period, maximum diameter of intra-ovarian, ripe oocytes was observed, while smaller oocytes were sampled during the new and full moon periods (Garcia, 1992). This species revealed a different kind of synchronization to the spring-neap cycle, probably with a different adaptive advantage.

TIME OF DAY

Is the time of day important for spawning? During the reproductive season and the specific lunar phase for a given species, fish may not continually release gametes during all day. Since fish synchronize reproduction to environmental factors in order to select the moment of the year and the moment of the lunar cycle, when best environmental conditions to the offspring can be obtained, it is reasonable to think that the best time of the day to lay their eggs. Furthermore, many fish species present well-known daily rhythms of behavior, such as feeding (Madrid *et al.*, 2001), what reinforces the concept of daily rhythmicity in most fish activities, including reproduction rhythms. Indeed, recent studies suggest the existence of daily reproduction rhythms in the gilthead sea bream, European sea bass, Senegal sole and zebrafish.

In the case of the gilthead sea bream, fish exposed to an artificial photoperiod of 9L:15D showed a daily spawning rhythm with spawning beginning in the afternoon and the acrophase around 18:00 h, during both winter and spring. Nevertheless, in fish exposed to a natural photoperiod of 12L:12D in spring, the acrophase of the rhythm was recorded later, at 21:28 h, in accordance with the latest beginning of the dark phase. The daily spawning rhythm in this species was capable to resynchronize when the LD cycle (9L:15D) was shifted by 12 h, resuming a stable phase-relationship after 4–5 transient days, which is characteristic of a endogenous circadian rhythm (Fig. 9.4) (Meseguer *et al.*, 2008). Furthermore, in the same species, the plasma levels of maturational GtH-II have also been observed to fluctuate throughout the day, reaching a peak at 8 h before spawning. This surge was accompanied by an increase in the plasma levels of MIS and E_2, as well as highest levels of GnRH-encoding mRNAs. Most of the morphological and hormonal changes, as well as shifts in gene expression, occurred daily 12-4 h before spawning (Gothilf *et al.*, 1997). In the European sea bass, *Dicentrarchus labrax*, although this species is thought to spawn during the night there are no studies evidencing a daily spawning rhythm; nevertheless daily variations in pituitary sbGnRH content, pituitary and plasma LH, and plasma T concentrations were observed. Pituitary sbGnRH profile

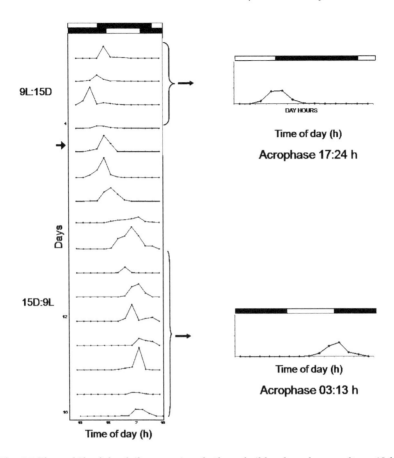

Fig. 9.4 Phase shift of the daily spawning rhythm of gilthead sea bream after a 12 h shift in the LD cycle. The black arrow shown on the left of the actogram, on day 5 indicates the shift of the LD cycle. The graphs on the right represent the average waveform from days 1 to 4 (top graph) and days 9 to 16 (bottom graph), inclusively (Meseguer *et al.*, 2008).

showed the lowest levels during the dark phase, while plasma LH peaked during the same period and T showed the highest levels around sunrise (Bayarri *et al.*, 2004c).

In the case of Senegal sole, the daily spawning rhythm observed appeared to be similar during both spawning seasons, and likely to be synchronized with the beginning of the night. During spring, spawning started around 21:00 h while in autumn it started earlier, around 19:00 h, but in this season photoperiod is shorter and sunset occurs also earlier. This fact suggested that the beginning of spawning was synchronized with the beginning of the night. Actually, in both spawning periods, the acrophase was very similar (23:07 h in spring and 23:25 h in autumn), which reinforces the accuracy of the rhythm, independently of the season. When the

Table 9.1. Summary of lunar and daily related reproductive activities.

Species	Lunar rhythms	Daily rhythms	Reference
Siganus guttatus	Spawning at the first quarter moon Peaks of GSI and plasma levels of sex steroids and VTG prior to spawning		Rahman *et al.* (2000a,b)
S. doliatus	Spawning at the first quarter moon coinciding with a peak in GSI		Park *et al.* (2006b)
S. spinus	Spawning at full moon coinciding with higher GSI sperm motility and quality		Harahap *et al.* (2001, 2002) Park *et al.* (2006a)
S. canaliculatus	Spawning at full moon coinciding with a peak in GSI Plasma VTG peak during pre-spawning		Hoque *et al.* (1999)
S. argenteus	Spawning at last quarter moon coinciding with higher sperm motility and quality		Rahman *et al.* (2003a,b,c) Park *et al.* (2006a)
Epinephelus guttatus and *E. striatus*	Aggregation and Spawning at full moon	Spawning occurs shortly before sunset	Colin *et al.* (1987)
E. merra	Spawning at full moon coinciding with higher GSI		Lee *et al.* (2002)
Plectropomus leopardus	Aggregation and Spawning at new moon	Spawning confined to a 22-min period on sunset	Samoilys and Squire (1994)
Neolamprologus morii	Spawning at full moon		Rossiter (1991)
Lepidiolamprologus profundicola	Spawning at first quarter moon		Watanabe (2000a)

Species	Description	Reference
Cyprichromis leptosoma	Spawning at first quarter moon synchronized with the *L. profundicola* lunar spawning rhythm	Watanabe (2000a,b)
Sparus aurata	Major incidence of spawning during full moon	Saavedra and Pousão-Ferreira (2006)
	Spawning in the afternoon plasma levels GtH-II, MIS and E_2 peaked 8 h before spawning	Meseguer et al. (2008) Gothilf et al. (1997)
Solea senegalensis	Spawning around last quarter and new moon	
	Spawning during the first half of the night, the beginning synchronized with the sunset.	Oliveira et al. (2009a)
Dicentrarchus labrax	Higher levels of sex steroids during full moon	Oliveira et al. (2009b) and Oliveira et al. (2009c)
	Higher levels of sexsteroids E_2 and T in the evening	
	Daily rhythms of pituitary sb GnRH and LH and T	Bayarri et al. (2004)
Takifugu niphobles	Semi-lunar spawning rhythm, with eggs laid during the high spring tides	
	Daily spawning rhythm synchronizedwith both tidal (high tides) and light–dark cycle (sunset).	Yamahira (2004)
Fundulus hetelocritus and *F. grandis*	Semi-lunar spawning rhythm, with eggs laid during the high spring tides	Hsiao et al. (1989)
Lates calcarifer	Semi-lunar spawning rhythm, with eggs laid during the low neap tides (quarter moons)	
	Spawning from the evening until dawn (from 19.00 to 05.00 h).	Garcia (1992)
Pleuronectes platessa	90% of the spawning occurred during the night	Nichols (1989)
Lutjanus campechanus	Peak of spawning at 16:00 h and ovulation 5 hours after complete the hydration of the oocyte	Jackson et al. (2006)

Table 9.1 contd.....

Table 9.1 contd....

Species	Lunar rhythms	Daily rhythms	Reference
Pagrus major		Plasma E_2 peak at 07:00 h, maturity at 10:00 h, ovulation at 13:00 h and spawning between 18:00 h and 19:00 h	Matsuyama *et al.* (1988)
Pseudolabrus japonicus		Diurnal oocyte maturation cycle: ovulation and spawning between 6:00 and 9:00 h	Matsuyama *et al.* (1998)
Halicoeres trimaculatus		Tidal spawning rhythms, with eggs being laid during or after high tide	Takemura *et al.* (2008)
Tharassoma duperrey		Tidal spawning rhythms, with eggs being laid around high tide	Ross (1983)
T. lucasanum		Tidal spawning rhythms, with eggs being laid around high tide	Warner (1982)
Labroides dimidiatus	Spawning at high tide, when between noon and evening (coinciding with quarter to new or full moon). Spawning during late afternoon independent of the state of the tide at cycle.the time of day, during the rest of the lunar		Sakai and Kohda (2001)

photoperiod was artificially extended by 4 hours the daily spawning rhythm was also capable to resynchronize, delaying the beginning of spawning, as well as the peak, while re-entraining the rhythm to the new LD cycle. This resynchronization to the new LD cycle, and especially to the new "sunset", suggested the role of the onset of darkness as a trigger to spawning in Senegal sole. Furthermore, when sole were maintained under conditions of constant light for two days, the spawning rhythm persisted, suggesting the existence of an endogenous timing mechanism controlling the circadian spawning time in sole (Oliveira *et al.*, 2009a). In accordance with the daily spawning rhythm observed in this species is the fact that also a daily rhythm of sex steroids has been observed in females during spring, with an acrophase at 20:00 h to E_2 and at 21:00 h to T (Oliveira *et al.*, 2009c), both coinciding with the beginning of the night, and preceding spawning.

The preference of Senegal sole for spawning during the night is in accordance with the fact that this species showed clear nocturnal activity rhythms (Bayarri *et al.*, 2004b). In fact, in gilthead sea bream as in sole, the daily spawning rhythm overlapped with the behavioural rhythms, since this species showed maximum feeding activity in the afternoon, during winter (Velázquez *et al.*, 2004). The same phenomenon was observed in zebrafish, *Danio rerio,* as this species showed a spawning peak in the morning, when the locomotor activity was also high (unpublished observations). This suggests that in each species, reproduction and behavioural rhythms (locomotor and feeding) are strongly related with each other.

Field observations have also pointed to the existence of daily spawning rhythms in several fish species. For instance, in several coral reef fishes, the existence of a daily periodicity in the spawning pattern has been described (Sancho *et al.*, 2000) and in the case of plaice (*Pleuronectes platessa* L.), plankton samples collected at different times of the day revealed that approximately 90% of the spawning in this species occurred during the night (Nichols, 1989). In the case of the coral trout, *Plectropomus leopardus,* courtship displays occurred at all times of the day. However, towards dusk small numbers of males established territories in which they courted and spawned with females. Trout spawned in pairs, exhibiting a rapid rush towards the surface presumably to release gametes, and spawning was only observed in a 22-min period on sunset (Samoilys *et al.*, 1994). In both groupers *Epinephelus guttatus* and *E. striatus* males patrolled territories containing several females and spawnings was observed low in the water column shortly before sunset (Colin *et al.*, 1987).

Fish gonads have also been seen to undergo daily maturation cycles, as in the red snapper (*Lutjanus campechanus*) in which spawning peaked at 16:00 h and ovulation 5 hours after complete hydration of the oocyte (Jackson *et al.*, 2006). In *Pagrus major*, spawning took place between 18:00 and 19:00 h, while the plasma levels of E_2 in females peaked at 07:00 h, maturity being reached at about 10:00 h and ovulation beginning at 13:00 h (Matsuyama *et al.*, 1988). A Japanese team (Kobayashi *et al.*, 1998) studied the *in vitro* response of oocytes to different concentrations of steroids and GtH at different times of the day in the kisu (*Sillago japonica*). The results indicated a daily rhythm in oocyte development similar to that observed in the bambooleaf *Pseudolabrus japonicus* (Matsuyama *et al.*, 1998), which ovulated and spawned between 6:00 and 9:00 h. These evidences further support the idea that fish in general possess a daily rhythm of gonadal maturation.

Besides the daily synchronization of fish to the light/dark cycles, in some species reproduction can also be synchronized to the daily tidal cycle. In the threespot wrasse, *Halicoeres trimaculatus*, collection of this species at a fixed time of day (09:00 h) revealed that ovarian histology sequentially changed according to the tidal cycle, suggesting that spawning occurred during or after the high tide. Authors noticed that tidal-related spawning was considerable in the morning and that most ovaries collected on the afternoon high tide exhibited post-spawning features (Takemura *et al.*, 2008). Two tropical wrasses, the *Tharassoma duperrey* and the *T. lucasanum* have been observed to spawn synchronously daily around the time of high tide, with little or no spawning occurring the days when the time of high tide was close to sunset (Warner, 1982; Ross, 1983; Hoffman and Grau, 1989).

It is hypothesized that synchronous spawning at the daytime high tide is an advantageous anti-predator tactic because it maximizes the distance between newly spawned eggs and the reef planktivore (Ross, 1983). In the case of the puffer *Takifugu niphobles* both the tidal cycle and the light–dark cycle influenced the timing of a day's spawning. However, the selection for synchronization with the tidal cycle is stronger than that for synchronization with the light–dark cycle (Yamahira, 2004). Another species, the bluestreak cleaner wrasse, *Labroides dimidiatus,* still presented a different strategy: when the high tide occurred between noon and evening (quarter moon to new or full moon), they spawned around the time of the tide providing fast offshore currents. For the rest of the lunar cycle, spawnings occurred during late afternoon independent of the state of the tide at the time of day (Sakai and Kohda, 2001). The onset of spawnings in the Asian sea bass, *Lates calcarifer* occurred during low tides in the evening until dawn (from 19.00 to 05.00 hours). Hormone-induced animals spawned any day of the semi-lunar cycle, but at the same time of the day, what suggested a tidal and diurnal cue entraining spawning of mature female sea bass with a possible prevalence of the daily rhythm on the lunar rhythm (Garcia, 1992).

CONCLUSIONS

As it has been discussed, each of the different environmental cycles play a distinct role in the synchronization of reproduction rhythms in fishes, and among species the synchronizing strategies vary greatly. Although just in a few species there studies pointing out evidences of synchronous reproduction rhythms with annual, lunar and daily environmental cycles, the evidences leads us to think that in many other the same pattern could be repeated. In the natural habitats all the environmental changes related with geophysical cycles are constantly present and influencing biological rhythms. To depict the different levels of influence of these environmental cyclical changes on the reproduction rhythms, we are going to use the Senegal sole, due to the availability of studies (Fig. 9.5). Briefly, Senegal sole reproduces once a year around spring, with spawning being located between April, May and June. Within this period, this species does not lay its eggs equally at any time of the lunar cycle; the darkest nights of the last quarter

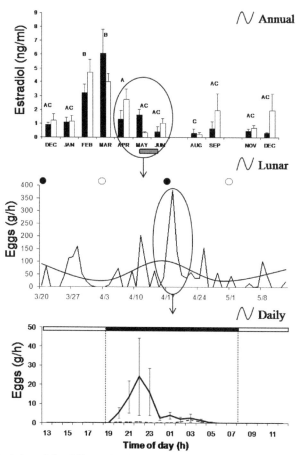

Fig. 9.5 Depiction of the different levels of the environmental influence on the reproduction rhythms in sole: annual rhythms of estradiol and spawning, lunar spawning rhythm and daily spawning rhythm.

and new moon are the selected for spawning. Finally, during this moon phase, Senegal sole selected the night to spawn.

To conclude, the results presented here strongly suggest that fish rely on the pineal, through hormonal (melatonin) and neural outputs, to transduce the cyclic changes in the environment to synchronize their reproduction rhythms, and that synchronizers with different periodicity (seasonal, lunar and daily) interact with each other to finely time reproduction in fish.

Acknowledgements

This work was supported by the Spanish Ministry of Science and Innovation by projects *"Circasole"* with ref AGL2007-66507-C02-02 and AQUAGENOMICS 28502 (Consolider-Ingenio Program), and by the Seneca Foundation by project *"Circabass"* 05690/PI/07.

References

Aguilar-Perera, A. and W. Aguilar-Dávila. 1996. A spawning aggregation of Nassau grouper *Epinephelus striatus* (Pisces: Serranidae) in the Mexican Caribbean. *Environmental Biology of Fishes* 45: 351–361.

Agulleiro, M.J., V. Anguis, J.P. Cañavate, G. Martínez-Rodríguez, C.C. Mylonas and J. Cerdá. 2006. Induction of spawning of captive reared Senegal sole (*Solea senegalensis*) using different administration methods for gonadotropin-releasing hormone agonist. *Aquaculture* 257: 511–524.

Ali, M.A. 1992. *Rhythms in Fishes*. Plenum Press, New York.

Amano, M., M. Iigo, K. Ikuta, S. Kitamura, H. Yamada and K. Yamamori. 2000. Roles of melatonin in gonadal maturation of underyearling precocious male masu salmon. *General and Comparative Endocrinology* 120: 190–197.

Anguis, V. and J.P. Cañavate. 2005. Spawning of captive Senegal sole (*Solea senegalensis*) under a naturally fluctuating temperature regime. *Aquaculture* 243: 133–145.

Aschoff, J. 1981. *Handbook of Behavioural Neurobiology*, Vol. 4: *Biological Rhythms*. Plenum Press, New York.

Asturiano, J.F., L.A. Sorbera, J. Ramos, D.E. Kime, M. Carrillo and S. Zanuy. 2000. Hormonal regulation of the European sea bass reproductive cycle: an individualized female approach. *Journal of Fish Biology* 56: 1155–1172.

Bayarri, M.J., J.A. Madrid and F.J. Sánchez-Vázquez. 2002. Influence of light intensity, spectrum and orientation on sea bass plasma and ocular melatonin. *Journal of Pineal Research* 32: 34–40.

Bayarri, M.J., R. García-Allegue, J.F. López-Olmeda, J.A. Madrid and F.J. Sánchez-Vázquez. 2004a. Circadian melatonin release *in vitro* by European sea bass pineal. *Fish Physiology and Biochemistry* 30: 87–89.

Bayarri, M.J., J.A. Muñoz-Cueto, J.F. López-Olmeda, L.M. Vera, M.A. Rol de Lama, J.A. Madrid and F.J. Sánchez-Vázquez. 2004b. Daily locomotor activity and melatonin rhythms in Senegal sole (*Solea senegalensis*). *Physiology and Behaviour* 81: 577–583.

Bayarri, M.J., L. Rodríguez, S. Zanuy, J.A. Madrid, F.J. Sánchez-Vázquez, H. Kagawa, K. Okuzawa and M. Carrillo. 2004c. Effect of photoperiod manipulation on the daily rhythms of melatonin and reproductive hormones in caged European sea bass (*Dicentrarchus labrax*). *General and Comparative Endocrinology* 136: 72–81.

Bolliet, V., M.A. Ali, F.J. Lapointe and J. Falcón. 1996. Rhythmic melatonin secretion in different teleost species: an *in vitro* study. *Journal of Comparative Physiology* B165: 677–683.

Bolliet, V., V. Bégay, J.P. Ravault, M.A. Ali, J.P. Collin and J. Falcón. 1994. Multiple circadian oscillators in the photosensitive pike pineal gland: A study using organ and cell culture. *Journal of Pineal Research* 16: 77–84.

Bromage, N., M. Porter and C. Randall. 2001. The environmental regulation of maturation in farmed finfish with special reference to the role of photoperiod and melatonin. *Aquaculture* 197: 63–98.

Cabral, H. 2000. Distribution and abundance patterns of flatfishes in the Sado estuary, Portugal. *Estuaries and Coasts* 23: 351–358.

Cabral, H. and M.J. Costa. 1999. Differential use of nursery areas within the Tagus Estuary by sympatric soles, *Solea solea* and *Solea senegalensis*. *Environmental Biology of Fishes* 56: 389–397.

Cahill, G.M. 1996. Circadian regulation of melatonin production in cultured zebrafish pineal and retina. *Brain Research* 708: 177–181.

Carr, A.-J., T.K. Tamai, L. Young, V. Ferrer, M. Dekens and D. Whitmore. 2006. Light reaches the very heart of the zebrafish Clock. *Chronobiology International* 23: 91–100.

Cerdá-Reverter, J.M., S. Zanuy, M. Carrillo and J.A. Madrid. 1998. Time-course studies on plasma glucose, insulin, and cortisol in sea bass (*Dicentrarchus labrax*) held under different photoperiodic regimes. *Physiology & Behavior* 64: 245–250.

Clark, D.S. and J.M. Green. 1990. Activity and movement patterns of juvenile Atlantic cod, *Gadus morhua*, in Conception Bay, Newfoundland, as determined by sonic telemetry. *Canadian Journal of Zoology* 68: 1434–1442.

Clark, R.W., A. Henderson-Arzapalo and C.V. Sullivan. 2005. Disparate effects of constant and annually-cycling daylength and water temperature on reproductive maturation of striped bass (*Morone saxatilis*). *Aquaculture* 249: 497–513.

Colin, P.L., D.Y. Shapiro and D. Weiler. 1987. Aspects of the reproduction of two groupers, *Epinephelus guttatus* and *E. striatus* in the West Indies. *Bulletin of Marine Science* 40: 220–230.

Collin, J.P., P. Brisson, J. Falcón, J. Guerlotté and J.P. Faure. 1986. Multiple cell types in the pineal: functional aspects. In: *Pineal and Retinal Relationships*, P.J. O'Brien and D.C. Klein (eds), Academic Press, Orlando, pp. 15–32.

Collin, J.P., P. Voisin, J. Falcón, P. Brisson and J.R. Defaye. 1989. Pineal transducers in the course of evolution: Molecular organization, rhythmic metabolic activity and role. *Archives of Histology and Cytology* 52: 441–449.

Cook, M.F. and E.P. Bergersen. 1988. Movements, habitat selection, and activity periods of northern pike in eleven mile reservoir, Colorado. *Transactions of the American Fisheries Society* 117: 495–502.

Dahle, R., G.L. Taranger, Ï. Karlsen, O.S. Kjesbu and B. Norberg. 2003. Gonadal development and associated changes in liver size and sex steroids during the reproductive cycle of captive male and female Atlantic cod (*Gadus morhua* L.). *Comparative Biochemistry and Physiology—Part A* 136: 641–653.

Duston, J. and N. Bromage. 1991. Circannual rhythms of gonadal maturation in female rainbow trout (*Oncorhynchus mykiss*). *Journal of Biological Rhythms* 6: 49–53.

Falcón, J. 1999. Cellular circadian clocks in the pineal. *Progress in Neurobiology* 58: 121–162.

Falcón, J., L. Besseau, S. Sauzet and G. Boef. 2007. Melatonin effects on the hypothalamo-pituitary axis in fish. *Trends in Endocrinology and Metabolism* 18: 81–88.

Falcón, J., C. Thibault, V. Bégay, A. Zachmann and J.P. Collin. 1992. Regulation of the rhythmic melatonin secretion by fish pineal photoreceptor cells. In: *Rhythms in Fishes*, M.A. Ali (ed), Plenum Press, New York, pp. 167–198.

Falcón, J., V. Bolliet, J.P. Ravault, D. Chesneau, M.A. Ali and J.P. Collin. 1994. Rhythmic secretion of melatonin by the superfused pike pineal organ: thermo- and photoperiod interaction. *Neuroendocrinology* 60: 535–543.

Garcia, L.Ma.B. 1992. Lunar synchronization of spawning in sea bass, *Lates calcarifer* (Bloch): effect of luteinizing hormone-releasing hormone analogue (LHRHa) treatment. *Journal of Fish Biology* 40: 359–370.

García-Allegue, R., J.A. Madrid and F.J. Sánchez-Vázquez. 2001. Melatonin rhythms in European sea bass plasma and eye: influence of seasonal photoperiod and water temperature. *Journal of Pineal Research* 31: 68–75.

García-López, A., V. Anguis, E. Couto, A.V.M. Canario, J.P. Cañavate, C. Sarasquete and G. Martínez-Rodríguez. 2006a. Non-invasive assessment of reproductive status and cycle of sex steroid levels in a captive wild broodstock of Senegalese sole *Solea senegalensis* (Kaup). *Aquaculture* 254: 583–593.

García-López, A., E. Pascual, C. Sarasquete and G. Martínez-Rodríguez. 2006b. Disruption of gonadal maturation in cultured Senegalese sole *Solea senegalensis* Kaup by continuous light and/or constant temperature regimes. *Aquaculture* 261: 789–798.

Gern, W., W.W. Dickhoff and L.C. Folmar. 1984. Increases in plasma melatonin titers accompanying seawater adaptation of coho salmon (*Oncorhynchus kisutch*). *General and Comparative Endocrinology* 55: 458–462.

Gibson, R.N. 1992. Tidally-synchronized behaviour in marine fishes. In: *Rhythms in Fishes*, M.A. Ali (ed.), Plenum Press, New York, pp. 63–81.

Gothilf, Y., I. Meiri, A. Elizur and Y. Zohar. 1997. Preovulatory changes in the levels of three gonadotropin-releasing hormone-encoding messenger ribonucleic acids (mRNAs), gonadotropin beta-subunit mRNAs, plasma gonadotropin, and steroids in the female gilthead seabream, *Sparus aurata*. *Biology of Reproduction* 57: 1145–1154.

Guzmán, J.M., B. Norberg, J. Ramos, C.C. Mylonas and E. Mañanós. 2008. Vitellogenin, steroid plasma levels and spawning performance of cultured female Senegalese sole (*Solea senegalensis*). *General and Comparative Endocrinology* 156: 285–297.

Hansen, L.P., N. Jonsson and B. Jonsson. 1993. Oceanic migration in homing Atlantic salmon. *Animal Behaviour* 45: 927–941.

Hansen, T., Ï. Karlsen, G.L. Taranger, G.I. Hemre, J.C. Holm and O.S. Kjesbu. 2001. Growth, gonadal development and spawning time of Atlantic cod (*Gadus morhua*) reared under different photoperiods. *Aquaculture* 203: 51–67.

Harahap, A.P., A. Takemura, S. Nakamura, M.D.S. Rahman and K. Takano. 2001. Histological evidence of lunar-synchronized ovarian development and spawning in the spiny rabbitfish *Siganus spinus* (Linnaeus) around the Ryukyus. *Fisheries Science* 67: 888–893.

Harahap, A.P., A. Takemura, M.D.S. Rahman, S. Nakamura and K. Takano. 2002. Lunar synchronization of sperm motility in the spiny rabbitfish *Siganus spinus* (Linnaeus). *Fisheries Science* 68: 706–708.

Hardeland, R., S.R. Pandi-Perumal and D.P. Cardinali. 2006. Melatonin. *The International Journal of Biochemistry and Cell Biology* 38: 313–316.

Herrero, M.J., J.A. Madrid and F.J. Sanchez-Vazquez. 2003. Entrainment to light of circadian activity rhythms in tench (*Tinca tinca*). *Chronobiology International* 20: 1001–1017.

Hoffman, K.S. and E.G. Grau. 1989. Daytime changes in oocyte development with relation to the tide for the Hawaiian saddleback wrasse, *Thalassoma duperrey*. *Journal of Fish Biology* 34: 529–546.

Holm, M., J.C. Holst and L.P. Hansen. 2000. Spatial and temporal distribution of post-smolts of Atlantic salmon (*Salmo salar* L.) in the Norwegian Sea and adjacent areas. *ICES Journal of Marine Science: Journal du Conseil* 57: 955–964.

Hoque, M.M., A. Takemura, M. Matsuyama, S. Matsuura and K. Takano. 1999. Lunar spawning in *Siganus canaliculatus*. *Journal of Fish Biology* 55: 1213–1222.

Hsiao, S.-M.and A.H. Meier. 1989. Comparison of semilunar cycles of spawning activity in *Fundulus grandis* and *F. heteroclitus* held under constant laboratory conditions. *Journal of Experimental Zoology* 252: 213–218.

Iigo, M. and K. Aida. 1995. Effects of season, temperature and photoperiod on plasma melatonin rhythms in the goldfish, *Carassius auratus*. *Journal of Pineal Research* 18: 62–68.

Iigo, M., H. Kezuka, K. Aida and I. Hanyu. 1991. Circadian rhythms of melatonin secretion from superfused goldfish (*Carassius auratus*) pineal glands *in vitro*. *General and Comparative Endocrinology* 83: 152–158.

Iigo, M., K. Mizusawa, M. Yokosuka, M. Hara, R. Ohtani-Kaneko, M. Tabata, K. Aida and K. Hirata. 2003. *In vitro* photic entrainment of the circadian rhythm in melatonin release from the pineal organ of a teleost, ayu (*Plecoglossus altivelis*) in flow-through culture. *Brain Research* 982: 131–135.

Iigo, M., Y. Fujimoto, M. Gunji-Suzuki, M. Yokosuka, M. Hara, R. Ohtani-Kaneko, M. Tabata, K. Aida and K. Hirata. 2004. Circadian rhythms of melatonin release from the photoreceptive pineal organ of a teleost, Ayu (*Plecoglossus altivelis*) in flow-thorough culture. *Journal of Neuroendocrinology* 16: 45–51.

Iigo, M., T. Abe, S. Kambayashi, K. Oikawa, T. Masuda, K. Mizusawa, S. Kitamura, T. Azuma, Y. Takagi, K. Aida and T. Yamamoto. 2007. Lack of circadian regulation of *in vitro* melatonin release from the pineal organ of salmonid teleosts. *General and Comparative Endocrinology* 154: 91–97.

Jackson, M.W., D.L. Nieland and J.H. Cowan. 2006. Diel spawning periodicity of red snapper *Lutjanus campechanus* in the northern Gulf of Mexico. *Journal of Fish Biology* 68: 695–706.

Jerez, S., C. Rodríguez, J.R. Cejas, A. Bolaños and A. Lorenzo. 2006. Lipid dynamics and plasma level changes of 17[beta]-estradiol and testosterone during the spawning season of gilthead seabream (*Sparus aurata*) females of different ages. *Comparative Biochemistry and Physiology Part B* 143: 180–189.

Kjesbu, O.S. 1994. Time of start of spawning in Atlantic cod (*Gadus morhua*) females in relation to vitellogenic oocyte diameter, temperature, fish length and condition. *Journal of Fish Biology* 45: 719–735.

Kleszczynska, A., L. Vargas-Chacoff, M. Gozdowska, H. Kalamarz, G. Martínez-Rodríguez, J.M. Mancera and E. Kulczykowska. 2006. Arginine vasotocin, isotocin and melatonin responses following acclimation of gilthead sea bream (*Sparus aurata*) to different environmental salinities. *Comparative Biochemistry and Physiology* A 145: 268–273.

Kobayashi, M., K. Aida, K. Furukawa, Y.K. Law, T. Moriwaki and I. Hanyu. 1998. Development of sensitivity to maturation-inducing steroids in the oocytes of the daily spawning teleost, the kisu *Sillago japonica*. *General and Comparative Endocrinology* 72: 264–271.

Kulczykowska, E. 2002. A review of the multifunctional hormone melatonin and a new hypothesis involving osmoregulation. *Reviews in Fish Biology and Fisheries* 11: 321–330.

Lacroix, G.L., B.J. Galloway, D. Knox and D. MacLatchy. 1997. Absence of seasonal changes in reproductive function of cultured Atlantic salmon migrating into a Canadian river. *Ices Journal of Marine Science* 54: 1086–1091.

Leatherland, J.F., K.J. Farbridge and T. Boujard. 1992. Lunar and semi-lunar rhythms in fishes. In: *Rhythms in Fishes*, M.A. Ali (ed.), Plenum Press, New York, pp. 83–107.

Lee, Y.D., S.H. Park, A. Takemura and K. Takano. 2002. Histological observations of seasonal reproductive and lunar-related spawning cycles in the female honeycomb grouper *Epinephelus merra* in Okinawan waters. *Fisheries Science* 68: 872–877.

López-Olmeda, J.F., C. Oliveira, H. Kalamarz, E. Kulczykowska, M.J. Delgado and F.J. Sánchez-Vázquez. 2009. Effects of water salinity on melatonin levels in plasma and peripheral tissues and on melatonin binding sites in European sea bass (*Dicentrarchus labrax*). *Comparative Biochemistry and Physiology* A 152: 486–490.

Madrid, J.A., T. Boujard and F.J. Sánchez-Vázquez. 2001. Feeding rhythms. In: *Food Intake in Fish*, D. Houlihan, T. Boujard and M. Jobling (eds), Blakwell Science, Oxford, pp. 189–215.

Maitra, S. and A. Chattoraj. 2007. Role of photoperiod and melatonin in the regulation of ovarian functions in Indian carp *Catla catla*: basic information for future application. *Fish Physiology and Biochemistry* 33: 367–382.

Mañanós, E., S. Zanuy and M. Carrillo. 1997. Photoperiodic manipulations of the reproductive cycle of sea bass (*Dicentrarchus labrax*) and their effects on gonadal development, and plasma 17β-estradiol and vitellogenin levels. *Fish Physiology and Biochemistry* 16: 211–222.

Mañanós, E., J. Núñez, S. Zanuy, M. Carrillo and F. Le Menn. 1994. Sea bass (*Dicentrarchus labrax* L.) vitellogenin. II—Validation of an enzyme-linked immunosorbent assay (ELISA). *Comparative Biochemistry and Physiology—Part B* 107: 217–223.

Matsuyama, M., S. Adachi, Y. Nagahama and S. Matsuura. 1988. Diurnal rhythms of oocyte development and plasma steroid hormone levels in the female red sea bream, *Pagrus major*, during the spawning season. *Aquaculture* 73: 357–372.

Matsuyama, M., S. Morita, T. Nasu and M. Kashiwagi. 1998. Daily spawning and development of sensitivity to gonadotropin and maturation-inducing steroid in the oocytes of the bambooleaf wrasse, *Pseudolabrus japonicus*. *Environmental Biology of Fishes* 52: 281–290.

Mayer, I., C. Bornestaf and B. Borg. 1997. Melatonin in non-mammalian vertebrates: Physiological role in reproduction? *Comparative Biochemistry and Physiology* A 118: 515–531.

Meseguer, C., J. Ramos, M.J. Bayarri, C. Oliveira and F.J. Sánchez-Vázquez. 2008. Light synchronization of the daily spawning rhythms of gilthead seabream (*Sparus aurata* L.) kept under different photoperiod and after shifting the LD cycle. *Chronobiology International* 25: 666–679.

Migaud, H., A. Davie, C.C. Martinez Chavez and S. Al-Khamees. 2007. Evidence for differential photic regulation of pineal melatonin synthesis in teleosts. *Journal of Pineal Research* 43: 327–335.

Mosconi, G., O. Carnevali, R. Carletta, M. Nabissi and A.M. Polzonetti-Magni. 1998. Gilthead seabream (*Sparus aurata*) vitellogenin: purification, partial characterization, and validation of an enzyme-linked Immunosorbent Assay (ELISA). *General and Comparative Endocrinology* 110: 252–261.

Mosconi, G., O. Carnevali, H.R. Habibi, R. Sanyal and A.M. Polzonetti-Magni. 2002. Hormonal mechanisms regulating hepatic vitellogenin synthesis in the gilthead seabream, *Sparus aurata*. *AJP—Cell Physiology* 283: C673–C678.

Nichols, J.H. 1989. The diurnal rhythm in spawning of plaice (*Pleuronectes platessa* L.) in the southern North Sea. *Journal du Conseil International pour l'Exploration de la Mer* 45: 277–283.

Nordqvist, O. 1924. Times of entering of the Atlantic salmon (*Salmo salar* L.) in the rivers. *Rapport et Procès-verbaux des Réunions, CIESM* 33: 1–58.

Oliveira, C., A. Ortega, J.F. López-Olmeda, L.M. Vera and F.J. Sánchez-Vázquez. 2007. Inflence of constant light and darkness, light intensity, and light spectrum on plasma melatonin rhythms in senegal sole. *Chronobiology International* 24: 615–627.

Oliveira, C., M.T. Dinis, F. Soares, E. Cabrita, P. Pousão-Ferreira and F.J. Sánchez-Vázquez. 2009a. Lunar and daily spawning rhythms of Senegal sole (*Solea senegalensis*). *Journal of Fish Biology* (In Press).

Oliveira, C., D.J. Neil, P. Pousão-Ferreira and F.J. Sánchez-Vázquez. 2009b. Influence of the lunar cycle on plasma melatonin, vitellogenin and sex steroids rhythms in Senegal sole *Solea senegalensis*. *Aquaculture* (In revision).

Oliveira, C., L.M. Vera, J.F. López-Olmeda, J.M. Guzmán, E. Mañanós, J. Ramos and F.J. Sanchez-Vazquez. 2009c. Monthly day/night changes and seasonal daily rhythms of sex steroids in Senegal sole (*Solea senegalensis*) under natural fluctuating or controlled environmental conditions. *Comparative Biochemistry and Physiology* A 152: 168–175.

Oliveira, C., E.M. Garcia, J.F. López-Olmeda and F.J. Sánchez-Vázquez. 2009d. Daily and circadian melatonin release in vitro by the pineal organ of two nocturnal teleost species: Senegal sole (*Solea senegalensis*) and tench (*Tinca tinca*). *Comparative Biochemistry and Physiology—Part A* (In Press).

Park, Y.-J., A. Takemura and Y.-D. Lee. 2006a. Annual and lunar-synchronized ovarian activity in two rabbitfish species in the Chuuk lagoon, Micronesia. *Fisheries Science* 72: 166–172.

Park, Y.J., A. Takemura and Y.D. Lee. 2006b. Lunar-synchronized reproductive activity in the pencil-streaked rabbitfish *Siganus doliatus* in the Chuuk Lagoon, Micronesia. *Ichthyological Research* 53: 179–181.

Pedersen, T. 1984. Variation of peak spawning of Arcto-Norwegian cod (*Gadus morhua* L.) during the time period 1929–1982 based on indices estimated from fishery statistics. In: *The Propagation of Cod Gadus morhua L. An International Symposium*, E. Dahl, D.S. Danielsen, E. Moksness and P. Solemdal (eds.), Flødevigen Rapportserie, Arendal 14–17 June 1983, pp. 301–316.

Pickering, A.D. and T.G. Pottinger. 1983. Seasonal and diel changes in plasma cortisol levels of the brown trout, *Salmo trutta* L. *General and Comparative Endocrinology* 49: 232–239.

Poeggeler, B., S. Thuermann, A. Dose, M. Schoenke, S. Burkhard and R. Hardeland. 2002. Melatonin's unique radical scavenging properties. Role of its functional substituents as revealed by a comparison with its structural analogs. *Journal of Pineal Research* 33: 20–30.

Prat, F., S. Zanuy, N. Bromage and M. Carrillo. 1999. Effects of constant short and long photoperiod regimes on the spawning performance and sex steroid levels of female and male sea bass. *Journal of Fish Biology* 54: 125–137.

Prat, F., S. Zanuy, M. Carrillo, A. De Mones and A. Fostier. 1990. Seasonal changes in plasma levels of gonadal steroids of sea bass, *Dicentrarchus labrax* L. *General and Comparative Endocrinology* 78: 361–373.

Rahman, M.D.S., A. Takemura and K. Takano. 2000a. Lunar synchronization of testicular development and plasma steroid hormone profiles in the golden rabbitfish. *Journal of Fish Biology* 57: 1065–1074.

Rahman, M., A. Takemura and K. Takano. 2000b. Correlation between plasma steroid hormones and vitellogenin profiles and lunar periodicity in the female golden rabbitfish, *Siganus guttatus* (Bloch). *Comparative Biochemistry and Physiology* B 127: 113–122.

Rahman, M.D.S., A. Takemura and K. Takano. 2001. Lunar synchronization of testicular development and steroidogenesis in rabbitfish. *Comparative Biochemestry and Physiology—Part B* 129: 367–373.

Rahman, M.D.S., A. Takemura and K. Takano. 2002. Lunar synchronization of in vitro steroidogenesis in ovaries of the Golden Rabbitfish *Siganus guttatus* (Bloch). *General and Comparative Endocrinology* 125: 1–8.

Rahman, M.D.S., M. Morita, A. Takemura and K. Takano. 2003a. Hormonal changes in relation to lunar periodicity in the testis of the forktail rabbitfish, *Siganus argenteus*. *General and Comparative Endocrinology* 131: 302–309.

Rahman, M.D.S., A. Takemura, S. Nakamura and K. Takano. 2003b. Rhythmic changes in testicular activiy with lunar cycle in the forktail rabbitfish. *Journal of Fish Biology* 62: 495–499.

Rahman, M.S., A. Takemura, Y.J. Park and K. Takano. 2003c. Lunar cycle in the reproductive activity in the forktail rabbitfish. *Fish Physiology and Biochemistry* 28: 443–444.

Rahman, S., B.H. Kim, A. Takemura, C.B. Park and Y.D. Lee. 2004a. Effects of moonlight exposure on plasma melatonin rhythms in the seagrass rabbitfish, *Siganus canaliculatus*. *Journal of Biological Rhythms* 19: 325–334.

Rahman, S., B.H. Kim, A. Takemura, C.B. Park and Y.D. Lee. 2004b. Influence of light-dark and lunar cycles on the ocular melatonin rhythms in the seagrass rabbitfish, a lunar-synchronized spawner. *Journal of Pineal Research* 37: 122–128.

Randall, C.F., N.R. Bromage, J.E. Thorpe, M.S. Miles and J.S. Muir. 1995. Melatonin rhythms in atlantic salmon (*Salmo salar*) maintained under natural and out-of-phase photoperiods. *General and Comparative Endocrinology* 98: 73–86.

Reiter, R.J. 1993. The melatonin rhythm: both a clock and a calendar. *Experientia* 49: 654–664.

Rensing, L.and P. Ruoff. 2002. Temperature effect on entrainment, phase shifting, and amplitude of circadian clocks and its molecular bases. *Chronobiology International* 19: 807–864.

Ross, R.M. 1983. Annual, semilunar, and diel reproductive rhythms in the Hawaiian labrid *Thalassoma duperrey*. *Marine Biology* 72: 311–318.

Rossiter, A. 1991. Lunar spawning synchroneity in a freshwater fish. *Naturwissenschaften* 78: 182–184.

Saavedra, M. and P. Pousão-Ferreira. 2006. A preliminary study on the effect of lunar cycles on the spawning behaviour of the gilt-head sea bream *Sparus aurata*. *Journal of the Marine Biological Association of the United Kingdom* 86: 899–901.

Sakai, Y. and M. Kohda. 2001. Spawning timing of the cleaner wrasse, *Labroides dimidiatus*, on a warm temperate rocky shore. *Ichthyological Research* 48: 23–30.

Samejima, M., S. Shavali, S. Tamotsu, K. Uchida, Y. Morita and A. Fukuda. 2000. Light-and temperature-dependence of the melatonin secretion rhythm in the pineal organ of the lamprey, *Lampetra japonica*. *Japanese Journal of Physiology* 50: 497–442.

Samoilys, M.A. and L.C. Squire. 1994. Preliminary observations on the spawning behavior of coral trout, *Plectropomus leopardus* (Pisces: Serranidae), on the Great Barrier Reef. *Bulletin of Marine Science* 54: 332–342.

Sánchez-Vázquez, F.J. and J.A. Madrid. 2001. Feeding anticipatory activity in fish. In: *Food Intake in Fish*, D.F. Houlihan, T. Boujard and M. Jobling (eds.), Blackwell Science, Oxford, pp. 216–232.

Sancho, G., A. Solow and P.S. Lobel. 2000. Environmental influences on the diel timing of spawning in coral reef fishes. *Marine Ecology Progress Series* 206: 193–212.

Servili, A., P. Herrera, J.A. Muñoz-Cueto, J.F. López-Olmeda and F.J. Sánchez-Vázquez. 2008. Interaction between gonadotropin-releasing-hormone and melatoninergic systems in the European sea bass. *Comparative Biochemistry and Physiology* A 151: S12.

Silversand, C., S.J. Hyllner and C. Haux. 1993. Isolation, immunochemical detection, and observations of the instability of vitellogenin from four teleosts. *Journal of Experimental Zoology* 267: 587–597.

Sugama, N., J.G. Park, Y.J. Park, Y. Takeuchi, S.J. Kim and A. Takemura. 2008. Moonlight affects nocturnal Period2 transcript levels in the pineal gland of the reef fish *Siganus guttatus*. *Journal of Pineal Research* 45: 133–141.

Sugden, D., M.A.A. Namboodiri, J.L. Weller and D.C. Klein. 1983. Melatonin synthesizing enzymes: serotonin N-acetyl-transferase and hydroxyindole-O-methyltransferase. In: *Methods in Biogenic Amine Research*, S. Parvez, T. Nagatsu, I. Nagatsu and H. Parvez (eds.), Elsevier, Amsterdam, pp. 567–572.

Sundararaj, B., S. Vasal and F. Halberg. 1982. Circannual rhythmic ovarian recrudescence in the catfish. *Advance Bioscience* 41: 319–337.

Takemura, A., M.D.S. Rahman, S. Nakamura, Y. Ju Park and K. Takano. 2004a. Lunar cycles and reproductive activity in reef fishes with particular attention to rabbitfishes. *Fish and Fisheries* 5: 317–328.

Takemura, A., E.S. Susilo, M.D.S. Rahman and M. Morita. 2004b. Perception and possible utilization of moonlight intensity for reproductive activities in lunar-synchronized spawner. *Journal of Experimental Biology* A301: 844–851.

Takemura, A., S. Ueda, N. Hiyakawa and Y. Nikaido. 2006. A direct influence of moonlight intensity on changes in melatonin production by cultured pineal glands on the golden rabbitfish *Siganus guttatus*. *Journal of Pineal Research* 40: 236–241.

Takemura, A., R. Oya, Y. Shibata, Y. Enomoto, M. Uchimura and S. Nakamura. 2008. Role of the tidal cycle in the gonadal development and spawning of the tropical wrasse *Halichoeres trimaculatus*. *Zoological Science* 25: 572–579.

Takemura, A., M.D.S. Rahman and Y.J. Park. 2009. External and internal controls of lunar-related reproductive activities in fish. *Journal of Fish Biology* (In press).

Thorpe, J.E. and L. Stradmeyer. 1995. The Atlantic Salmon. In: *Conservation of Fish and Shellfish Resources*, J.E. Thorpe, G.A.E. Gall, J.E. Lannan, C.E. Nash (eds.), Academic Press, London, pp. 79–114.

Velázquez, M., S. Zamora and F.J. Martínez. 2004. Influence of environmental conditions on demand-feeding behaviour of gilthead seabream (*Sparus aurata*). *Journal of Applied Ichthyology* 20: 536–541.

Vera, L.M., J.F. López-Olmeda, M.J. Bayarri, J.A. Madrid and F.J. Sánchez-Vázquez. 2005. Influence of light intensity on plasma melatonin and locomotor activity rhythms in tench. *Chronobiology International* 22: 67–78.

Vera, L.M., C. Oliveira, J.F. López-Olmeda, J. Ramos, E. Mañanós, J.A. Madrid and F.J. Sánchez-Vázquez. 2007. Seasonal and daily plasma melatonin rhythms and reproduction in Senegal sole kept under natural photoperiod and natural or controlled water temperature. *Journal of Pineal Research* 43: 50–55.

Warner, R.R. 1982. Mating systems, sex change and sexual demography in the rainbow wrasse, *Thalassoma lucasanum*. *Copeia* 3: 653–661.

Watanabe, T. 2000a. The nesting site of a piscivorous cichlid *Lepidiolamprologus profundicola* as a safety zone for juveniles of a zooplanktivorous cichlid *Cyprichromis leptosoma* in Lake Tanganyika. *Environmental Biology of Fishes* 57: 171–177.

Watanabe, T. 2000b. Lunar cyclic spawning of a mouthbrooding cichlid, *Cyprichromis leptosoma*, in Lake Tanganyika. *Ichthyological Research* 47: 307–310.

Yamahira, K. 2004. How do multiple environmental cycles in combination determine reproductive timing in marine organisms? A model and test. *Functional Ecology* 18: 4–15.

Ziv, L. and Y. Gothilf. 2006. *Period2* expression pattern and its role in the development of the pineal circadian Clock in zebrafish. *Chronobiology International* 23: 101–112.

Zohar, Y. and H. Gordin. 1979. Spawning kinetics in the gilthead sea-bream, *Sparus aurata* L. after low doses of human chronic gonadotropin. *Journal of Fish Biology* 15: 665–670.

Zohar, Y., R. Billard and C. Weil. 1984. La reproduction de la daurade (*Sparus aurata*) et du bar (*Dicentrarchus labrax*). In: *Connaissance du Cycle Sexuel et Controle de la Gametogenese et de la ponte*, G. Barnane and R. Billard (eds.), L'Aquaculture de bar et des Sparides, Paris, pp. 3–24.

10

INFLUENCE OF MELATONIN ON HYPOTHALAMO-PITUITARY-OVARIAN AXIS IN CARP; MECHANISM AND ACTION

Włodzimierz Popek,[CA] Ewa Drąg-Kozak and *Ewa Łuszczek-Trojnar*

INTRODUCTION

Maturation and reproduction processes in fish are considerably influenced by environmental factors. Information on the changing light period reaches the body through the pineal gland, mainly because of regular changes in blood melatonin concentration. During the night hours, melatonin synthesis and release send an endocrine signal to the central nervous system proportionately to the dark period (night) (Reiter, 1991a). This is particularly important for seasonally breeding species. At the same time, by acting on the hypothalamic-pituitary-gonadal axis, melatonin synchronizes animals with their reproductive cycles (Reiter, 1991b; Zachmann *et al.*, 1991). Through rhythmic changes in synthesis and secretion, melatonin also plays a very important role in regulating circadian changes in the concentration of other hormones, especially those related to sexual maturation and reproduction.

In fish, like in other vertebrates, melatonin affects pituitary activity (Fenwick, 1970a) and regulates gonadal function (Peter, 1968, Fenwick, 1970b; Urosaki, 1972a; De Vlaming, 1974; Popek *et al.*, 1994a, 1997a,b, 2000, 2006). In teleosts, the role of the pineal gland and melatonin has not been conclusively resolved to date. Melatonin can inhibit or stimulate gonadal

Department of Ichthyobiology and Fisheries, University of Agriculture in Krakow, Krakow, Poland.
[CA]Corresponding author: rzpopek@cyf-kr.edu.pl

maturation depending on many factors such as season of the year, light conditions, and water temperature (Urosaki, 1972b; De Vlaming, 1974). A study with *Clarias batrachus* (Nayak and Singh, 1987) has shown that melatonin inhibits gonadal development by reducing the synthesis of sex steroids. However, the authors did not specify the stage of maturity which the experimental fish were in. In contrast, evidence that melatonin stimulates 17β-estradiol secretion in mature carp females during the prespawning period was obtained by Popek (Popek *et al.*, 1997a). Furthermore, melatonin injections in mature carp females in the middle of the dark phase significantly increase LH levels (Breton et al., 1993). In crucian carp, the absence of the pineal gland (pinealectomy) inhibits or stimulates LH release depending on photoperiod (Hontela and Peter, 1980). The pineal gland of *Notemigonus crysoleucas* can also stimulate or inhibit sexual maturation depending on day length and water temperature (De Vlaming, 1975).

Autoradiographic research conducted in the late 1980s was aimed to localize specific binding sites, namely specific melatonin receptors in cerebral tissue. This research showed that mammalian brain contains only three such sites: suprachiasmatic nucleus, median eminence and posterodorsal part of the fourth cerebral ventricle (Vaneček, 1988). The greatest density of melatonin receptors is located in the median eminence adjacent to the arcuate nucleus (Weaver *et al.*, 1989). In mammals, this site comprises axon terminals from LHRH-producing cells as well as dopamine-containing neurons of the tuberoinfundibular system. At this site, dopamine inhibits the release of LHRH into the pituitary portal circulation (Wuttke *et al.*, 1971), thus controlling LH and FSH secretion from the pituitary (Gallo, 1980a,b; 1981a,b). The localization of melatonin in these sites is ample evidence of its effect on reproduction in mammals. Much more convincing evidence about the role of dopamine in the mammalian hypothalamus was provided by Zisapel and Laudon (1983) and Zisapel *et al.* (1985). Using prepared hypothalamic cells of the rat, they conclusively showed that melatonin inhibits the release of dopamine. What is more, even picomolar concentrations of melatonin block dopamine release from the retina of ox, rabbit and poultry (Cardinali *et al.*, 1979; Dubocovich, 1985).

In fish, there was initially no evidence for the role of melatonin in the hypothalamic dopamine activity and only comprehensive studies showed that a similar mechanism may also exist in this class of vertebrates. It was observed that intracerebral microinjections of melatonin increase fluorescence intensity in aminergic nuclei of the hypothalamus (Popek and Epler, 1999), indicating that melatonin inhibits the secretion of catecholamines in the carp hypothalamus. Another evidence supporting the effect of the pineal gland on the level of hypothalamic catecholamines in carp was the interrelationship between endogenous rhythms of melatonin and catecholamines. These authors showed that blood melatonin level in fish peaks during the night and falls during the day (Popek *et al.*, 1997a). Therefore, the level of hypothalamic biogenic amines should show an inverse tendency. Indeed, studies on the circadian rhythm of catecholamine levels in carp hypothalami have confirmed this idea (Popek *et al.*, 1994b; Popek and Epler, 1999).

The proximity of melatonin receptors and dopamine and LHRH production sites in mammals suggests that dopamine may provide a link between melatonin and hypothalamic LHRH. It was also hypothesized that melatonin together with gonadal steroids (in feedback) may affect seasonal changes in LHRH secretion in the hypothalamus (Maywood *et al.*, 1996). Meanwhile, studies concerning the effect of 17β-estradiol on the pineal gland and melatonin secretion in mammals have shown that 17β-estradiol may directly influence melatonin synthesis and secretion from pineal cells. This may take place through intracellular receptors or indirectly by modulating the activity of the web of synaptic nerves (Yuwiler, 1989; Hernandez *et al.*, 1990).

All the studies discussed above put forward a hypothesis that in the hypothalamic-pituitary system of carp, melatonin most probably plays a role through the hypothalamic aminergic system, mainly through dopamine (as also suggested in mammals). In addition to the hypothalamic factor stimulating the pituitary secretion of LH (GnRH), this catecholamine is the second hypothalamic factor controlling secretory activity of the pituitary in fish (Chang and Peter, 1983). It inhibits the release of LH into the blood by blocking the formation of adenosine monophosphate (cAMP) in pituitary gonadotrophs. It is now known that the inhibitory effect of dopamine often overrides the stimulating effect of GnRH in LH release. This was used to develop a method for artificial stimulation of ovulation in fish, in which dopamine receptor blockers were used on pituitary gonadotrophs in addition to superactive LHRH analogues. Thus, the different pituitary responses to pinealectomy, light manipulation, temperature and melatonin injection, observed in the experiments discussed above, may be due to changes in the activity of the hypothalamic dopaminergic system.

To finally prove this hypothesis, *in vitro* experiments were conducted to determine whether during the spawning period of carp, melatonin will show the ability to inhibit dopaminergic activity in direct contact with the hypothalamus.

Using a kit for dynamic perifusion of tissue (see Fig. 6.2 Chapter 6), hypothalami harvested during summer (spawning period) from mature carp females were placed into perifusion columns. One pituitary gland or one hypothalamus was placed in biogel in each of five perifusion columns (2 ml for pituitary and 4 ml for hypothalami). The perifusion medium, pumped through Teflon tubes, flowed through the columns and was then collected using a fraction collector into Eppendorf tubes.

They were then perifused with mineral medium in separate groups. Three pineal glands as well as medium in which melatonin was dissolved at a concentration of 300 pg/ml, were implanted in each hypothalamus. Perifusion was held at 20°C. During the first hour of perifusion, columns were illuminated with white light at 2000 lx. During the second and third hour, columns were darkened (0 lx). Medium flowing from columns was collected individually at 15-minute intervals. Dopamine concentration in the collected samples was determined using radioenzymatic kits (Catechola). Mean dopamine concentration ranged from 0.4 to 1.6 pmol/ml, with the highest concentration found in the group in which just hypothalamus was perifused. In the group in which pineal glands were added to the hypothalamus the concentration of dopamine decreased but only in samples collected from columns during the darkness period. Continuous perifusion of hypothalami with medium and melatonin caused dopamine concentration in filtrates to be the lowest during the entire experiment (Fig. 10.1).

Earlier studies (Popek *et al.* 1991, 1997 a,b) have demonstrated that the pineal gland and melatonin only affect the hypothalamic-pituitary-gonadal

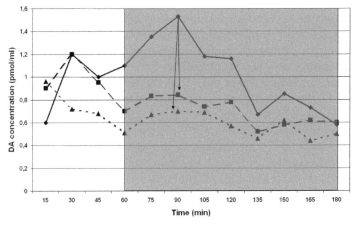

Fig. 10.1 Mean dopamine concentration in medium (summer period) during perifusion of hypothalami harvested from mature carp. Continuous line—hypothalami alone, broken line—hypothalami with implanted pineal glands, dotted line—hypothalami in the presence of melatonin (300 pg/ml). Bright area (0–60 min.) 2000 lx. Dark area (60–180 min.) 0 lx. Arrows indicate significant means (p<0.01) differing from others (90 min of perifusion).

system during the spawning period. To confirm these findings and to show if there is a seasonal rhythm in the dopaminergic activity of the hypothalamus and a seasonal rhythm of hypothalamic sensitivity to melatonin, hypothalami of mature carp females were perifused during the winter period (L:D=8:16) at 5°C using the identical design. This experiment clearly demonstrated that the mean concentration of dopamine in filtrates was low in all the groups (no significant differences between the means) (Fig. 10.2).

The results obtained support the hypothesis about the role of melatonin in hypothalamic control of carp reproduction in summer. During the

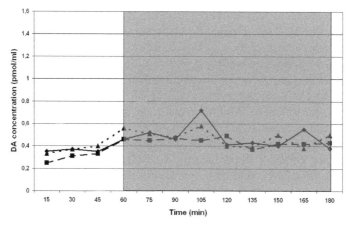

Fig. 10.2 Mean dopamine concentration in medium (winter period) during perifusion of hypothalami harvested from mature carp. Continuous line—hypothalami alone, broken line—hypothalami with implanted pineal glands, dotted line—hypothalami in the presence of melatonin (300 pg/ml). Bright area (0–60 min.) 2000 lx. Dark area (60–180 min.) 0 lx.

spawning season, which is characterized by long day and optimum temperature for carp reproduction, melatonin, by inhibiting the secretion of hypothalamic dopamine can additionally stimulate the gonadotropic activity of the pituitary. During winter, dopaminergic activity of the hypothalamus in mature carp females is low and melatonin has no effect on dopamine synthesis and/or secretion.

These findings and the results of earlier experiments lead us to outline the mechanism of melatonin's action on hypothalamic control of sexual maturation in carp. The pineal gland and melatonin have no significant effect on changes in blood LH concentration nor modify the circadian rhythm and seasonal LH concentration in blood. Likewise, *in vitro* studies showed no direct influence of this hormone on LH secretion from the pituitary. This is probably because no specific melatonin receptors have been found in teleost pituitaries to date. That proximal pituitary cells are not sensitive to melatonin was also confirmed by an *in vitro* study in which pituitaries were perifused in the presence of pineal glands or melatonin (Popek *et al.*, 2000). A considerable amount of their receptors was found in the thalamic region, pretectal area and optic tectum of the rainbow trout (*Oncorhynchus mykiss*) (Mazurais *et al.*, 1999).

Other experiments, which showed that melatonin has no effect on steroidogenesis or carp oocyte maturation *in vitro* (Popek *et al.*, 1996) lead to us to suggest that this hormone has no direct influence on gonadal maturation and oocyte ovulation. Thus, the hypothalamus is one of the possible sites of melatonin action in the hypothalamic-pituitary-gonadal axis.

The carp hypothalamus, which forms a ventral part of the diencephalon, has two main neurohormones that control LH secretion: gonadoliberin

(GnRH), which stimulates LH release from the pituitary (Chang and Peter, 1983; Sokołowska *et al.*, 1985) and catecholamine—dopamine which inhibits the synthesis and secretion of both GnRH (Yu and Peter, 1992) and LH (Peter *et al.*, 1986). The results of these studies indicate that melatonin in carp has an indirect effect on sexual maturation and reproduction through the hypothalamic dopaminergic system. During the spawning period of carp, melatonin has an inhibitory effect on the secretion of hypothalamic dopamine, thus reducing the inhibitory effect of the hypothalamus on GnRH synthesis and LH secretion from the pituitary gland. This enhances the stimulatory effect of hypothalamic GnRH and leads to a considerable increase in blood LH concentration. The modulating action of melatonin in carp shows a clear seasonal activity, which also extends to the activity of hypothalamic dopamine. This activity peaks during the long day and optimum temperature, coinciding with the reproductive period during the sexual cycle of the carp. During winter gonadal regression, the activity of melatonin and hypothalamic DA is low.

It is also worth noting the role of gonadal steroids, which regulate LH secretion depending on the phase of sexual cycle and degree of gonadal maturity. The main steroid hormone of ovaries (17β-estradiol) and testosterone in males relay information between the gonads and the hypothalamus based on the mechanism of positive and negative feedback.

In vitro research on steroidogenesis in many fish species showed that in adult fish the blood level of steroids varies during the annual cycle according to ovarian maturity. In teleosts, testosterone and 17β-estradiol inhibit LH release, most probably through indirect effect on GnRH (Peter *et al.*, 1991). A study which investigated the effect of this hormone on brain dopamine concentration showed that a negative feedback of steroids that affects LH secretion occurs by increasing the inhibitory effect of dopamine on GnRH release (Kah *et al.*, 1992).

During vitellogenesis, 17β-estradiol induces a negative feedback and inhibits LH secretion. In mature fish, both a negative and positive feedback were found to coexist, with possible inhibition or stimulation of LH depending on the stage of the sexual cycle. In the final stage of the sexual cycle, the blood concentration of 17β-estradiol falls before the last stage of oocyte maturation and ovulation, which was reported in many fish species, including salmonids (Kah *et al.*, 1997) and Cyprinidae (Trudeau, 1997). During this time, the estrogen-induced negative feedback that blocks LH secretion is suppressed. Another study reported that the increasing blood level of 17β-estradiol increases the activity of hypothalamic dopamine, leading to inhibition of LH secretion (Saligaut *et al.*, 1998). Giving 17β-estradiol to ovariectomized fish produces a rapid increase in the hypothalamic concentration of dopamine. Further studies with other fish species demonstrated that a decrease in dopamine activity, i.e., a weakening of its inhibitory effect on LH secretion from the pituitary, is paralleled by a decrease in the concentration of 17β-estradiol towards the end of vitellogenesis (Joy *et al.*, 1998; Hernandez-Rauda *et al.*, 1999). Sites with receptors specific to 17β-estradiol were also identified using an autoradiographic method. In crucian carp and in other cyprinid fish, the highest concentration of these receptors was detected in the preoptic area (NPO and NPP), tuberal area (NLT) and ventral part of the hypothalamus (NRL and NRP), where GnRH and dopamine are synthesized (Kim *et al.*, 1978).

In summary, it is concluded that in addition to melatonin, which relays chronobiological information on environmental conditions such as photoperiod and temperature, the mechanism that synchronizes and modulates LH secretion should also include gonadal steroids, which carry information about the current status of gonadal maturity. By acting through the dopaminergic system, these two hormones synchronize the full maturity of fish gonads with optimum spawning time.

ANTIGONADAL AND PROGONADAL ACTION OF MELATONIN IN CARP; THE ROLE OF GONADAL STEROIDS

The succession of day and night affects all live organisms to the same extent, forcing them to follow a strictly regulated lifestyle. The succession of seasons forces fish to know the yearly time scale and to adjust their sexual cycles and reproductive strategies accordingly. This is because melatonin synchronizes physiological processes and behavioural changes in certain seasons of the year, adjusting gonadal development and reproduction time to a period that assures best conditions for their offspring to develop and survive. Research with mammals proved that in immature animals, melatonin inhibits LH secretion, thus protecting the organism from premature puberty. This inhibiting effect of melatonin on sexual maturation occurs at the hypothalamic and pituitary levels (Cassone *et al.*, 1987). In mature animals, melatonin inhibits or stimulates maturation depending on the length of sexual cycles and reproduction date.

As noted above, many studies have shown that melatonin in fish may stimulate or inhibit gonadal development, but the effect of melatonin varies according to species, stage of sexual development and temperature and light conditions (Popek *et al.*, 1991, 2000, 2005, 2006; Popek and Epler, 1999). In fish, the inhibiting effect of melatonin on gonadal development was found in the catfish *Clarias batrachus*, in which melatonin was shown to reduce the synthesis of sex steroids (Nayak and Singh, 1987). A study with mature carp females demonstrated that in the prespawning period melatonin stimulates estradiol synthesis (Popek *et al.*, 1997b). Melatonin injected in mature carp females in the middle of the dark phase significantly increases

LH concentration. It was also shown that melatonin does not directly affect gonads or the pituitary, but probably modulates the activity of the hypothalamic dopaminergic system (Popek *et al.*, 1991, 1994b).

The inhibitory or stimulatory effect of melatonin on the hypothalamic-pituitary-gonadal axis depends largely on the degree of gonadal maturity. To investigate the effect of melatonin on the activity of hypothalamic catecholamines in immature carp females, melatonin (1µg/1µl/1kg) was microinjected into the third cerebroventricle. After 15 minutes, fish were decapitated and hypothalamic dopamine concentration was determined radioimmunologically. Melatonin microinjections caused a rapid and significant (p<0.01) decrease in the concentration of hypothalamic dopamine in relation to the control fish, whose ventricles were injected with physiological saline (Fig. 10.3).

As is evident from Figure 10.3, the presence of melatonin in the hypothalamus caused the concentration of hypothalamic dopamine to decrease almost two-fold in immature fish. Therefore it may be suspected that melatonin increased the secretion of this catecholamine or even partly flushed it from the dopaminergic neurons of the hypothalamus. This entails certain consequences, because dopamine is the main hormone inhibiting LH release into the blood. It is worth noting that similar findings were obtained in immature carp females studied using the radioenzymatic method.

In the experiment in which melatonin was injected into the brain of live fish, we cannot exclude an interaction of other systems, neurohormones or feedbacks in the hypothalamus, the effect of which could not be eliminated.

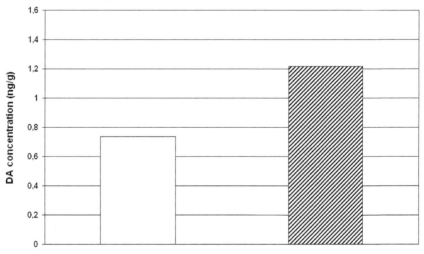

Fig. 10.3 Mean melatonin concentration in the hypothalamus of immature carp females given intraventricular injections. Empty bar—microinjections of melatonin (1µg/1µl/1kg), hatched bar—microinjections of physiological salt. Means are significantly different (p<0.01).

Thus, the observed changes in hypothalamic dopamine concentration could result from the action of other factors. To make sure that melatonin has a direct effect on dopamine secretion in the hypothalamus, parallel *"in vitro"* experiments were conducted outside the body. Prepared hypothalami harvested from immature carp females were placed in perifusion columns through which medium with dissolved melatonin (300pg/ml) flowed.

The results obtained using the *"in vitro"* method have completely confirmed the earlier findings. Similarly to the *"in vivo"* conditions, in the hypothalamus deprived of body control, exogenous melatonin increased dopamine secretion from the hypothalamus. In the experimental group (melatonin medium), it was found that in the samples of medium flowing from perifusion columns, the mean concentration of dopamine during 120 minutes of perifusion was higher than in the control group (Fig. 10.4).

The highest mean concentration of dopamine was found at 30 minutes and was significantly higher in relation to the control group (p<0.01). These results indicate that melatonin injected into the third cerebroventricle accelerated dopamine secretion from the aminergic nuclei of the hypothalamus, whereas dopamine added to the perifusion medium flushed this hormone from hypothalamus fragments. It can therefore be assumed that in immature fish, melatonin accelerates dopamine metabolism in the hypothalamus. These results confirm our earlier experiment in which fluorescent methods were used to investigate circadian and seasonal changes

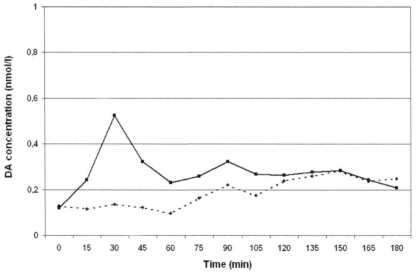

Fig. 10.4 Mean dopamine concentration in medium during perifusion of hypothalami of immature carp females. Continuous line—perifusion in the presence of melatonin, dotted line—pure medium. At 30 min of perifusion, dopamine concentration in medium was significantly (p<0.01) higher than control means.

in the rate of catecholamine metabolism in the hypothalamus of immature and mature carp (Popek *et al.*, 1994b). The results of this study showed that fluorescence intensity in the aminergic nuclei of the hypothalamus is much lower in immature than in mature fish, which is probably due to the high activity of the dopaminergic system. At the same time it is known that the inhibitory activity of hypothalamic dopamine often overrides the stimulating effect of GnRH in LH release from the pituitary gland. In the case of immature fish in whose ovaries oogonia and oocytes undergoing proptoplasmatic growth predominate, melatonin can control their development through dopamine. In addition, in immature fish in which the activity of dopaminergic processes in the hypothalamus is highest (Hernandez-Rauda *et al.*, 1999), melatonin, similarly as in mammals, may slow LH secretion to prevent premature puberty.

In light of this evidence, it can be hypothesized that melatonin is a hormone with antigonadal action in the hypothalamus of immature carp females. Rather than being direct, this action involves stimulating the activity of the hypothalamic dopaminergic system, and the released dopamine is directly involved in the inhibition of LH secretion from the pituitary (Fig. 10.5).

The response of the dopaminergic system is rapid and occurs within 15 to 30 minutes of the appearance of melatonin in the hypothalamus. It should be noted that in immature mammals, the mechanism of melatonin action is similar (Karasek, 1997).

Further experiments to show the mechanism of melatonin action on the hypothalamic-pituitary system were conducted with 3-year-old maturing carp females, whose ovaries began trophoplasmatic growth and oocytes

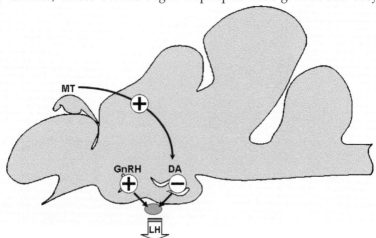

Fig. 10.5 Diagram of probable mechanism of melatonin action on hypothalamic dopaminergic system in immature carp females.
Abbreviations: MT—melatonin, GnRH—gonadotropin-releasing hormone, DA—dopamine, LH—luteinizing hormone. ⊕ sign designates stimulation, ⊖ sign stands for inhibition.

started to store reserve substances when entering the vitellogenesis period. During this process, some material found in yolk (vitellogenin) is synthesized in oocyte plasma. The rest of vitellogenin is produced in liver hepatocytes, from where it is transported to the ovary via blood circulation. There it is incorporated into oocytes through micropinocytosis. This period of oocyte growth and maturation is largely controlled by 17β-estradiol and LH. In pinealectomized fish, yolk-containing oocytes undergo atresion, but administration of 17β-estradiol arrests this process. In both hypophisectomized and control fish, the injections of 17β-estradiol trigger hepatic biosynthesis of proteins intended for yolk and help to incorporate them into the oocyte. What is more, 17β-estradiol only stimulates vitellogenesis when working together with LH.

To identify the possible mechanism of these processes, experiments were conducted to determine if melatonin and 17β-estradiol given to maturing carp females stimulate or inhibit the release of hypothalamic dopamine. As a result of intraventricular microinjections of melatonin, 17β-estradiol, or melatonin and 17β-estradiol the dopamine level was found to decrease within 15 minutes of the appearance of these hormones in the hypothalamus. Thus, both melatonin and 17β-estradiol caused a release of dopamine from the dopaminergic nuclei of the hypothalamus. Joint microinjection of these hormones caused dopamine concentration in the hypothalamus of maturing fish to decrease significantly ($p < 0.01$) almost twofold (Fig. 10.6).

The synergistic action of melatonin and 17β-estradiol caused the strongest reaction of the dopaminergic system, leading to a rapid release of dopamine from the aminergic nuclei of the hypothalamus.

In further research, in which isolated hypothalami were placed into perifusion columns, the strongest reaction was found during perifusion of the hypothalamus with melatonin medium. Dopamine concentration in the filtered medium increased gradually to peak at 90 minutes of the experiment, which was followed by a steady decrease in the concentration of this hormone in medium. Similar though weaker changes in dopamine concentrations in filtrates were observed in the other experimental groups. In the control group (pure medium), dopamine concentration in the filtrate was low throughout perifusion (Fig. 10.7).

The above results indicate that melatonin as well as 17β-estradiol stimulate the release or flushing of dopamine from hypothalamic aminergic nuclei of maturing fish. It should be noted that the response of the hypothalamic dopaminergic system to melatonin is delayed by 60 minutes in relation to the reaction rate in immature fish. It can therefore be assumed that in maturing carp females, the increasing level of 17β-estradiol during vitellogenesis has a stimulating effect on the release of hypothalamic dopamine, which is known to serve as a hypothalamic inhibitor of LH secretion. Most probably, this effect is part of the mechanism of negative

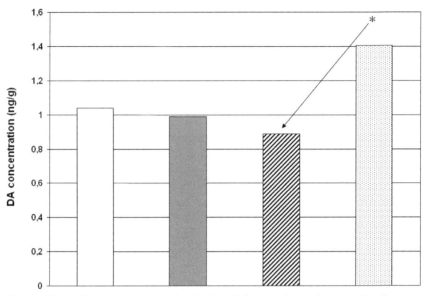

Fig. 10.6 Mean dopamine concentration in hypothalamus of maturing carp females given intraventricular microinjections of melatonin (1µg/1µl/1kg)—empty bar, 17β-estradiol (1µg/1µl/1kg)—grey bar, melatonin and 17β-estradiol—hatched bar, physiological salt—dotted bar. Arrow and asterisk (*) indicate means differing significantly (p<0.01).

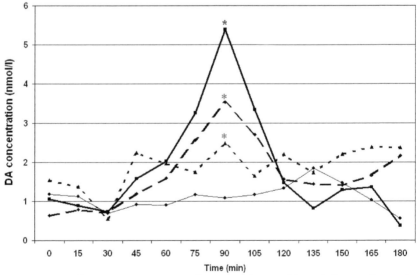

Fig. 10.7 Mean dopamine concentration during perifusion of hypothalamus of maturing carp females. Hypothalami perifused in the presence of melatonin (300 pg/ml)—bold line, in the presence of 17β-estradiol (10 ng/ml)—dotted line, in the presence of melatonin and 17β-estradiol—broken line. Control hypothalami—thin line. Asterisks (*) indicate means differing significantly (p<0.01) from the mean for the control group, at 90 min of perifusion.

feedback between gonadal steroids (17β-estradiol) and the central nervous system. This process occurs both in the hypothalamus and pituitary gland and leads to a decrease in blood LH level.

Similarly to *"in vivo"* conditions, in the hypothalamus of maturing carp females deprived of the body control, the effect of exogenous melatonin was stronger than that of 17β-estradiol. It is interesting to note, however, that the joint microinjections of melatonin and 17β-estradiol into the third cerebroventricle caused the strongest reaction in the form of dopamine release/flushing from hypothalamic cells. Perhaps this outcome resulted from the interaction of other factors in fish brains in addition to hypothalamic factors. It can therefore be hypothesized that in maturing fish, the inhibiting effect of melatonin on LH release is additionally strengthened by the blood concentration of 17β-estradiol, which increases during vitellogenesis, and by the contribution of other extrahypothalamic factors.

In summary, the results obtained indicate a probable mechanism by which the pineal gland and melatonin serve their chronobiological and physiological functions in immature and mature fish. At this developmental phase, ovaries are in maturity stages II and III, which are characterized by ovarian growth and incorporation of reserve materials during vitellogenesis. The activity of gonadal steroids, mainly 17β-estradiol, is rapidly growing. The high level of 17β-estradiol persists throughout vitellogenesis. At this developmental stage of gonads, the role of estrogens is mainly to stimulate vitellogenin production in the liver. The high level of 17β-estradiol in the blood also affects the brain (mainly the hypothalamus) in a negative feedback with LH. LH synthesis and secretion is inhibited by hypothalamic dopamine, which in turn is controlled by melatonin. This was shown by the results of the above experiments, where in maturing carp females whose ovaries were at different stages of vitellogenesis, melatonin caused a rapid release of this catecholamine from dopaminergic nuclei of the hypothalamus (Fig. 10.8).

Thus, in the initial period of ovarian growth in carp, the pineal gland and its hormone melatonin stimulate dopamine secretion through dopaminergic nuclei of the hypothalamus. The process, which also occurs during later stages of maturation, inhibits LH synthesis and secretion. This delays the maturation of ovaries, and melatonin protects the organism from premature puberty. The mechanism is supported by a negative feedback of 17β-estradiol, whose blood concentration rapidly increases during vitellogenesis. By limiting the growth rate of gonads, this process completely synchronizes sexual maturation with the development of other systems, the somatic development and above all the rhythmically changing environment, which enables the final stages of maturity to be reached at the optimum moment. This especially concerns fish living in medium latitudes, where seasons of the year and thus the reproductive season are strongly influenced by cyclically changing environmental conditions.

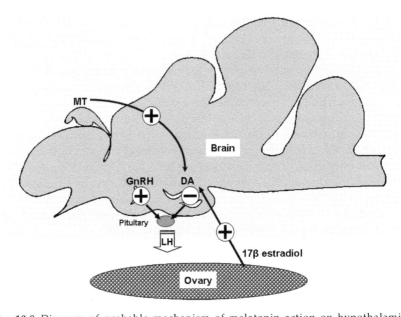

Fig. 10.8 Diagram of probable mechanism of melatonin action on hypothalamic dopaminergic system in maturing carp females.
Abbreviations: MT—melatonin, GnRH—gonadotropin-releasing hormone, DA—dopamine, LH—luteinizing hormone.⊕sign designates stimulation,⊖sign stands for inhibition.

Vitellogenesis is completed when fish enter reproduction. The ovaries of mature carp are characterized by fairly rapid and significant changes inside oocytes such as migration of the nucleus, germinal vesicle breakdown (GVBD), breakdown of the follicular envelope and ovulation. The level of 17β-estradiol falls, and the negative feedback of this steroid with the hypothalamus becomes positive. During this time the whole hormonal system is oriented almost entirely towards stimulating LH secretion. The role of the pineal gland and melatonin during this period is to maintain this tendency and to ensure that these processes are completed at the appropriate time.

All the experiments discussed in previous chapters, based on the histochemical, fluorescent and radioenzymatic study of the hypothalamic dopaminergic activity, melatonin microinjections into the third cerebroventricle and pinealectomy, as well as the *"in vitro"* research (perifusions) were carried out with mature carp females. The results obtained led us to conclude that melatonin stimulates sexual maturation and reproduction through the dopaminergic system of the hypothalamus. During the spawning period of the carp (L:D=16:8, water temperature 20°C), melatonin inhibits the secretion of hypothalamic dopamine, thus reducing its inhibitory effect on GnRH synthesis and LH secretion from the pituitary gland. This increases the stimulatory effect of hypothalamic GnRH and other hypothalamic neurohormones on pituitary gonadotrophs, leading to the ovulatory wave of LH in blood, ovulation of mature oocytes and spawning of fish (Fig. 10.9).

It is notable that both melatonin activity and the hypothalamic dopamine activity in mature carp show clear seasonality. This activity is the highest during the long day and at optimum temperature, when reproduction occurs

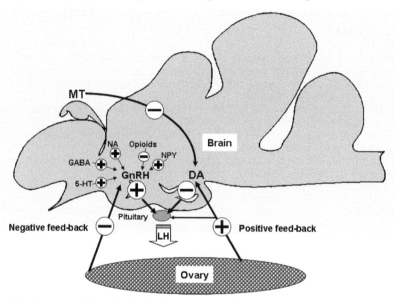

Fig. 10.9 Diagram of hypothalamic-pituitary control of carp reproduction and probable mechanism of melatonin action on hypothalamic dopaminergic system in mature carp females. LH—luteinizing hormone, GABA—gamma-aminobutyric acid, NPY—neuropeptide Y, 5-HT—serotonin, NA—noradrenalin. ⊕ sign designates stimulation,⊖sign stands for inhibition.

in the sexual cycle of carp. In winter, the activity of melatonin and hypothalamic dopamine is low and this system is inactive.

When observing the relationships among melatonin, dopamine, 17β-estradiol and the pituitary of immature and mature fish, it is easy to notice a change in the direction of melatonin action. Initially it stimulates the hypothalamic dopaminergic system by synchronizing the rate of sexual maturation with somatic maturity and season of the year. During this period, melatonin inhibits LH secretion and thus modulates the rate of growth and the duration of vitellogenesis. The change in the direction of 17β-estradiol and LH feedback from negative to positive also alters the way in which melatonin affects the secretion of hypothalamic dopamine. In mature fish during the reproductive period, melatonin inhibits dopamine secretion. In this way it indirectly stimulates GnRH synthesis and LH secretion from the pituitary. The change in the direction of melatonin action probably occurs when the negative feedback of 17β-estradiol with the hypothamalus and pituitary gland is suppressed, this steroid being most probably key to this mechanism. During spawning the pineal gland, which is also sensitive to temperature in this period, indirectly contributes to stimulating LH secretion by synchronizing ovulation with that time of day, night and season which provides optimum environmental parameters for a given species of fish.

References

Breton, B., T. Mikołajczyk and W. Popek. 1993. The neuroendocrine control of the gonadotropin (GtH2) secretion in teleost fish. In: *Aquaculture: Fundamental and Applied Research*. B. Lahlou and P. Vitiello (eds). *American Geophysical Union USA*, pp. 199–215.

Cardinali, D.P., M.I. Vacas and E.E. Boyer. 1979. Specific binding sites of melatonin in bovine brain. *Endocrinology* 105: 437–441.

Cassone, V.M., M.H. Roberts and R.Y. Moore. 1987. Melatonin inhibits metabolic activity in the rat suprachiasmatic nuclei. *Neuroscience Letters* 81: 29–34.

Chang, J.P. and R.E. Peter. 1983. Effects of pimozide and des Gly10,[D-Ala6] luteinizing hormone-releasing hormone ethylamide on serum gonadotropin concentrations, germinal vesicle migration and ovulation in female goldfish, *Carassius auratus*. *General and Comparative Endocrinology* 52: 30–37.

De Vlaming, V.L. 1974. Environmental and endocrine control of teleost reproduction. In: *Control of Sex in Fishes* C.B. Schreck (ed), Virgina Polytechnic Institute and State University, Blacksburg, VA, pp. 13–83.

De Vlaming, V.L. 1975. Effects of pinealectomy on gonadal activity in the cyprinid teleost, *Notemigonus cryoleucas*. *General and Comparative Endocrinology* 26: 36–49.

Drąg-Kozak, E., W. Popek, E. Łuszczek-Trojnar, D. Fortuna-Wrońska and P. Epler. 2004. The influence of melatonin on noradrenaline and adrenaline release in maturing carp females (*Cyprinus carpio* L.). *Scientific Messenger of Lviv National Academy of Veterinary Medicine named after S.Z. Gzhytskyj* 6: 64–70.

Dubocovich, M.L. 1985. Characterization of a retinal melatonin receptor. *The Journal of Pharmacology and Experimental Therapeutics* 234: 395–401.

Fenwick, J.C. 1970a. The pineal organ: Photoperiod and reproductive cycles in the goldfish, *Carassius auratus* L. *Journal of Endocrinology* 46: 101–111.

Fenwick, J.C. 1970b. Demonstration and effect of melatonin in fish. *General and Comparative Endocrinology* 14: 86–97.

Gallo, R.V. 1980a. Effects of manipulation of brain dopaminergic or serotonergic systems on basal pulsatile LH release and perisuprachiasmatic-induced suppression of pulsatile LH release in ovariectomized rats. *Neuroendocrinology* 31: 161–167.

Gallo, R.V. 1980b. Neuroendocrine regulation of pulsatile luteinizing hormone release in the rat. *Neuroendocrinology* 30: 122–131.

Gallo, R.V. 1981a. Pulsatile LH release during the ovulatory LH surge on proestrus in the rat. *Biology of Reproduction* 24: 100–104.

Gallo, R.V. 1981b. Further studies on dopamine-induced suppresion of pulsatile LH release in ovariectomized rats. *Neuroendocrinology* 32: 187–192.

Hernandez, G., P. Abreu, R. Alonso, C. Santana, F. Moujir and C.H. Calzadilla. 1990. Castration reduces the nocturnal rise of pineal melatonin levels in the male rat by impairing its noradrenergic input. *Journal of Neuroendocrinology* 2: 777–782.

Hernandez-Rauda, R., G. Rozas, P. Rey, J. Otero and M. Aldegunde. 1999. Changes in the pituitary metabolism of monoamines (dopamine, norepinephrine, and serotonin) in female and male rainbow trout (*Oncorhynchus mykiss*) during gonadal recrudescence. *Physiological and Biochemical Zoology* 72: 352–359.

Hontela, A. and R.E. Peter. 1980. Effects of pinealectomy, binding and sexual condition on serum gonadotropin levels in the goldfish. *General and Comparative Endocrinology* 40: 168–179.

Joy, K.P., B. Senthilkumaran and C.C. Sudhakumari. 1998. Periovulatory changes in hypothalamic and pituitary monoamines following GnRH analogue treatment in the catfish, *Heteropneustes fossilis*: a study correlating changes in plasma hormone profiles. *Journal of Endocrinology* 156: 365–372.

Kah, O., I. Anglade, B. Linard, F. Pakdel, G. Salbert, T. Bailhache, B. Ducoutet, C. Saligaut, P. Le Goff, Y. Valotaire and P. Jego. 1997. Estrogen receptors in the brain—pituitary complex and the neuroendocrine control of gonadotropin release in rainbow trout. *Fish Physiology and Biochemistry* 17: 53–62.

Kah, O., V.L. Trudeau, B.D. Sloley, J.P. Chang, P. Dubourg, K.L. Yu and R.E. Peter. 1992. Influence of GABA on gonadotropin release in the goldfish. *Neuroendocrinology* 55: 396–404.

Karasek, M. 1997. Pineal and melatonin. 1997. *Polskie Wydawnictwo Naukowe, Warszawa* (In Polish).

Kim, Y.S., W.E. Stumpf and M. Sar. 1978. Topography of estradiol target cells in the forebrain of goldfish, *Carassius auratus*. *The Journal of Comparative Neurology* 182: 611–620.

Maywood, E.S., E.L. Bittman and M.H. Hastings. 1996. Lesions of the melatonin- and androgen-responsive tissue of the dorsomedial nucleus of the hypothalamus block the gonadal response of male Syrian hamsters to programmed infusions of melatonin. *Biological of Reproduction* 54: 470–477.

Mazurais, D., I. Brierley, I. Anglade, J. Drew, C. Randall, N. Bromage, D. Michel, O. Kah, and L.M. Williams. 1999. Central melatonin receptors in rainbow trout: Comparative distribution of ligand binding and gene expression. *Journal of Comparative Neurology* 409: 313–324.

Nayak, P.K. and T.P. Singh. 1987. Effect of melatonin and 5-methoxytryptamine on the sex steroids and thyroid hormones during the prespawning phase of the annual reproductive cycle in the freshwater teleost, *Clarias batrachus*. *Journal of Pineal Research* 4: 377–386.

Peter, R.E. 1968. Failure to detect an effect of pinealectomy in goldfish. *General and Comparative Endocrinology* 10: 443–449.

Peter, R.E., V.L. Trudeau and B. D. Sloley. 1991. Brain regulation of reproduction in teleosts. *Bulletin of the Institute of Zoology, Academia Sinica, Monograph* 16: 89-118.

Peter, R.E., J. Chang, C.S. Nahorniak, R.J. Omeljaniuk, M. Sokołowska, S.H. Shih and R. Billard. 1986. Interactions of catecholamines and GnRH in regulation of gonadotropin secretion in teleost fish. *Recent Progress in Hormone Research* 42: 513–548.

Popek, W. and P. Epler. 1999. Effects of intraventricular melatonin microinjections on hypothalamic catecholamine activity in carp females during a year. *Electronic Journal of Polish Agricultural Universities* (www.ejpau.media.pl), 2(1) *Fisheries.*

Popek, W., K. Bieniarz and P. Epler. 1991. Role of the pineal gland in sexual cycle in common carp. In: *Chronobiology and Chronomedicine*, J. Surowiak and M.H. Lewandowski (eds). *Verlag Peter Lang, Frankfurt*, pp. 99–102.

Popek, W., P. Epler and M. Sokołowska-Mikołajczyk. 1996. Melatonin does not affect steroidogenesis or maturation of carp oocytes *in vitro* during the prespawning period. *Polish Archives of Hydrobiology* 43: 379–385.

Popek, W., E. Łuszczek-Trojnar and P. Epler. 2000. Effects of pineal gland and melatonin on maturation gonadotropin (GtH2) secretion from perifused pituitary glands of mature carp during spawning. *Electronic Journal of Polish Agricultural Universities* (www.ejpau.media.pl), 3(1) *Fisheries.*

Popek, W., P. Epler, K. Bieniarz and M. Sokołowska-Mikołajczyk. 1997a. Contribution of factors regulating melatonin release from pineal gland of carp (*Cyprinus carpio* L.) in normal and in polluted environments. *Archives of Polish Fisheries* 5: 59–75.

Popek, W., J. Galas and P. Epler. 1997b. The role of pineal gland in seasonal changes of blood estradiol level in immature and mature carp females. *Archives of Polish Fisheries* 5: 259–265.

Popek, W., B. Breton, W. Piotrowski, K. Bieniarz and P. Epler. 1994a. The role of the pineal gland in the control of a circadian pituitary gonadotropin release rhythmicity in mature female carp. *Neuroendocrinology Letters* 16: 183–193.

Popek, W., H. Natanek, K. Bieniarz and P. Epler. 1994b. Seasonal changes in the catecholamine levels of the hypothalamus in mature and immature carp (*Cyprinus carpio* L.). *Polish Archives of Hydrobiology* 41: 227–236.

Popek, W., E. Łuszczek-Trojnar, E. Drąg-Kozak, D. Fortuna-Wrońska and P. Epler. 2005. Effect of the pineal gland and melatonin on dopamine release from perifused hypothalamus of mature female carp during spawning and winter regression. *Acta Ichthyologica and Piscatoria* 35: 65–72.

Popek, W., Łuszczek-Trojnar, E., Drąg-Kozak, E., Rz¹sa, J. and P. Epler. 2006. Effect of melatonin on dopamine secretion in the hypothalamus of mature female common carp *Cyprinus carpio* L. *Acta Ichthyologica and Piscatoria* 36: 135–141.

Reiter, R.J. 1991a. Melatonin: The chemical expression of darkness. *Molecular and Cellular Endocrinology* 79: 153–158.

Reiter, R.J. 1991b. Pineal melatonin: Cell biology of its synthesis and of its physiological interactions. *Endocrine Reviews* 12: 151–180.

Saligaut, C., B. Linard, E. Mananos, O. Kah, B. Breton and M. Govoroun. 1998. Release of pituitary gonadotropins GtHI and GtHII in the rainbow trout (*Oncorhynchus mykiss*): modulation by estradiol and catecholamines. *General and Comparative Endocrinology* 109: 302–309.

Sokołowska, M., R.E. Peter, C.S. Nahorniak and J.P. Chang. 1985. Seasonal effects of pimozide and des Gly10 [D-Ala6] LH-RH ethylamide on gonadotrophin secretion in goldfish. *General and Comparative Endocrinology* 57: 472–479.

Trudeau, V.L. 1997. Neuroendocrine regulation of gonadotropin II release and gonadal growth in the goldfish, *Carassius auratus*. *Reviews of Reproduction* 2: 55–68.

Urosaki, H. 1972a. Effects of pinealectomy on gonadal development in the Japanese killifish (medaka) *Oryzias latipes*. *Annotationes Zoologicae Japonenses* 45: 10–15.

Urosaki, H. 1972b. Role of the pineal gland in gonadal development in the fish, *Oryzias latipes*. *Annotationes Zoologicae Japonenses* 45: 152–158

Vaneček, J. 1988. The melatonin receptors in rat ontogenesis. *Neuroendocrinology* 48: 201–203.

Weaver, D.R., S.A. Rivkees and S.M. Reppert. 1989. Localization and characterization of melatonin receptors in rodent brain *in vitro*. *The Journal of Neuroscience* 9: 2581–2590.

Wuttke, W., E. Cassel and J. Meites. 1971. Effects of ergocornine on serum prolactine and LH, and on hypothalamic contents of PIF and LRF. *Endocrinology* 88: 737–741.

Yu, K.L. and R.E. Peter. 1992. Adrenergic and dopaminergic regulation of gonadotropin-releasing hormone release from goldfish preoptic-anterior hypothalamus and pituitary *in vitro*. *General and Comparative Endocrinology* 85: 138–146.

Yuwiler, A. 1989. Effects of steroids on serotonin-N-acetyltransferase activity of pineals in organ culture. *Journal of Neurochemistry* 52: 46–53.

Zachmann, A., S.C.M. Knijf, V. Bolliet and M.A. Ali. 1991. Effects of temperature cycles and photoperiod on rhythmic melatonin secretion from the pineal organ of the white sucker (*Catostomus commersoni*) *in vitro*. *Neuroendocrinology Letters* 13: 325–330.

Zisapel, N. and M. Laudon. 1983. Inhibition by melatonin of dopamine release from rat hypothalamus: regulation of calcium entry. *Brain Research* 272: 378–381.

Zisapel, N., Y. Egozi and M. Laudon. 1985. Circadian variations in the inhibition of dopamine release from adult and newborn rat hypothalamus by melatonin. *Neuroendocrinology* 40: 102–108.

11

CIRCADIAN AND SEASONAL RHYTHM IN SECRETION OF MELATONIN, DOPAMINE AND LH IN CARP

Włodzimierz Popek,[CA] Hanna Natanek and *Ewa Łuszczek-Trojnar*

INTRODUCTION

Seasonal changes in the concentration of melatonin and hormones that control the development and maturation of gonads in vertebrates are a basic indicator in the annual reproductive cycle of seasonally breeding animals. By acting on the hypothalamic-pituitary-gonadal axis, these hormones synchronize animals with the environment by determining phases of the sexual cycle and the optimum time for breeding.

CIRCADIAN AND SEASONAL RHYTHM IN SECRETION OF MELATONIN IN CARP

The release of melatonin is a complex process, but the main controlling factor is light. In mammals, light information reaches the pineal gland via neural pathway, which begins in retina and involves the optic nerve, the suprachiasmatic nucleus of the hypothalamus (SCN), the paraventricular nucleus (PVN), the medial forebrain bundle, the diencephalon cover, the intermediolateral nucleus *of the spinal cord and the superior cervical ganglion whose postganglionic fibres innervating the* pineal gland *are the final element of*

Department of Ichthyobiology and Fisheries, University of Agriculture in Krakow, Krakow, Poland.
[CA]Corresponding author: rzpopek@cyf-kr.edu.pl

this pathway (Klein, 1993). In mammals and in all other vertebrates, light inhibits melatonin synthesis and lack of light stimulates melatonin production and release into the blood and cerebrospinal fluid. The biological half-life of melatonin ranges from several to several dozen minutes (Reiter, 1986).

Pineal and blood melatonin levels are subject to rhythmic circadian changes, with low levels during the day and high levels during the night. Therefore, the succession of day and night forces the characteristic biological rhythm of blood melatonin concentration, known as the circadian rhythm. The rhythm of melatonin release is controlled by the biological clock located in the hypothalamic SCN (Reiter, 1987). The clock is driven by rhythmic changes in activity of the enzyme N-acetyltransferase (NAT), generated in the cervical nerve ganglion by 24-hour changes in light intensity (Klein *et al.*, 1971). As noted above, the characteristic nocturnal pattern of melatonin release is omnipresent in all animals, from mammals to insects and even unicellular ciliates (*Gonyaulax polyedra*) (Poeggeler *et al.*, 1991). However, the pathway of light to the pineal gland of non-mammals is different. The pineal gland of birds, amphibians, reptiles and fish contains photosensitive cells known as photoreceptors. In fish, the exposed front of the pineal gland is located under partly depigmented and transparent calvaria (Fig. 11.1). This area is so characteristic as to be termed pineal window in many species (Holmgren, 1965).

Because of the extracerebral location of the pineal gland, its photoreceptors respond directly to changes in light intensity by controlling melatonin synthesis and release (Meissl and Brandstätter, 1992). Melatonin biosynthesis in fish follows the same course as in mammals and no other

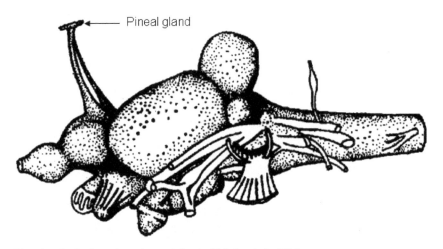

Fig. 11.1 Sagittal section of perch brain (Falcón *et al.*, 1994).

formation mechanisms are known. This hormone is produced from tryptophan. The pineal gland contains all the substrates and enzymes necessary for production of melatonin from tryptophan. Tryptophan is taken up from the blood by pinealocytes and hydroxylated into 5-hydroxy-tryptophan by 5-tryptophan-hydroxylase. In turn, 5-hydroxy-tryptophan is decarboxylated to serotonin (5-hydroxytryptamine) by aromatic amino acid decarboxylase. Under the influence of serotonin N-acetyltransferase, serotonin is acetylated to N-acetylserotonin. The last step in melatonin synthesis is methylation of N-acetylserotonin by hydroxyindole-O-methyltransferase (Karasek, 1997).

Research on changes in melatonin concentration in the blood of carp showed that similarly to higher vertebrates, the lowest blood concentration of this hormone, which practically does not exceed 60 pg/ml, occurs during the entire light phase (under natural conditions—during the day). The concentration of this hormone begins to increase immediately after the beginning of the dark phase (night). It peaks (160 pg/ml on average) in the second part of the dark phase, 2 h before its end (2:00 am). When the next light cycle begins, melatonin concentration returns to its initial value (Popek *et al.*, 1997) (Fig. 11.2).

This profile of changes in melatonin concentration in carp blood, characterized by a peak occurring towards the end of the dark phase, is indicative of a type A profile (Reiter, 1982). Melatonin concentrations in the blood of carp were studied during the summer period and fish were exposed to natural photoperiod (L:D=16:8) and temperature (20°C).

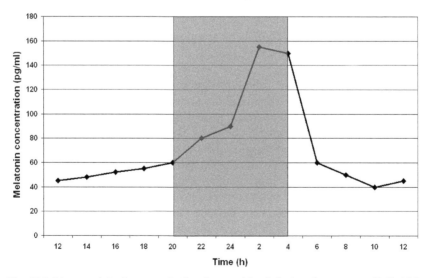

Fig. 11.2 Mean melatonin concentration in carp blood during the summer (L:D=16:8; pond water temperature 20°C). Dark area—dark phase (night).

Further experiments were aimed to determine changing trends resulting from the change of photoperiod length. Because all observations were made under natural conditions, the next sampling of fish blood was conducted in the autumn (L:D=12:12, 15°C). The longer exposure of fish to darkness increased the duration of elevated concentration of melatonin in the blood. At the same time, maximum melatonin concentration, which occurred at 4:00 am during this period, was lower (150 pg/ml) (Fig. 11.3).

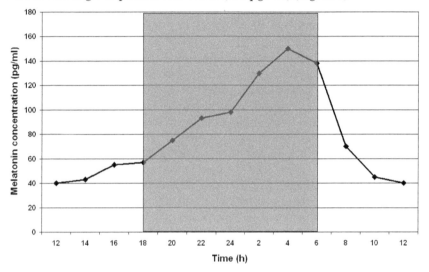

Fig. 11.3 Mean melatonin concentration in carp blood during the autumn (L:D=12:12; pond water temperature 15°C). Dark area—dark phase (night).

As indicated by the figure, the acrophase was moved by 2 hours and occurred, like in the summer period, 2 hours before the end of the dark phase (4:00 am).

The next experiments involved the same group of fish in the winter period. At medium latitudes, the shortest day of the year is towards the end of December. The light phase (day) to dark phase (night) length ratio is 8:16, with water temperature of 5°C. Under these conditions, the profile of 24-hour changes in blood melatonin concentration is variable. The highest melatonin concentration occurs at 8:00 am. The acrophase continues to be moved towards the end of the dark phase, but like in the summer and autumn period, it occurs 2 hours before dawn. The amplitude is flatter because the highest value of melatonin concentration is 140 pg/ml (Fig. 11.4).

Analysis of changes in the concentration of melatonin in carp blood during different photoperiods shows that the changes are dynamic and it can be assumed that the melatonin profile is different every day. In this way we can compare the role of melatonin to the hands of a 24-hour biological clock (pineal gland). The clock is very accurate because it is synchronized

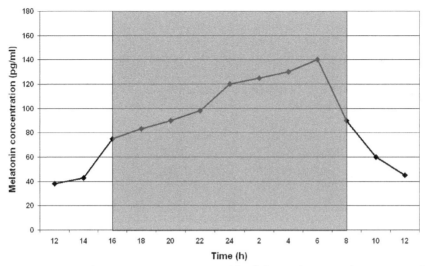

Fig. 11.4 Mean melatonin concentration in carp blood during the winter (L:D=8:16; pond water temperature 5°C). Dark area—dark phase (night).

with the Sun. Also changes that occur in the long term, especially the lengthened duration of elevated melatonin concentration in blood in proportion to dark phase (night) length, can be an important indicator of the succession of seasons. What is more, although the maximum concentration (acrophase) of melatonin from summer to winter assumes an increasingly low value (amplitude), the period of elevated melatonin concentration during 24 hours increases. As a result, the total amount of melatonin produced over 24 hours increases. In the above experiments, the total amount of melatonin release during 24 hours was 930 pg/ml in summer, 1034 pg/ml in autumn, and 1137 pg/ml in winter.

As can be seen from these experiments, the pineal gland and melatonin, in addition to functioning as a 24-hour biological clock, serve as a fairly accurate biological calendar by synchronizing birds for vital life functions.

CIRCADIAN AND SEASONAL RHYTHM IN SECRETION OF DOPAMINE IN CARP

Rhythmic circadian changes in melatonin levels synchronize the circadian rhythms of many different hormones, including hormones responsible for sexual maturation and reproduction in mammals (Reiter, 1991).

The discovery and cloning of melatonin receptors present on the membranes of the hypothalamic neurosecretory cells and secretory cells of the anterior pituitary provided insights into the role of these receptors in the reproductive physiology processes involving melatonin. Because the mammalian brain contains a relatively small number of melatonin receptors, which are much more numerous in the hypothalamus and anterior pituitary, it is conjectured that the hypothalamus most probably acts as a functional link between melatonin and the endocrine system. The mechanism of melatonin action in the hypothalamus has not been elucidated to date. It is suggested that together with gonadal steroids (steroid feedback), melatonin can affect seasonal changes in LH-RH secretion in the hypothalamus, or hypothalamic dopaminergic pathways take part in the photoperiod effect being transferred by melatonin to reproductive functions, while dopamine modifies melatonin action in the hypothalamus.

It has been known since the 1970s that in fish, especially in Cyprinidae, the pituitary gland is strongly influenced by the LH-release inhibiting factor (Peter, 1970). Because of its action, this factor was initially called gonadotropin release inhibitory factor (GRIF). Further experiments revealed that this neurohormone originates from the *anteroventral* preoptic region of the hypothalamus. In this area, neural fibres along which GRIF runs, enter the pituitary peduncle and then the gland itself. It was also conclusively shown that GRIF is one of the catecholamines, namely dopamine (Chang and Peter, 1983; Chang *et al.*, 1983).

In fish, hypothalamic dopamine originates from cell clusters that form two aminergic nuclei: the nucleus recessus lateralis (NRL) located in the medial part of the hypothalamus, and the nucleus recessus posterioris (NRP) located in the posterior part (Popek, 1988; Popek *et al.*, 1994). Both these nuclei send fibres along the ventral part of the hypothalamus towards the preoptic area. There, after making a loop, most of these enter pituitary peduncle through the nucleus lateral tuberis (NLT), reaching LH-producing cells in the glandular (proximal) part of the pituitary.

Studies with crucian carp have shown an inhibiting effect of dopamine on GnRH synthesis and secretion. These results were confirmed by an *in vitro* study, which showed that GnRH release is probably inhibited by neuronal projections of dopaminergic fibres and GnRH fibres that are adjacent to each other (Yu and Peter, 1992). Research involving tilapia, African catfish and carp has also confirmed that dopamine inhibits LH secretion from the pituitary gland.

It was also shown that dopamine can decrease the sensitivity of gonadotropic cells to GnRH stimulation by reducing the number of GnRH receptors in the pituitary. Specific dopamine antagonists were used to prove that dopamine inhibits LH secretion by activating D-2 dopamine receptors found in gonadotropic cells of the pituitary (Peter *et al.*, 1986). This was also supported by radioreceptor studies. Other research also showed that gonadotropic cells of the crucian carp are directly innervated by dopaminergic fibres (Kah *et al.*, 1986).

Dopamine appears to play a key role in the hypothalamus, especially in the context of the environmental signal transmitted by melatonin. It would therefore be interesting to determine whether dopamine shows a similar circadian fluctuation pattern in the hypothalamus to melatonin. This study used sexually immature and mature carp whose brains were harvested at 6-hour intervals during summer (L:D=16:8; water temperature 20°C) and during winter gonadal regression (L:D=8:16; water temperature 5°C). Then, using the *Falck-Hillarp* fluorescence *histochemical* technique (Falck *et al.*, 1962; Björklund *et al.*, 1972), dopamine-containing NRL and NRP regions were localized in the hypothalamus and dopamine activity was estimated based on fluorescence intensity (Popek *et al.*, 1994). This was possible because many experiments have shown that the activity of the aminergic system is always inversely proportional to fluorescence intensity. Thus, the low intensity of nerve tissue fluorescence observed under the fluorescence microscope corresponds to high activity of the system. It was shown that in immature carp there are no cyclic changes in aminergic activity during both the summer and winter periods. In sexually mature fish in the summer, the aminergic system is active in the light phase (day) and peaks at noon (Fig. 11.5).

Similar data were obtained during winter with the difference that fluorescence intensity in the fish hypothalami was always high (above 3 on a 5-point scale), possibly indicating lower activity of the system than in summer (Fig. 11.6).

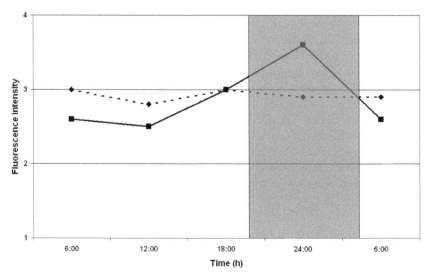

Fig. 11.5 Circadian changes in fluorescence intensity in aminergic nuclei of carp hypothalamus during the summer (L:D=16:8; water temperature 20°C). Continuous line—mature fish, broken line—immature fish. Dark area—dark phase (night).

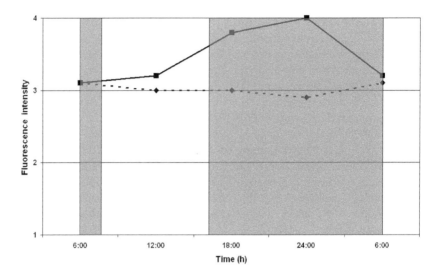

Fig. 11.6 Circadian changes in fluorescence intensity in aminergic nuclei of carp hypothalamus during winter gonadal regression (L:D=8:16; water temperature 5°C). Continuous line—mature fish, broken line—immature fish. Dark area—dark phase (night).

These data indicate that the activity of the aminergic system (dopamine) may be related to the hormonal processes controlling sexual maturity. What is more, the highest aminergic activity of the hypothalamus in mature fish occurs during the day (6:00 am–noon), and the lowest in the middle of the dark phase (night). Assuming that catecholamines, especially dopamine, are a link in relaying information between the pineal gland and the pituitary and that the above results can be associated with dopamine (70% of hypothalamic catecholamines), it should be stated that the circadian rhythm of melatonin synchronizes the circadian rhythm of dopaminergic activity in the hypothalamus. Thus, the high nocturnal concentration of melatonin in the blood inhibits the activity of the dopaminergic system and, conversely, the dopaminergic system is active during the light phase (day) when blood melatonin concentration is low. These statements suggest that melatonin can inhibit the secretion of hypothalamic dopamine and thus regulate its activity.

Similar observations were previously made about changes in fluorescence intensity in the hypothalamus of European eels (*Anguilla anguilla*) during summer and winter (Popek, 1988). These experiments clearly showed that low fluorescence intensity in hypothalamic aminergic nuclei corresponds to high activity of the system. In eels, which are typical nocturnal fish, this situation occurred during the night hours. Fluorescence intensity was always high during the day and clear circadian and seasonal rhythms were also observed (Figs. 11.7 and 11.8).

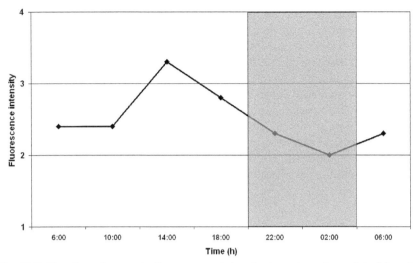

Fig. 11.7 Circadian changes in fluorescence intensity in aminergic nuclei of female eel hypothalamus during the summer (L:D=16:8; water temperature 20°C). Dark area—dark phase (night).

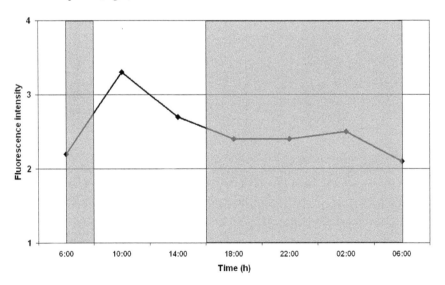

Fig. 11.8 Circadian changes in fluorescence intensity in aminergic nuclei of female eel hypothalamus during the winter (L:D=8:16; water temperature 5°C). Dark area—dark phase (night).

An indirect proof of the link between melatonin and the aminergic system were the results of an experiment which studied fluorescence intensity in hypothalami of fish that were pinealectomized (completely deprived of melatonin) and fish subjected to intracranial injection of melatonin (1µg/

1μL/1kg) directly into the third ventricle of the diencephalon. Injections were made in the middle of the dark phase (midnight) during summer (L:D=16:8, water temperature 20°C) and winter (L:D=8:16; water temperature 5°C) (Popek and Epler, 1999). The results obtained indicate that similarly to mammals, melatonin can act through the hypothalamic aminergic system in fish. The increased fluorescence intensity, observed in aminergic nuclei of the hypothalamus in summer (during the spawning period of carp) and resulting from intraventricular injections of melatonin, as well as the low fluorescence intensity in the group of pinealectomized fish (Fig. 11.9) clearly show that melatonin inhibits the secretion of catecholamines (mainly dopamine) in the carp hypothalamus.

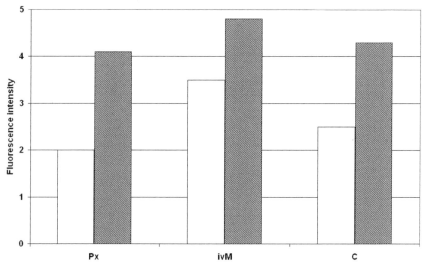

Fig. 11.9 Fluorescence intensity in aminergic nuclei of the hypothalamus of carp subjected to pinealectomy (Px) and melatonin microinjection into the third cerebroventricle (ivM), and in control fish (C). Light bars—summer period, hatched bars—winter period.

As noted above, this is evidence that melatonin can affect the hypothalamic dopaminergic system by synchronizing/modulating its activity.

In the present study, also time of the day had a decisive effect because earlier experiments did not show differences in the hypothalamic fluorescence intensity of carp that were decapitated at the beginning of the light phase. This effect was probably caused by variable amount of active melatonin receptors in the carp hypothalamus. This hypothesis is supported by the results of experiments with mammals, in which melatonin efficiency was dependent on the concentration of melatonin receptors in the nerve tissue, which was higher during the night than during the morning hours (Peschke *et al.*, 1988).

The number of active receptors in melatonin is also probably associated with the less variable reaction of the dopaminergic system to melatonin during the winter period (L:D=8:16; water temperature 5°C), where no differences in fluorescence intensity were found between the groups. Fluorescence was equally strong, being evidence of the low system activity.

A direct proof showing that melatonin can affect the hypothalamic dopaminergic system were the results of an experiment in which actual dopamine concentration in fish hypothalami was studied radioenzymatically (Catechola kit) using the same experimental design (Popek *et al.*, 2006). The experiment demonstrated that during the summer, the highest dopamine concentration (2.175 mg/g tissue) occurs in the hypothalamus of mature carp females given intraventricular microinjection of melatonin. Meanwhile, in the hypothalamus of pinealectomized fish, dopamine concentration was the lowest. What is more, differences between these means were statistically significant (p<0.01) (Fig. 11.10).

The results obtained in the experiments discussed above indicate a certain pattern. During the long photoperiod, the increase in melatonin concentration within the hypothalamus (direct microinjections) leads to increased fluorescence intensity and increased dopamine concentration in this part of diencephalon. Conversely, pinealectomy and thus clearance of melatonin from blood reduces fluorescence intensity and dopamine concentration in the hypothalamus. These findings support a hypothesis that melatonin can control (inhibit) the release of hypothalamic dopamine during summer.

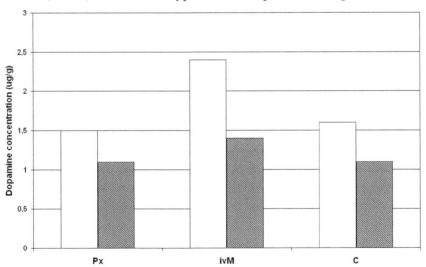

Fig. 11.10 Mean dopamine concentration in aminergic nuclei of the hypothalamus of carp subjected to pinealectomy (Px) and melatonin microinjection into the third cerebroventricle (ivM), and in control fish (C). Light bars—summer period, hatched bars—winter period.

The results obtained also point to the seasonal activity of the pineal-hypothalamic system, because during the short day (winter), when Cyprinidae fish undergo gonadal regression, no significant melatonin-induced differences were found in fluorescence intensity (activity of the aminergic system) and in the actual catecholamine content of the hypothalamus of mature carp females.

CIRCADIAN AND SEASONAL RHYTHM IN SECRETION OF LH IN CARP

The next step to show the relationship between circadian rhythm of melatonin and synchronization of the endocrine activity of pituitary was to determine if LH concentration in the blood of fish changes cyclically during 24 hours and if there are seasonal changes during the sexual cycle. This question is important because LH release from vertebrate pituitary is largely dependent on environmental factors such as temperature, light and stage of gonadal maturity. In immature fish, blood LH concentration remains low until the onset of gametogenesis. During oocyte growth and vitellogenesis, the level of this hormone increases. A further increase in LH concentration occurs during maturation of oocytes, migration of the nucleus in the oocyte, and ovulation. After ovulation and spawning, blood LH concentration decreases and remains low during regeneration of the ovary.

Experiments conducted in 4 seasons of the year (spring, summer, autumn, winter) in immature and mature carp females showed that in immature fish (early vitellogenesis) LH concentration ranged from 4.07 to 6.61 ng/ml regardless of the time of day and night and season of the year, and changes in this concentration were not rhythmic. Mature fish (completed vitellogenesis) showed clear circadian rhythm of LH concentration in blood. This rhythm only occurred in the summer, during the spawning period of the carp (Figs. 11.11 and 11.12) (Popek *et al.,* 1994).

The above data demonstrate that a clear rhythm in LH concentration in carp blood occurs only in sexually mature fish and only during the long day (breeding season). In the other seasons of the year and in immature fish, this cyclicity was not shown. To find if there is a direct relationship between the circadian rhythm of melatonin concentration in the blood of both immature and mature carp that occurs regardless of season of the year, blood LH concentration was analysed in an experiment with pinealectomized fish. Mature carp females during the spawning period were subjected to pinealectomy. It was found that the pinealectomy, which completely clears the blood of melatonin, had no effect on the concentration and circadian dynamic (amplitude, acrophase) of the profile of LH concentration in the blood of female carp (Popek *et al.,* 1994).

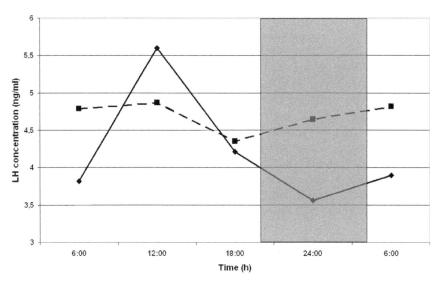

Fig. 11.11 Mean 24-hour concentration of LH in female carp blood during the summer (L:D=16:8; water temperature 20°C). Continuous line—mature fish, broken line—immature fish. Dark area—dark phase (night).

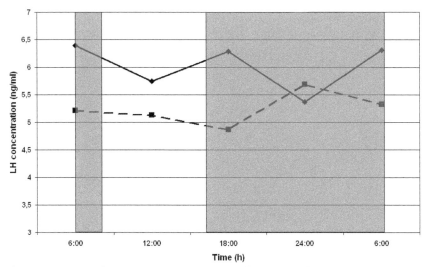

Fig. 11.12 Mean 24-hour concentration of LH in female carp blood during the winter (L:D=8:16; water temperature 5°C). Continuous line—mature fish, broken line—immature fish. Dark area—dark phase (night).

The lack of direct effect of melatonin on carp pituitary was also confirmed in *in vitro* experiments. As discussed above, because of photoreceptive cells, the pineal gland removed from the body reacts to changing light intensity by changes in melatonin concentration that are identical as under physiological conditions. Using tissue perifusion equipment described previously, pituitary glands were perifused with 3 implanted pineal glands and pituitary glands were perifused in the presence of melatonin (300 pg/ml). During the first hour of perifusion, columns were illuminated by white light at an intensity of 2000 lx. In the second and third hour, the columns were darkened. Thus, in the second (dark) stage of the experiment, pituitary glands were perifused in the presence of endogenous and synthetic melatonin. The experiments were performed during the summer and winter (Popek *et al.*, 2005). It was found that LH secretion from carp pituitaries in the presence of pineal glands, and also under the influence of exogenous melatonin, did not change significantly in the first (light) and second (dark) part of the experiment. This is evidence that melatonin does not affect LH synthesis and secretion from the carp pituitary under *in vitro* conditions (Figs. 11.13 and 11.14).

In a further search for the site of melatonin effect on synchronization of the sexual cycle in carp, the effect of this hormone on steroidogenesis and final stages of oocyte maturation *in vitro* was studied (Popek *et al.*, 1996). In this study, too, melatonin at different concentrations of Cortland's medium (1 to 10µg/ml) did not cause over a 24-hour period any changes in the concentration of testosterone, testosterone glucuronide, 17-hydroxyprogesterone and 17-alpha-20-beta-dihydroxyprogesterone

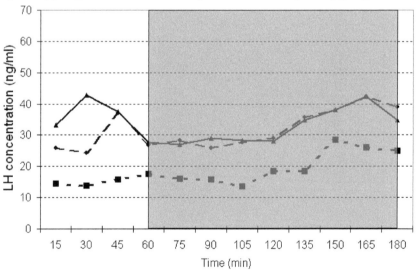

Fig. 11.13 Mean LH concentration in medium (summer period) during perifusion of pituitaries harvested from carp. Continuous line—pituitaries alone, broken line—pituitaries in the presence of 3 pineal glands, dotted line—pituitaries in the presence of melatonin (300 pg/ml). Light field (0–60 min.) 2000 lx. Dark field (60–180 min.) 0 lx.

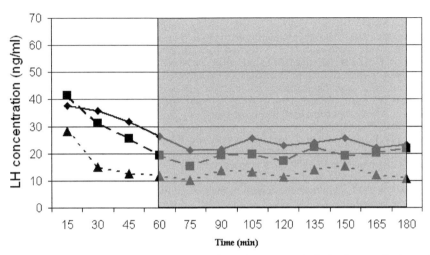

Fig. 11.14 Mean LH concentration in medium (winter period) during perifusion of pituitaries harvested from carp. Continuous line—pituitaries alone, broken line—pituitaries in the presence of 3 pineal glands, dotted line—pituitaries in the presence of melatonin (300 pg/ml). Light field (0–60 min.) 2000 lx. Dark field (60–180 min.) 0 lx.

(17α20βP). No changes were also found for the percentage of oocytes with different nucleus locations under the influence of melatonin.

In summary, melatonin in carp during the summer period does not directly affect the cells of the proximal pituitary in which LH is synthesized, nor does it directly affect steroidogenesis in the follicular envelope of oocytes and is not directly involved in the final stages of oocyte maturation, its role in synchronization of sexual maturation and reproduction being restricted mainly to the hypothalamus.

References

Chang, J.P. and R.E. Peter. 1983. Effects of dopamine on gonadotropin release in female goldfish, *Carassius auratus*. *Neuroendocrinology* 36: 351–357.

Chang, J.P., A.F. Cook and R.E. Peter. 1983. Influence of catecholamines on gonadotropin secretion in goldfish, *Carassius auratus*. *General and Comparative Endocrinology* 49: 22–31.

Falcón, J., V. Bolliet, J.P. Ravault, D. Chesneau, M.A. Ali and J.P. Collin. 1994. Rhythmic secretion of melatonin by the superfused pike pineal organ: thermo- and photoperiod interaction. *Neuroendocrinology* 60: 535–543.

Falck, B., N.A. Hillarp, G. Thieme and A. Torp. 1962. Fluorescence of catecholamines and related compounds condensed with formaldehyde. *Journal of Histochemistry and Cytochemistry* 10: 348–354.

Holmgren, U. 1965. On the ontogeny of the pineal organs in teleost fishes. *Progress in Brain Research* 10: 172–182.

Kah, O., B. Breton, J.G. Dulka, J. Nunez-Rodriguez, R.E. Peter, J.E. Rivier and W.W. Vale. 1986. A reinvestigation of Gn-RH (gonadotropin-releasing hormone) systems in the goldfish brain using antibodies to salmon Gn-RH. *Cell and Tissue Research* 244: 327–337.

Karasek, M. 1997. Pineal and melatonin. *Polskie Wydawnictwo Naukowe, Warszawa* (In Polish).

Klein, D.C. 1993. The mammalian melatonin rhythm-generating system. In: *Light and Biological Rhythms in Man*, L. Wetterberg (ed). Pergamon Press, Oxford, pp. 55–72.

Klein, D.C., J.L. Weller and R.Y. Moore. 1971. Melatonin metabolism neural regulation of pineal serotonin: acetyl coenzyme A N-acetyltransferase activity. *Proceedings of the National Academy of Sciences of the United States of America* 68: 3107–3110.

Meissl, H. and R. Brandstätter. 1992. Photoreceptive functions of the teleost pineal organ and their implications in biological rhythms. In: *Rhythms in Fishes*, M.A. Ali (ed). Plenum Press, New York, pp. 235–254.

Peschke, E., D. Peschke, J. Peil, C. Ruzsas and B.Mess. 1988. Changes in the dailly rhythm of serum testosterone levels following superior cervical sympathetic ganglionectomy in the cold-exposed rat: The role of the pineal. *Journal of Pineal Research* 5: 179–189.

Peter, R.E. 1970. Hypothalamic control of the thyroid gland activity and gonadal activity in the goldfish *Carassius auratus*. *General and Comparative Endocrinology* 14: 334–356.

Peter, R.E., J. Chang, C.S. Nahorniak, R.J. Omeljaniuk, M. Sokołowska, S.H. Shih and R. Billard. 1986. Interactions of catecholamines and GnRH in regulation of gonadotropin secretion in teleost fish. *Recent Progress in Hormone Research* 42: 513–548.

Popek, W. 1988. Effects of hormonal stimulation of sexual maturation on the hypothalamic catecholamine content in European eel (*Anguilla anguilla* L.). *Polish Archives of Hydrobiology* 35: 119–126.

Popek, W. and P. Epler. 1999. Effects of intraventricular melatonin microinjections on hypothalamic catecholamine activity in carp females during a year. *Electronic Journal of Polish Agricultural Universities* (*www.ejpau.media.pl*), 2 (1) *Fisheries*.

Popek, W., H. Natanek, K. Bieniarz and P. Epler. 1994. Seasonal changes in the catecholamine levels of the hypothalamus in mature and immature carp (*Cyprinus carpio* L.). *Polish Archives of Hydrobiology* 41: 227–236.

Popek, W., P. Epler, K. Bieniarz and M. Sokołowska-Mikołajczyk. 1997. Contribution of factors regulating melatonin release from pineal gland of carp (*Cyprinus carpio* L.) in normal and in polluted environments. *Archives of Polish Fisheries* 5: 59–75.

Popek, W., B. Breton, W. Piotrowski, K. Bieniarz and P. Epler. 1994. The role of the pineal gland in the control of a circadian pituitary gonadotropin release rhythmicity in mature female carp. *Neuroendocrinology Letters* 16: 183–193.

Popek, W., E. Łuszczek-Trojnar, E. Drąg-Kozak, D. Fortuna-Wrońska and P. Epler. 2005. Effect of the pineal gland and melatonin on dopamine release from perifused hypothalamus of mature female carp during spawning and winter regression. *Acta Ichthyologica et Piscatoria* 35: 65–72.

Popek, W., E. Łuszczek-Trojnar, E. Drąg-Kozak, J. Rząsa and P. Epler. 2006. Effect of melatonin on dopamine secretion in the hypothalamus of mature female common carp, *Cyprinus carpio* L. *Acta Ichthyologica et Piscatoria* 36: 135–141.

Reiter, R.J. 1982. Neuroendocrine effect of the pineal gland and melatonin. In: *Frontiers in Neuroendocrinology*, W.F. Ganong and L. Martini (eds.). Raven, New York 7: 287–316.

Reiter, R.J. 1986. Normal levels of melatonin in the pineal gland and body fluids in humans and experimental animals. *Journal of Neural Transmission* 21: 35–54.

Reiter, R.J. 1987. The melatonin message duration versus coincidence hypothesis. *Life Sciences* 40: 2119–2131.

Reiter, R.J. 1991. Pineal gland: Interface between the photoperiodic environment and the endocrine system. *Trends in Endocrinology and Metabolism* 1: 13–19.

Yu, K.L. and R.E. Peter. 1992. Adrenergic and dopaminergic regulation of gonadotropin-releasing hormone release from goldfish preoptic-anterior hypothalamus and pituitary *in vitro*. *General and Comparative Endocrinology* 85: 138–146.

12

CIRCADIAN CLOCKS IN RETINA OF GOLDFISH

P.M. Iuvone,[1,CA] E. Velarde,[2] M.J. Delgado,[2] A.L. Alonso-Gomez,[2] and R. Haque[1]

INTRODUCTION

The retinas of most vertebrate species contain circadian clocks that allow physiological anticipation of daily changes of light intensity as a function of the solar day-night cycle. These clocks regulate light and dark adaptation, turnover of photoreceptor outer segment membranes, gene expression, metabolism, and neurotransmitter release. The goldfish (*Carassius auratus*) has been an extremely useful model to study circadian mechanisms of adaptation. However, until recently the clock molecules responsible for generating these circadian rhythms in goldfish retina have not been studied. In this chapter we will review studies of circadian regulation of physiology and biochemistry as well as the cloning and rhythmic expression of circadian clock genes in the goldfish retina, with comparisons to other vertebrate species.

THE MOLECULAR BASIS OF CIRCADIAN CLOCKS

Circadian clocks are generally thought to be composed of interlocking transcription-translation feedback loops involving of a set of highly conserved clock genes and their protein products (Fig. 12.1). The proteins encoded by clock genes serve as transcription factors, and are either positive

[1]Departments of Ophthalmology and Pharmacology, Emory University School of Medicine, Atlanta, GA 30322, USA.
[2]Department of Physiology (Animal Physiology II), Faculty of Biology, Complutense University of Madrid, 28040 Madrid, Spain.
[CA]Corresponding author: miuvone@emory.edu

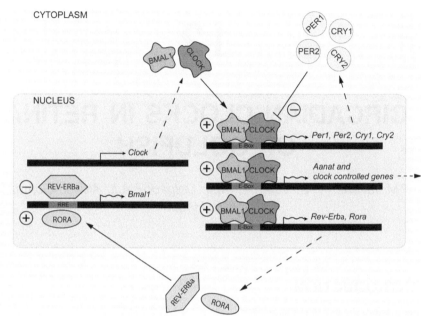

Fig. 12.1 The molecular basis of circadian clocks. See text for details. Reproduced from Tosini *et al.*, BioEssays 30: 624–633, 2008 © Wiley Periodicals, Inc.

or negative transcription regulators (Ko and Takahashi, 2006). The positive regulators include BMAL1, CLOCK, and NPAS2, which heterodimerize and bind to E box enhancer elements in the promoters of genes encoding the negative regulators, the PERIOD and CRYPTOCHROME proteins. BMAL1/ CLOCK or BMAL1/NPAS2 heterodimers activate transcription of the *Period* and *Cryptochrome* genes. In a negative feedback loop, the PERIOD and CRYPTOCHROME proteins block the action of BMAL1/CLOCK and thereby inhibit the transcription of their own genes. This feedback loop generates a daily rhythm of PERIOD and CRYPTOCHROME protein expression. In a second feedback loop, BMAL1/CLOCK activates the *Rev-erbá* and *Rora* genes. The protein products of these genes compete for the RRE element in the *Bmal1* gene, leading to rhythmic expression of BMAL1. Rhythmic expression of circadian clock gene proteins generates circadian clock outputs by regulating the expression of clock-controlled genes (CCGs). Some of the CCGs contain circadian E boxes that are activated rhythmically by heterodimers of BMAL1/CLOCK or BMAL1/NPAS2 (Chong *et al.*, 2000; Munoz and Baler, 2003). While most of the details of this mechanism have been elucidated in mammalian systems, there is compelling evidence that similar mechanisms operate in non-mammalian vertebrates, including fish (Cahill, 2002).

CIRCADIAN RHYTHMS OF MELATONIN AND DOPAMINE

Studies of the retinas of non-mammalian and mammalian species have established the role of a highly conserved reciprocal feedback loop involving two neuromodulators, melatonin and dopamine, that regulates daily rhythms of adaptation and physiology (Fig. 12.2; reviewed in Green and Besharse, 2004; Iuvone *et al.*, 2005; Tosini *et al.*, 2008). Melatonin is synthesized in photoreceptors under the influence of a circadian clock, such that melatonin release occurs predominantly at night in darkness. Melatonin diffuses throughout the retina to act on a family of G-protein coupled receptors (GPCRs)—the MT1 (Mel1a), MT2 (Mel1b), and Mel1c receptors, which are found on neurons and glia. Melatonin may also act on other binding sites, such as the *Mt3* binding site, although the affinity of melatonin for these other sites is much lower than that of the melatonin GPCRs. Melatonin functions to promote dark adaptation in the retinal

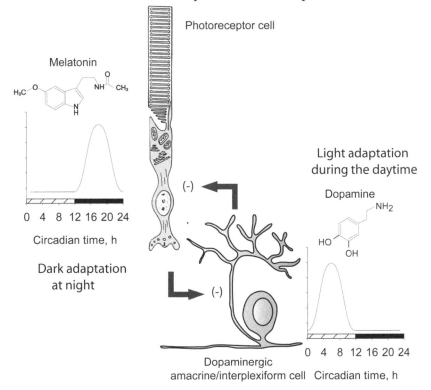

Fig. 12.2 Dopamine-melatonin feedback loop in retina regulates circadian adaptation. Melatonin promotes dark adaptation in part by inhibiting dopamine release from interplexiform cells at night, while dopamine promotes light adaptation and inhibits melatonin during the daytime. Modified from Tosini *et al.*, BioEssays 30: 624–633, 2008 © Wiley Periodicals, Inc.

network, in part by inhibition of dopamine release at night. In contrast to melatonin, dopamine release is stimulated by light and subject to regulation by circadian clocks such that release is high during the daytime. Dopamine, in turn, diffuses throughout the retina to act on GPCRs, the dopamine D1, D2, D4, and D5 receptors, which are found on photoreceptors, retinal pigment epithelial cells, and inner retinal neurons. Dopamine promotes light adaptive physiology and acts on D2-like (D2 or D4) receptors on photoreceptor cells to inhibit melatonin release during the daytime. Dopamine also sets the phase of the circadian clock in photoreceptors cells of some species.

The goldfish retina has been a useful model for elucidating some of the mechanisms controlling circadian adaptation by this dopamine-melatonin feedback loop. A circadian clock in goldfish retina controls melatonin content and release (Iigo *et al.*, 1997a). Melatonin content and release from goldfish retina is high during the subjective night in constant darkness and is inhibited by light. This regulation is similar to that observed in mammals, chicken, *Xenopus*, quail, pigeon, iguana, ugui, and zebrafish (Hamm and Menaker, 1980; Underwood *et al.*, 1990; Cahill and Besharse, 1995; Tosini and Menaker, 1996; Iigo *et al.*, 1997a). The regulation of melatonin in retinas of some other fish species may differ. For example, in the retinas of the European sea bass, melatonin levels appear to be higher during the daytime than at night and do not appear to be rhythmic in constant darkness (Iigo *et al.*, 1997b).

Dopamine release from isolated superfused goldfish retina also exhibits a circadian rhythm, which persists in constant darkness with higher levels of release during the subjective day (Ribelayga *et al.*, 2004). Interestingly, this rhythm of dopamine release appears to be driven by the inhibitory influence of melatonin. Dopamine release from retinas superfused in the continuous presence of melatonin becomes arrhythmic, with release rates tonically at the nighttime level. In contrast, continuous exposure to luzindole, a melatonin receptor blocker, results in arrthymic dopamine release with levels comparable to those during the daytime. Thus, the dopamine release rhythm appears to be generated by melatonin-mediated suppression of dopamine release at night. A similar mechanism has been proposed to explain the circadian release and metabolism of dopamine in pigeon, iguana, and mouse retinas (Adachi *et al.*, 1999; Doyle *et al.*, 2002; Bartell *et al.*, 2007).

A circadian clock in the goldfish retina profoundly affects adaptation in part by regulating the balance of rod and cone input to second order retinal neurons (Wang and Mangel, 1996). Following exposure to prolonged darkness during the night but not during the day, L-type cone horizontal cell responses to a flash of light resemble those of rod horizontal cells with respect to waveform, threshold, and spectral sensitivity, indicative of predominantly rod input during the night. In retinas maintained in constant darkness, cone horizontal cell responses to a bright flash of light are bigger during the day than during the night. The daytime responses show a spectral sensitivity similar to red cones while nighttime responses have a rod-like spectral sensitivity. Thus, a circadian clock imparts a cone dominant state during the daytime and rod-dominant state at night.

The circadian rhythm of rod-cone input in goldfish retina appears to be mediated by dopamine. Application of dopamine or quinpirole, a dopamine D2-like receptor agonist, to the retina during the subjective night, when endogenous dopamine release is low, reduces rod input to horizontal cells and increases cone input, mimicking the daytime situation (Ribelayga *et al.*, 2002). In contrast, spiperone, a D2-like receptor antagonist, decreases cone input and increases rod input during the subjective daytime, demonstrating that endogenous dopamine mediates the cone dominant daytime physiology.

These circadian changes in rod-cone input were observed in L-type cone horizontal cells, which receive direct synaptic contact from cones but not rods in the goldfish retina. This implies that the rod responses at night must be indirect. Recent evidence suggests that the circadian clock via dopamine controls rod input to cone horizontal cells by gating electrical synapses between rods and cones (Ribelayga *et al.*, 2008). Electrical and dye coupling between rods and cones via gap junctions is weak during the daytime, when cone input predominates. In contrast, coupling between rods and cones is pronounced at night. Thus, at night rods drive dim light cone synaptic input to the L-type horizontal cells. Consistent with observations of rod-cone input to horizontal cells, blocking D2-like receptors with spiperone increases rod-cone coupling during the daytime while activating the receptors with quinpirole uncouples the rods from the cones at night. This synaptic switching mechanism controlled by the circadian clock may allow the fish to see dim, large objects at night while having high acuity vision during the daytime.

RHYTHMIC EXPRESSION OF CLOCK GENES IN GOLDFISH RETINA

To begin to elucidate the molecular mechanism of circadian oscillators in goldfish retina, we have cloned partial cDNA fragments for six clock genes and studied their temporal expression patterns (Velarde *et al.*, 2009). The three *period* (*per1-3*) and three *cryptochrome* (*cry1-3*) transcript cDNAs are 84–92% identical to their zebrafish counterparts. All six transcripts are expressed in retina, as well as in brain, gut, heart and liver. The transcripts are rhythmically expressed in retina (Fig. 12.3). The three *cryptochrome* mRNAs show similar patterns of rhythmicity; *cry1* and *cry 2* transcript rhythms peaked at approximately zeitgeber time (ZT) 14, while *cry3* peaked later in the night at ~ZT 20. *Per1* and *per3* transcript rhythms also peak at ~ZT20, but *per2* mRNA is most highly expressed in the morning, ~ZT2.

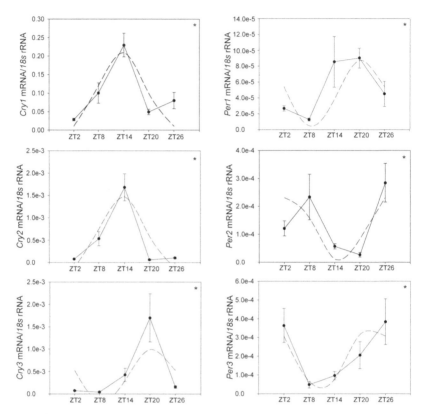

Fig. 12.3 Daily expression of cryptochrome and period transcripts in the goldfish retina. Relative mRNA expression for six clock gene transcripts is plotted against time of day (zeigeber time—ZT), with ZT 0 reflecting the time of light onset of the daily 12h-12h light-dark cycle. All six transcripts are rhythmically expressed in retina. Modified from Velarde *et al.*, 2009, Journal of Biological Rhythms 24: 104–113, 2009 © Sage Publications.

When compared to other vertebrate species, the temporal control of clock gene expression shows some similarities but many differences, indicative of interspecies differences in circadian organization of the vertebrate retina. The high nighttime expression pattern of the three goldfish *cryptochromes* is similar to the expression pattern of zebrafish *cry2* but differs from that of zebrafish *cry1* and *cry3*, which peak during the daytime (Kobayashi *et al.*, 2000). This pattern of expression of the goldfish *cryptochromes* is also distinct from daytime expression of *cry1* in quail and chicken retinas (Fu *et al.*, 2002; Haque *et al.*, 2002; Chaurasia *et al.*, 2006). Regarding the *period* transcripts, the rhythm of goldfish *per1* peaks at night. This pattern of expression is opposite to that observed in *Xenopus* retina, where *per1* peaks in the early daytime (Zhuang *et al.*, 2000). The goldfish *per2* rhythm shows a peak at mid- to late-day and resembles that of *Xenopus*, chick, quail, rat, and mouse

(Yoshimura *et al.*, 2000; Zhuang *et al.*, 2000; Kamphuis *et al.*, 2005; Chaurasia *et al.*, 2006; Ruan *et al.*, 2008). The goldfish *per3* rhythm, with high levels of mRNA late at night and early in the day, is similar to that in zebrafish, quail, and chick (Yoshimura *et al.*, 2000; Bailey *et al.*, 2004; Ruan *et al.*, 2008).

CONCLUSIONS

The goldfish retina has proven to be a powerful model system for exploring the circadian control of adaptation using both physiological and biochemical approaches. With the cloning of goldfish clock genes, this system will also be amenable to future molecular analysis of retinal circadian organization. The rhythmic expression of clock gene transcripts clearly demonstrates that the goldfish retina contains autonomous circadian clocks capable of driving rhythms of physiology.

Acknowledgements

Research in the authors' laboratories is funded by the US Public Health Service (NIH grants R01 EY004864 and P30 EY0063630), Research to Prevent Blindness, the Spanish MEC (AGL2007-65744C03-03), and the European Social Fund.

References

Adachi, A., Y. Suzuki, T. Nogi and S. Ebihara. 1999. The relationship between ocular melatonin and dopamine rhythms in the pigeon: effects of melatonin inhibition on dopamine release. *Brain Research* 815: 435–440.

Bailey, M.J., P.D. Beremand, R. Hammer, E. Reidel, T.L. Thomas and V.M. Cassone. 2004. Transcriptional profiling of circadian patterns of mRNA expression in the chick retina. *Journal of Biological Chemistry* 279: 52247–52254.

Bartell, P.A., M. Miranda-Anaya, W. McIvor and M. Menaker. 2007. Interactions between dopamine and melatonin organize circadian rhythmicity in the retina of the green iguana. *Journal of Biological Rhythms* 22: 515–523.

Cahill, G.M. 1996. Circadian regulation of melatonin production in cultured zebrafish pineal and retina. *Brain Research* 708: 177–181.

Cahill, G.M. 2002. Clock mechanisms in zebrafish. *Cell and Tissue Research* 309: 27–34.

Cahill, G.M. and J.C. Besharse. 1995. Circadian rhythmicity in vertebrate retinas: regulation by a photoreceptor oscillator. *Progress in Retinal and Eye Research* 14: 267–291.

Chaurasia, S.S., N. Pozdeyev, R. Haque, A. Visser, T.N. Ivanova and P.M. Iuvone. 2006. Circadian clockwork machinery in neural retina: evidence for the presence of functional clock components in photoreceptor-enriched chick retinal cell cultures. *Molecular Vision* 12: 215–223.

Chong, N.W., M. Bernard and D.C. Klein. 2000. Characterization of the chicken serotonin N-acetyltransferase gene. Activation *via* clock gene heterodimer/E box interaction. *Journal of Biological Chemistry* 275: 32991–32998.

Doyle, S.E., M.S. Grace, W. McIvor and M. Menaker. 2002. Circadian rhythms of dopamine in mouse retina: the role of melatonin. *Visual Neuroscience* 19: 593–601.

Fu, Z., M. Inaba, T. Noguchi and H. Kato. 2002. Molecular cloning and circadian regulation of cryptochrome genes in Japanese quail (*Coturnix coturnix japonica*). *Journal of Biological Rhythms* 17: 14–27.

Green, C.B. and J.C. Besharse. 2004. Retinal circadian clocks and control of retinal physiology. *Journal of Biological Rhythms* 19: 91–102.

Hamm, H.E. and M. Menaker. 1980. Retinal rhythms in chicks—circadian variation in melatonin and serotonin N-acetyltransferase. *Proceedings of the National Academy of Sciences of the United States of America* 77: 4998–5002.

Haque, R., S.S. Chaurasia, J.H. III. Wessel and P.M. Iuvone. 2002. Dual regulation of cryptochrome 1 mRNA expression in chicken retina by light and circadian oscillators. *Neuroreport* 13: 2247–2251.

Iigo, M., M. Hara, R. Ohtani-Kaneko, K. Hirata, M. Tabata and K. Aida. 1997a. Photic and circadian regulations of melatonin rhythms in fishes. *Biological Signals* 6: 225–232.

Iigo, M., F.J. Sanchez-Vazquez, J.A. Madrid, S. Zamora and M. Tabata. 1997b. Unusual responses to light and darkness of ocular melatonin in European sea bass. *Neuroreport* 8: 1631–1635.

Iuvone, P.M., G. Tosini, N. Pozdeyev, R. Haque, D.C. Klein and S.S. Chaurasia. 2005. Circadian clocks, clock networks, arylalkylamine N-acetyltransferase and melatonin in the retina. *Progress in Retinal and Eye Research* 24: 433–456.

Kamphuis, W., C. Cailotto, F. Dijk, A. Bergen and R.M. Buijs. 2005. Circadian expression of clock genes and clock-controlled genes in the rat retina. *Biochemical and Biophysical Research Communications* 330: 18–26.

Ko, C.H. and Takahashi, J.S. 2006. Molecular components of the mammalian circadian clock. *Human Molecular Genetics* 15 Spec No 2: R271–R277.

Kobayashi, Y., T. Ishikawa, J. Hirayama, H. Daiyasu, S. Kanai, H. Toh, I. Fukuda, T. Tsujimura, N. Terada, Y. Kamei, S. Yuba, S. Iwai and T. Todo. 2000. Molecular analysis of zebrafish photolyase/cryptochrome family: two types of cryptochromes present in zebrafish. *Genes Cells* 5: 725–738.

Munoz, E. and R. Baler, R. 2003. The circadian E-box: when perfect is not good enough. *Chronobiology International* 20: 371–388.

Ribelayga, C., Y. Wang and S.C. Mangel. 2002. Dopamine mediates circadian clock regulation of rod and cone input to fish retinal horizontal cells. *Journal of Physiology* 544: 801–816.

Ribelayga, C., Y. Wang and S.C. Mangel. 2004. A circadian clock in the fish retina regulates dopamine release via activation of melatonin receptors. *Journal of Physiology* 554: 467–482.

Ribelayga, C., Y. Cao and S.C. Mangel. 2008. The circadian clock in the retina controls rod-cone coupling. *Neuron* 59: 790–801.

Ruan, G.X., G.C. Allen, S. Yamazaki, D.G. McMahon. 2008. An autonomous circadian clock in the inner mouse retina regulated by dopamine and GABA. *PLoS Biology* 6: e249.

Tosini, G. and Menaker, M. 1996. Circadian rhythms in cultured mammalian retina. *Science* 272: 419–421.

Tosini, G., N. Pozdeyev, K. Sakamoto and P.M. Iuvone. 2008. The circadian clock system in the mammalian retina. *BioEssays* 30: 624–633.

Underwood, H., R.K. Barrett and T. Siopes. 1990. The quail's eye: a biological clock. *Journal of Biological Rhythms* 5: 257–265.

Velarde, E., R. Haque, P.M. Iuvone, C. Azpeleta, A.L. Alonso-Gomez and M.J. Delgado. 2009. Circadian clock genes of goldfish, *Carassius auratus*: cDNA cloning and rhythmic expression of period and cryptochrome transcripts in retina, liver and gut. *Journal of Biological Rhythms* 24: 104–113.

Wang, Y. and S.C. Mangel. 1996. A circadian clock regulates rod and cone input to fish retinal cone horizontal cells. *Proceedings of the National Academy of Sciences of the United States of America* 93: 4655–4660.

Yoshimura, T., Y. Suzuki, E. Makino, T. Suzuki, A. Kuroiwa, Y. Matsuda, T. Namikawa and S. Ebihara. 2000. Molecular analysis of avian circadian clock genes. *Brain Research. Molecular Brain Research* 78: 207–215.

Zhuang, M., Y. Wang, B.M. Steenhard and J.C. Besharse. 2000. Differential regulation of two period genes in the *Xenopus* eye. *Brain Research. Molecular Brain Research* 82: 52–64.

INDEX

ABOUT THE EDITORS

Ewa Kulczykowska, DSc, a graduate of the Biological Faculty, University of Warsaw, made her research career at the Polish Academy of Sciences, first at the Medical Research Centre, Warsaw, and then at the Institute of Oceanology, Sopot, where she heads the Department of Genetics and Marine Biotechnology, present focus being on fish endocrinology.

Wlodzimierz Popek is Professor at the Cracow University of Agriculture. For many years he has been engaged in studying chronobiology and physiology of reproduction in fish. He used light and temperature to synchronize the circadian rhythm of melatonin and basic blood hormone levels in carp. He also provided evidence that the pineal gland of fish may act as both a calendar and a daily biological clock.

B.G. Kapoor was formerly Professor and Head of Zoology in Jodhpur University (India). Dr. Kapoor has co-edited 20 books published by Science Publishers, Enfield, NH, USA. The most recent ones are Fish Defenses (Volume 1), 2009 (with Giacomo Zaccone, José Meseguer Peñalver and Alfonsa García-Ayala); The Biology of Blennies, 2009 (with Robert A. Patzner, Emanuel J. Gonçalves and Philip A. Hastings); Development of Non-Teleost Fishes, 2009 (with Yvette W. Kunz and Carl A. Luer); Fish Defenses (Volume 2), 2009 (with Giacomo Zaccone, C. Perrière and A. Mathis); Fish Locomotion: An Eco-ethological Perspective, 2010 (with Paolo Domenici), Biology of Subterranean Fishes, 2010 (with Eleonora Trajano and Maria Elina Bichuette). Dr. Kapoor has also co-edited The Senses of Fish: Adaptations for the Reception of Natural Stimuli, 2004 (with G. von der Emde and J. Mogdans); and co-authored Ichthyology Handbook, 2004 (with B. Khanna), both from Springer, Heidelberg. He has also been a contributor in books from Academic Press, London (1969, 1975 and 2001). His E-mail is: bhagatgopal.kapoor@rediffmail.com

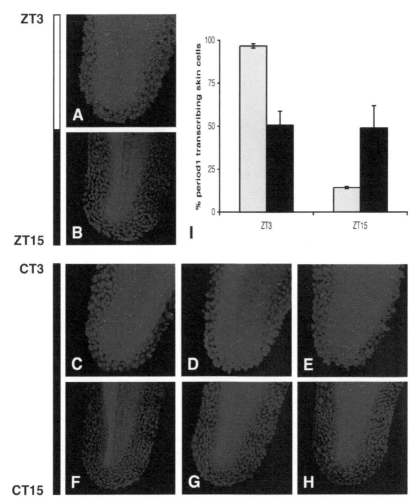

Fig. 7.4 Fluorescent *in situ* hybridization was performed in an attempt to examine gene expression changes within single cells of the developing embryo. All images of *per1* expression were made on the second day of development. It is clear, from panels A and B, that there is a dramatic rhythm in *per* expression between cells at ZT3 and ZT15 raised on a light dark cycle. Panels C to H show examples of expression for cells raised under constant dark conditions. If the clock had not started in the embryo one would predict all cells to be expressing *per1* at a mid-/average level. If, however, the clock has started, but is not synchronized between cells, then some cells will show high levels, and others low levels of expression of *per1*. Panel I shows a quantification of these data, and reveals an average number of cells expressing *per1*, supporting the idea of a population of de-synchronized clocks existing in the early embryo when raised in the dark. (From Dekens *et al.*, 2008).

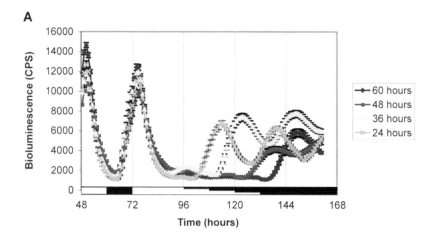

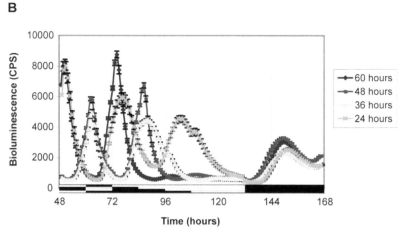

Fig. 7.9 To explore the light "stopping" action of sustained light exposure, *per1* expression was compared under a series of long-pulse experiments. The top trace shows a "forwards wedge", with light pulses beginning at the same circadian phase. The circadian oscillation in *per1* is clearly held motionless during the sustained light exposure, and begins again upon return to constant dark conditions. A "reverse wedge" (lower panel) shows that the pacemaker begins from approximately the same phase upon release into DD, about ZT12 or dusk. The clock appears to be held motionless by light pulses longer than 12 hours (From Tamai *et al.*, 2007).

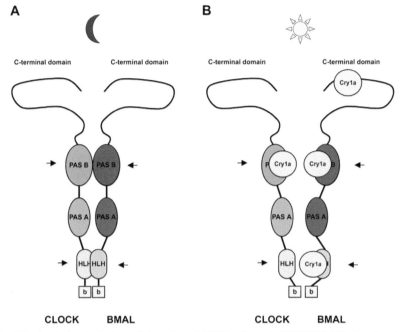

Fig. 7.10 A schematic for how light-induced CRY1a disrupts CLOCK:BMAL function in a light-dependent manner. CLOCK and BMAL have been shown to interact through their PAS B and bHLH domains. CRY1a in the light binds to these regions and prevents the formation of a transcriptionally active dimer. In this way, long light exposures can "stop" circadian pacemaker function, until the CRY1a protein is degraded (From Tamai *et al.*, 2007).

T - #0401 - 071024 - C3 - 229/152/12 - PB - 9780367383954 - Gloss Lamination